Soil Erosion Measurement Techniques and Field Experiments, 2nd Edition

Soil Erosion Measurement Techniques and Field Experiments, 2nd Edition

Guest Editors

Vito Ferro
Alessio Nicosia

MDPI

Basel • Beijing • Wuhan • Barcelona • Belgrade • Novi Sad • Cluj • Manchester

Guest Editors
Vito Ferro
Department of Agricultural,
Food and Forest Sciences
University of Palermo
Palermo
Italy

Alessio Nicosia
Department of Agricultural,
Food and Forest Sciences
University of Palermo
Palermo
Italy

Editorial Office
MDPI AG
Grosspeteranlage 5
4052 Basel, Switzerland

This is a reprint of the Special Issue, published open access by the journal *Water* (ISSN 2073-4441), freely accessible at: https://www.mdpi.com/journal/water/special_issues/47AZ4C142H.

For citation purposes, cite each article independently as indicated on the article page online and as indicated below:

Lastname, A.A.; Lastname, B.B. Article Title. *Journal Name* **Year**, *Volume Number*, Page Range.

ISBN 978-3-7258-8529-9 (Hbk)
ISBN 978-3-7258-8530-5 (PDF)
https://doi.org/10.3390/books978-3-7258-8530-5

Cover image courtesy of Alessio Nicosia

Contents

About the Editors

Vito Ferro

Vito Ferro is a Full Professor of hydraulics and hydrology, soil erosion and conservation, and environmental restoration at the University of Palermo. He has been involved in many projects, such as the National Biodiversity Future Center (NBFC). In 2007, he won the "Karl Emil Hilgard Hydraulic Prize".

Alessio Nicosia

Alessio Nicosia is an Assistant Professor at the University of Palermo. His research deals with hydraulics and hydrology, soil erosion and conservation, and environmental restoration. The main subject of his research is the study of flow resistance in rills.

water

MDPI

Editorial

Soil Erosion Measurement Techniques and Field Experiments, 2nd Edition

Vito Ferro [1,2] and Alessio Nicosia [1,*]

1 Department of Agricultural, Food and Forestry Sciences, University of Palermo, Viale Delle Scienze, Building 4, 90128 Palermo, Italy; vito.ferro@unipa.it
2 National Biodiversity Future Center (NBFC), 90133 Palermo, Italy
* Correspondence: alessio.nicosia@unipa.it

1. Introduction

The soil erosion processes involve the detachment of soil particles from the soil surface, followed by their transport by erosive agents, such as rainfall, overland flows, and channelized flows (rills, ephemeral gullies, and gullies) [1–3]. Soil erosion determines a decrease in soil organic matter and nutrients, with a consequent reduction in soil productivity in both natural and anthropogenic ecosystems [4,5]. Other on-site consequences of soil erosion include reduced cultivable soil layer depth and degradation of soil structure [6]. Erosion also determines off-site impacts caused by soil particles entering the water system, such as contamination from chemical pesticides and fertilizers, eutrophication of waterways, siltation of reservoirs with a consequent loss of storage, and sedimentation into channels [7].

To comprehend and implement accurate soil erosion models, precise and consistent measurements of erosion processes are necessary [8]. Experiments offer a chance to examine the degree to which the ideas employed in models accurately describe the erosion processes taking place.

The objective of the second edition of this Special Issue was to combine manuscripts concerning (1) experimental areas for soil erosion measurements at different spatial scales (plot, hillslope, and basin), (2) field experiments for studying the soil erosion in its different forms (interrill, rill, and gully erosion), and (3) innovative techniques and procedures (e.g., tracers, three-dimensional photo-reconstruction, and aerial and terrestrial acquisition platforms) for measuring soil erosion processes.

2. Main Contributions of the Special Issue

This Special Issue brings together nine papers (eight research articles and one review) that explore different aspects of soil erosion processes. The contributions examine cause–effect relationships, propose new predictive models, and analyze how factors such as rainfall and agricultural practices influence erosion. Both rill and gully erosion are addressed, alongside studies on interrill erosion and sediment deposition.

The importance of measuring soil erosion at the field scale is underlined by the only review included in this Special Issue. This review by Nicosia et al. [9] reports recent scientific developments on the measurement of rainfall erosivity, soil loss at the plot scale, and rill and gully erosion using close-range photogrammetry. A critical analysis of all the literature papers leads to the conclusion that the available techniques can be considered a solid basis, which, however, could still be improved.

Among the research papers, those by Čupić et al. [10] and Ghaderi Dehkordi et al. [11] deal with erosion and sedimentation tracing by using the radioisotope cesium-137 (^{137}Cs).

In particular, the first [10] assessed soil erosion intensity and soil properties across the Crveni Potok catchment in Serbia, based on ^{137}Cs activities and the profile distribution model. These authors found a significant negative correlation between erosion rates, soil organic matter, and indicators of soil structural stability, and compared the applied model with the RUSLE (Revised Universal Soil Loss Equation) one, finding similar mean erosion rates but significantly different median values. The second [11], instead, employed geochemical tracers and ^{137}Cs to assess sediment contributions in a small sub-basin located in western Iran to quantify how various land uses (i.e., rangeland, rainfed agriculture, irrigated agriculture, and orchards) contribute to sediment yield over time. The findings revealed significant changes in sediment yield contributions over the past 60 years in dependence of land use.

Only the paper by Chen and Wu et al. [12] deals with interrill erosion, studying the impact of antecedent soil moisture content on the loss of clay soil through two-year runoff plot experiments under natural rainfall. The main result of this study showed that the impact of antecedent soil moisture on interrill erosion is conditional, and the impact only exists in erosion events with a low rainfall–runoff erosivity index.

The article by Li et al. [13] studied the sediment deposition characteristic of ecological riverbanks associated with vegetation restoration in the deep waterway regulation scheme of Yangtze River, by using two types of typical ecological riverbanks (i.e., lattice gabion ballasted vegetation mat riverbank LGBVR and mesh grid riverbank MGR) and a traditional riprap riverbank (TRR). The findings demonstrated that the ecological restoration effect of the LGBVR was better than that of the mesh grid riverbank (MGR), while that of the TRR was relatively poor.

The paper by Todisco et al. [14] studied the spatio-temporal dynamics of rainfall erosivity in the Umbria region (Italy), using a wide database of 30 min precipitation records from 54 stations collected for 20 years. In particular, using the RUSLE2 framework, the authors evaluated models of varying complexity to estimate the R-factor (the original model, or models based solely on event rainfall depth or daily rainfall depth). All the models showed consistency in the spatial and temporal patterns of the R-factor, revealing that most of the annual erosivity is concentrated in the summer because of the high average intensity of rainfall events.

Rainfalls were also studied from another point of view in the paper by Caruso et al. [15], who characterized a rainfall simulator originally designed to assess the erosive effects of precipitation on heritage surfaces. Using volumetric methods and disdrometric techniques with a Parsivel2 optical disdrometer, the study confirmed that the simulator generates rainfall velocities comparable to natural rainfall. The authors highlighted the value of combining multiple measurement approaches to improve accuracy and reliability.

Lu et al. [16] investigated the spatial coupling between geohazard susceptibility and soil conservation capacity in Guangdong Province, China, using integrated spatial analysis and machine learning approaches. Through kernel density estimation, hotspot analysis, principal component analysis, and t-SNE clustering applied to 11,252 geohazard records and nine soil conservation factors, they identified some critical mechanisms.

Finally, Rodrigo-Comino et al. [17] combined controlled field experiments with proximal remote sensing-derived geomorphometric variables and machine learning to identify key factors of erosion in a Mediterranean citrus orchard located close to Seville and the National Park of Doñana (Southern Spain) on Gleyic Regosols (clayic, arenic). Results showed that soil loss was most strongly linked to maximum compaction and hydraulic conductivity, while sediment concentration increased with surface compaction but had weak correlations with topography. Runoff decreased with average infiltration, which is tied to compaction. The study concluded that integrating standardized field protocols with

 https://doi.org/10.3390/w18020202

proximal morphometrics and machine learning is an efficient way to prioritize explanatory variables, reducing experimental effort while maintaining explanatory power.

3. Concluding Remarks and Future Perspectives

The papers collected in this Special Issue highlight the complexity of soil erosion processes and the diverse methodologies available to study them, ranging from isotopic tracing and rainfall simulation to machine learning and spatial analysis. Together, they demonstrate that integrating traditional field measurements with innovative technologies can significantly improve our understanding of erosion dynamics across scales and land uses. The novel and significant results of these studies will aid in the investigation of soil erosion and the comprehension of the dynamics of certain processes related to this subject. The selected articles contribute to our understanding of how water soil erosion can be managed and minimized. They may also provide more insights into future research directions and the primary obstacles in this area of study. Looking ahead, future research should focus on refining measurement techniques, enhancing predictive models through interdisciplinary approaches, and promoting the application of advanced tools such as remote sensing and artificial intelligence. Strengthening the link between scientific advances and practical land management strategies will be essential to mitigate erosion risks, safeguard ecosystems, and ensure sustainable agricultural and environmental practices.

Author Contributions: Writing—original draft preparation, V.F. and A.N.; writing—review and editing, V.F. and A.N. All authors have read and agreed to the published version of the manuscript.

Funding: This research received no external funding.

Conflicts of Interest: The authors declare no conflicts of interest.

References

1. Li, Z.W.; Zhang, G.H.; Geng, R.; Wang, H.; Zhang, X.C. Land use impacts on soil detachment capacity by overland flow in the Loess Plateau, China. *Catena* **2015**, *124*, 9–17. [CrossRef]
2. Di Stefano, C.; Nicosia, A.; Palmeri, V.; Pampalone, V.; Ferro, V. Estimating flow resistance in steep slope rills. *Hydrol. Process.* **2021**, *35*, e14296. [CrossRef]
3. Ferro, V.; Nicosia, A. Comment on "Rill erosion processes on steep colluvial deposit slope under heavy rainfall in flume experiments with artificial rain by F. Jiang et al.". *Catena* **2020**, *185*, 103793. [CrossRef]
4. Panagos, P.; Borrelli, P.; Ballabio, C.; Lugato, E.; Meusburger, K.; Montanarella, L.; Alewell, C. The new assessment of soil loss by water erosion in Europe. *Environ. Sci. Pol.* **2015**, *54*, 438–447. [CrossRef]
5. Carollo, F.G.; Di Stefano, C.; Nicosia, A.; Palmeri, V.; Pampalone, V.; Ferro, V. A new strategy to assure compliance with soil loss tolerance at a regional scale. *Catena* **2023**, *223*, 106945. [CrossRef]
6. Food and Agricultural Organization (FAO). *Soil Erosion: The Greatest Challenge to Sustainable Soil Management*; Food and Agricultural Organization (FAO): Rome, Italy, 2019; 100p. Available online: https://www.fao.org/documents/card/en/c/ca4395en (accessed on 2 December 2025).
7. Mullan, D. Soil erosion under the impacts of future climate change: Assessing the statistical significance of future changes and the potential on-site and off-site problems. *Catena* **2013**, *109*, 234–246. [CrossRef]
8. Pampalone, V.; Carollo, F.G.; Nicosia, A.; Palmeri, V.; Di Stefano, C.; Bagarello, V.; Ferro, V. Measurement of Water Soil Erosion at Sparacia Experimental Area (Southern Italy): A Summary of More than Twenty Years of Scientific Activity. *Water* **2022**, *14*, 1881. [CrossRef]
9. Nicosia, A.; Carollo, F.G.; Di Stefano, C.; Palmeri, V.; Pampalone, V.; Serio, M.A.; Bagarello, V.; Ferro, V. The Importance of Measuring Soil Erosion by Water at the Field Scale: A Review. *Water* **2024**, *16*, 3427. [CrossRef]
10. Čupić, A.; Smičiklas, I.; Manić, M.; Đokić, M.; Dragović, R.; Đorđević, M.; Gocić, M.; Jović, M.; Topalović, D.; Dragović, S. ^{137}Cs-Based Assessment of Soil Erosion Rates in a Morphologically Diverse Catchment with Varying Soil Types and Vegetation Cover: Relationship with Soil Properties and RUSLE Model Predictions. *Water* **2025**, *17*, 526. [CrossRef]
11. Ghaderi Dehkordi, N.; Khaledi Darvishan, A.; Zare, M.R.; Porto, P. Temporal changes in the average contribution of land uses in sediment yield using the ^{137}Cs method and geochemical tracers. *Water* **2024**, *17*, 73. [CrossRef]

12. Chen, Y.; Wu, C. The Contradictory Issue of the Impact of Antecedent Soil Moisture to Interrill Erosion in Clay Soil: A Two-Year Field Study. *Water* **2024**, *16*, 2076. [CrossRef]
13. Li, J.; Wang, X.; Wu, L. Research on Sediment Deposition Characteristics and the Vegetation Restoration of Ecological Riverbanks in the Deep Waterway Regulation Scheme of Yangtze River. *Water* **2024**, *16*, 2350. [CrossRef]
14. Todisco, F.; Massimi Alunno, A.; Vergni, L. Spatial Distribution, Temporal Behaviour, and Trends of Rainfall Erosivity in Central Italy Using Coarse Data. *Water* **2025**, *17*, 801. [CrossRef]
15. Caruso, R.; Serio, M.A.; Búrdalo-Salcedo, G.; Carollo, F.G.; Ortiz-Marqués, A.; Ferro, V.; Fernández-Raga, M. Integrated Approach to Assess Simulated Rainfall Uniformity and Energy-Related Parameters for Erosion Studies. *Water* **2025**, *17*, 3429. [CrossRef]
16. Lu, Y.; Fu, J.; Tang, L. Correlation Analysis of Geological Disaster Density and Soil and Water Conservation Prevention and Control Capacity: A Case Study of Guangdong Province. *Water* **2025**, *17*, 2527. [CrossRef]
17. Rodrigo-Comino, J.; Cambronero-Ruiz, L.; Moreno-Cuenca, L.; González-Vivar, J.; González-Moreno, M.T.; Rodríguez-Galiano, V. Integrating UAV-LiDAR and Field Experiments to Survey Soil Erosion Drivers in Citrus Orchards Using an Exploratory Machine Learning Approach. *Water* **2025**, *17*, 3541. [CrossRef]

https://doi.org/10.3390/w18020202

water

MDPI

Review

The Importance of Measuring Soil Erosion by Water at the Field Scale: A Review

Alessio Nicosia [1], Francesco Giuseppe Carollo [1], Costanza Di Stefano [1], Vincenzo Palmeri [1,2], Vincenzo Pampalone [1], Maria Angela Serio [1], Vincenzo Bagarello [1] and Vito Ferro [1,2,*]

1 Department of Agricultural, Food and Forest Sciences, University of Palermo, 90128 Palermo, Italy; alessio.nicosia@unipa.it (A.N.); francescogiuseppe.carollo@unipa.it (F.G.C.); costanza.distefano@unipa.it (C.D.S.); vincenzo.palmeri02@unipa.it (V.P.); vincenzo.pampalone@unipa.it (V.P.); mariaangela.serio@unipa.it (M.A.S.); vincenzo.bagarello@unipa.it (V.B.)

2 National Biodiversity Future Center (NBFC), 90133 Palermo, Italy

* Correspondence: vito.ferro@unipa.it

Abstract: Water erosion is a significant global threat due to the high soil loss rate and all its consequent implications. Technologies to predict erosion are strongly related to measurements and vice versa. Measurements can simply provide empirical evidence of the erosion process and are hard to extrapolate in time and space. Measurements were used to develop some erosion models, such as the Universal Soil Loss Equation (USLE), and also for their calibration and validation. Several measurement techniques are used to collect soil erosion data at different spatial and temporal scales, but they cannot be considered fully accurate in any experimental condition. Each technique exhibits advantages and disadvantages, so extensive knowledge of their feasibility, accuracy, and limitations is required to correctly plan experiments and use the performed measurements. In this paper, recent scientific developments on the measurement of rainfall erosivity, soil loss at the plot scale, and rill and gully erosion using close-range photogrammetry are presented. Further considerations are made on the quality of soil erosion measurements and the usefulness and importance of measuring plot soil loss. Our critical analysis highlighted that the techniques reported in the literature are a solid basis, which, however, should be developed to improve their range of applicability and data quality.

Keywords: soil erosion; soil loss; rainfall erosivity; rills; gullies

1. Introduction

The European Union (EU) soil thematic strategy [1] defines soil as "*essentially a non-renewable resource and a very dynamic system, which performs many functions and delivers services vital to human activities and ecosystem survival*". This definition is based on the fact that soil formation is a slow process, as hundreds to thousands of years are required to form a thin layer of topsoil when standard agricultural conditions are maintained [2,3]. On the contrary, its degradation or destruction attributable to erosion processes is fast and is accelerated by anthropogenic activities [4–9].

Erosion by water is a significant threat to soil as, in the European Union's lands (forests, semi-natural, and agricultural areas), the mean soil loss rate is 2.46 t ha^{-1} yr^{-1} [10], while the mean rate of soil formation is 1.4 t ha^{-1} yr^{-1}.

Impacts can be grave, not only because of land degradation and fertility loss but also due to different off-site effects, such as the sedimentation, eutrophication, and siltation of waterways, or enhanced flooding [11]. In Europe, Panagos et al. [12] estimated the economic loss associated with soil loss due to erosion processes, hypothesizing a restoration cost of USD 20 t^{-1}, roughly equal to USD 20 billion yr^{-1}. Panagos et al. [12] also estimated that a European land area of 12 million ha is affected by soil erosion processes, with an estimated loss of crop productivity of EUR 1.25 billion.

Soil conservation strategies are required to alleviate agricultural productivity losses in areas that suffer from soil loss ("on-site" effects) and also to reduce "off-site" effects related to sediment transport along the flow paths developing on the hillslopes and in the river network. A recent study on the costs of sediment removal from reservoirs in the EU and United Kingdom (UK) [13] demonstrated that each year, 135 million m^3 of sediment due to water erosion is potentially transported to the nearest river networks, and the associated sediment removal cost exceeds EUR 2.3 billion per year.

Technologies to predict erosion are strongly related to measurements and vice versa. Measurements can simply provide empirical evidence of the erosion process and are hard to extrapolate in time and space. Measurements were used to develop some erosion models, such as the Universal Soil Loss Equation (USLE) [14,15], and also for their calibration and validation [16]. These models are also applied to design soil conservation strategies.

According to Toy et al. [17], soil loss measurements are useful for (1) developing scientific erosion research, (2) developing erosion prediction technology, (3) developing erosion conservation strategies, (4) evaluating the environmental impact of erosion and conservation practices, and (5) establishing the allocation of conservation regulations, policies, and programs.

Several measurement techniques can be used to collect soil erosion data at different spatial and temporal scales [18], but no method has been proven to be fully accurate in any experimental condition [19]. As remarked by Raza et al. [20], an important aspect of measuring soil erosion is spatio-temporal soil heterogeneity. In fact, the accuracy of erosion rate measurements depends on the considered spatial dimension, as soil heterogeneity has an increasing effect with the analyzed spatial scale. Moreover, each developed method exhibits both advantages and disadvantages, so extensive knowledge of its feasibility and limitations is required to plan correct experiments and use the obtained measurements correctly [21].

As an example, erosion pins are the simplest tool for obtaining local measurements of soil erosion [22–24]. The pin, which can be implemented with different materials [25,26], is inserted vertically into the soil, and the exposed height, generally equal to 5 cm, is measured after insertion as a reference. Pins also have the following advantages: (1) they are little, have reduced surface exposure, and do not significantly modify the surrounding soil surface; and (2) they are cheap, lightweight, and can be installed with little effort. However, when deposition processes are relevant, pins can be buried by deposition, and during the measurement time, erosion processes can be balanced or exceeded by deposition. Moreover, human and animal activity can disturb pin measurements [27], and the small size and surface exposure of pins can hinder their field visibility, especially when vegetation cover is present [25].

The measurement technique also depends on the study's aims and the type of experimental data to be obtained. As an example, erosion due to rainfall impact is measured by small-scale devices, such as splash cups, that differ appreciably from the plots in which interrill and rill erosion are simultaneously measured [18]. At the plot scale, soil loss is due to interrill and rill erosion processes, and the rill component can be directly measured by measuring the geometry of the rill channels [28] or can be obtained by calculating the difference between the measured total soil loss (interrill plus rill) and the interrill component contemporaneously measured by small plots [17,29,30].

Measurements of rill or gully erosion can be performed by direct or remote sensing methods (image-based and laser scanners) [31,32]. The direct method is based on the survey of some cross-sections along the channel. Cross-sections can be measured using a pin profiler (profilometer) (Figure 1) made of stainless steel pins with transverse constant spacing which is positioned perpendicularly to the channel axis. The direct method is simple, low-cost, and widely used [28]. However, it has the disadvantage of being time-consuming, and, in a rilled area, the operator may cause a possible modification of the plot surface when entering the area. Close-range and low-altitude aerial photogrammetry can be efficiently applied for measuring soil erosion by the Structure from Motion (SfM) technique. This technique, using oblique images taken from uncalibrated and non-metric cameras coupled with photogrammetric software packages [33], allows the creation of accurate three-dimensional (3D) (Figure 2) and 2.5D models (Digital Elevation Models, DEMs) of the surveyed surface [34–37].

Figure 1. View of a pin profiler (profilometer) (modified from Di Stefano et al. [35]).

Figure 2. Example of 3D Digital Terrain Model (3D-DTM) of a plot area surveyed for measuring soil erosion (modified from Di Stefano et al. [38]).

The measurement spatial scale (cup, microplot, plot, hillslope, basin) and technique (global, distributed, direct, remote) depend on the investigated variable (soil loss, sediment yield) and the involved erosion sub-processes (impact, interrill, rill, gully erosion). Stroosnijder [16] suggested the following five spatial scales for human-induced water erosion at agricultural scales: (i) the point (1 m^2) scale for interrill (splash) erosion; (ii) the plot (<100 m^2) scale for rill erosion; (iii) the hillslope (<500 m^2) scale for sediment deposition; (iv) the field (<1 ha) scale for channels; and (v) the small basin (<50 ha) scale for spatial interaction effects.

The total soil loss can be measured by two opposed approaches: (1) direct survey (in-site measurement) of the investigated area at fixed time intervals (e.g., event, month, year); or (2) measurement of the magnitude of soil loss from the area and accumulated in a storage device, such as a sedimentation tank or a sediment-collecting system (out-site measurement) (Figure 3). The choice of a specific technique is related to the requested spatial resolution and measurement accuracy. Soil loss in a known temporal interval (e.g., event, month, year) at the plot scale can be detected by using erosion pins distributed on the plot area or measuring the weight of sediments reaching the downstream end of the plot and stored in a collecting system. These two approaches are quite different, and the results in terms of data quality and type of collected information (distributed, lumped) depend on the technological and procedural matter of the used technique.

Figure 3. Views of sedimentation tanks used to collect soil loss from a plot.

The selected condition (laboratory or field) of the investigation affects the results of the erosion research. Laboratory measurements are useful for investigating specific erosion sub-processes (e.g., interrill erosion) in highly controlled conditions even if the natural variability of these sub-processes due to environmental factors is hindered. Laboratory experiments can be used to represent small-scale studies of field conditions and have the advantage of being fast and repeatable [17]. Particular attention merits the preparation of soil samples, which risk being unrepresentative of field conditions.

Field data are strongly related to the characteristics (climate, soil, morphology, and land use) of the investigated area. The main advantage of field investigations is the opportunity to carry out measurements at a real scale, considering the actual soil, the existing crop cover, and the temporal variation in the environmental variables affecting the studied process [16]. When field measurements are collected in a short time interval, corresponding to a few erosive events, they can be unrepresentative of field conditions. In other words, the collected data may not be representative of average or extreme conditions for the investigated area, which require longer experiments. Furthermore, the practical difficulties related to field activities make these data less accurate than those collected in laboratory-controlled conditions.

Another problem is related to some specific erosion sub-processes, which require indirect measurement of the considered variable (e.g., soil erodibility). When the investigated variable is not representative of an intrinsic soil property, its indirect measurement requires the choice of a mathematical model representing the investigated sub-process. The main limitation of this type of measurement is that the measured non-intrinsic soil property varies with the equation used by the simulation model.

In any condition (laboratory, field) and for each investigated property (intrinsic or non-intrinsic), the main objective is to obtain precise and unbiased measurements [17], i.e., with a small standard deviation in the same condition and without systematic errors in the measured value. In any case, statistical data analysis cannot replace the ability of the researcher to both detect and explain unexpected results. After about twenty years, the scientific and technological progress in the field of soil erosion measurement allows us to take a second look at the conclusion of Stroosnijder [16] that "*there is a crisis in erosion measurements because there are insufficient empirical data of adequate quality, a lack of funds to improve that situation, a lack of development of new technologies and equipment, and a lack of skilled personnel*". The importance of measuring soil loss is related to the design of erosion prediction technology as well as its calibration and performance evaluation. Furthermore, the measurement provides empirical evidence of the investigated erosion processes and is required in modeling approaches for its extrapolation in time and space.

Considering the current state of the art of water soil erosion measurements, the aim of this paper is to present recent scientific developments on the measurement of rainfall erosivity, soil loss at the plot scale, and rill and gully erosion using close-range photogrammetry. Further considerations concern the quality, usefulness, and importance of soil erosion measurements.

This paper deals with measurements of soil erosion processes at the field scale and its structure reflects the distinction between the measurement techniques of each erosion process (interrill, rill, and gully). This choice depends on the circumstance that the adopted measurement technique is strictly related to the relevant variables of the specific soil erosion process. The usefulness of soil loss measurements is discussed, and research needs and future perspectives are also presented.

2. Measuring Rainfall Erosivity

Rainfall erosivity is a property expressing the climatic forcing of the erosion process. Following the definition of Wischmeier and Smith [39], rainfall erosivity is obtained as the product of the rainfall energy E (MJ ha^{-1}) and the maximum continuous 30 min intensity I_{30} (mm h^{-1}) during a single rainfall event [40]. At the plot and field scales, soil loss by water erosion is related to different rainfall properties [41], such as rainfall intensity, I (mm h^{-1}); rainfall momentum, RM (N m^{-2}); and rainfall kinetic energy per unit time and area, named kinetic power, P_n (J m^{-2} h^{-1}) [42–46].

The fact that devices measuring rainfall erosivity were not widespread favored the application of different empirical P_n-I relationships within empirical soil erosion prediction models [42,47]. Salles et al. [47], performing an overview of several empirical P_n-I relationships, showed that, for a fixed rainfall intensity, these relationships show great variability in the estimation of kinetic power. Like Parsons and Gadian [48], Salles et al. [47] concluded that a global parameter such as rainfall intensity is not sufficient to determine P_n because it also depends on other parameters (rain type, altitude, climate, and measurement method). Carollo et al. [44], using raindrop size distribution (DSD) measurements performed with different techniques and in different geographical areas, confirmed that rainfall intensity is insufficient to determine rainfall kinetic power, and moreover, they detected that the most commonly used empirical P_n-I relationships [47] are site-specific and thus need to be calibrated.

The rainfall kinetic power or momentum can also be indirectly measured by knowing the DSD and the raindrop falling velocity. For each raindrop of a given volume and mass, raindrop velocity can be measured or calculated by a literature relationship that uses the drop diameter [43,45,46,49–52].

For a drop with a diameter equal to D, falling from a height H, the raindrop velocity [49,53–63] increases with H up to a threshold value of 20 m, after which it becomes constant (terminal velocity).

At present, few terminal velocity measurements are available in the literature [64,65], and many relationships describing raindrop terminal velocity–diameter have been obtained by using measurements of simulated single raindrops that fall in still air [50,55,64–70].

Many scholars have proposed relationships to determine raindrop falling velocity as a function of the drop diameter and the atmosphere density and viscosity, which depend on the fall height [53–63,71]. To describe the variation in terminal velocity with the diameter of the drop, which in nature can change its spherical shape while falling through the air, Best [56] used the measurements of terminal velocity by Spilhaus [54] and proposed three empirical relationships to estimate raindrop falling velocity.

Assouline et al. [72] evaluated the raindrop velocity at the soil surface as a function of H, considering fall velocity measurements by Laws [68] and Wolfs [73]. For simulated rainfalls, they stated that the drop velocity at the soil surface is comparable to the terminal velocity of equivalent diameters for falling heights higher than 5.5 m.

Considering that the terminal velocity increases with D for $D < 0.56$ cm and then becomes constant for the highest D values, Ferro [49] deduced the following relationship by using the following available literature measurement [54,64,65,68–70]:

$$V(D) = V_{max}\,[1 - exp(-a_n\,D)] \tag{1}$$

in which V_{max} (m s^{-1}) and a_n (cm^{-1}) are two parameters that vary with H (m). The parameters $V_{max} = 9.5$ m s^{-1} and $a_n = 6$ cm^{-1} are used to calculate the terminal velocity. This equation was also used by [43,44,74] for estimating the kinetic power by DSD in Italy.

DSD is the result of some physical phenomena (i.e., raindrop collision, coalescence, break up) that influence raindrop formation and size changes during the falling process [42,75,76].

Numerous techniques (optical disdrometers, photographic method, acoustic disdrometers, flour pellet method, stain method, oil immersion method, raindrop spectrographs, micro rain radars) have been tested for DSD measurements on the ground and for assessing and monitoring rainfall erosivity [43,45,52,68,77–93]. Kathiravelu et al. [93], performing an overview of raindrop measurement techniques, concluded that manual raindrop measurement techniques are (1) generally unsuitable for the measurement of smaller and larger drop sizes outside the normal range (0.3–6.0 mm), (2) not capable of accurate drop counts, and (3) unsuitable for continuous rainfall monitoring investigations, especially during intense and windy storm conditions. Moreover, all of these measurement methods cannot be used to measure or report drop velocity.

Improvements for reproducing DSD and sampling continuously over long durations have been made by introducing automated measurement techniques, i.e., impact, acoustic, and optical disdrometers.

As the disdrometer is a widely applied instrument for measuring DSD, many studies have been developed with the aims of evaluating the quality of these data [94–96] and identifying the factors affecting the measurement quality, such as the simultaneous arrival of multiple drops, which can be confused with a single larger drop, and the passage of the drops to the side of the detector that inhibits their detection [97]. The disdrometer measurement technique presents some disadvantages (Table 1): (1) the management of a high amount of numerical data, (2) the high costs of installation and management, and (3) use limited to experimental installations for scientific research and not suitable for DSD detection at a large spatial scale. Moreover, disdrometers do not allow us to make direct measurements of rainfall variables such as P_n, RM, and I for quantifying the aggressiveness of precipitation.

Table 1. Characteristics of the presented measurement techniques for the relevant variables of rainfall erosivity, interrill, rill, and gully erosion at the field scale.

Process	Variable	Measurement Technique	Limits	Advantages	Disadvantages
Soil detachment due to rainfall	Drop size distribution Raindrop velocity	Manual raindrop measurements	- Suitable only for the normal drop range (0.3–6.0 mm)	- Simple - Applicable for laboratory studies	- Unsuitable for continuous rainfall monitoring - Raindrop velocity is not measured
		Disdrometer	- Simultaneous arrival of multiple drops	- Useful in experimental installations - Temporal continuous measurements - High-quality data	- Management of many numerical data - High costs of installation and management - Unsuitable for regional investigation - Indirect measurement of P_n and RM
	Rainfall kinetic power P_n and rainfall momentum RM	Impact meters	- Sensor dimension - Signal/noise ratio	- Direct measurement of P_n or RM - Low costs - Useful for continuous measurements - Suitable for regional investigation	- Need for a signal–rainfall variable (P_n or RM) calibration curve
		New patented device (https://www.uibm.gov.it/bancadati/Number_search/type_url?type=wpn; Patent Number: 102018000010691 (accessed on 3 September 2024)	- Sensor dimension - Signal/noise ratio - Indirect measurement of P_n or RM	- Only measures rainfall intensity I and number of raindrops N - Simplicity of the measurement technique - Cheapness of the device - Coupling with existing recording rain gauges	- Prototype is required
Interrill and rill erosion at field scale	Plot soil loss	Bounded field plot	- Plot size dependent on the investigated process (interrill or interrill+rill) - Scale effects detected on runoff, sediment concentration, and soil loss - Uncertainties and source of errors due to the sediment sampling technique	- Temporal continuous measurements of soil loss - Measurement at rainfall event scale - High-quality data - Narrow plots can be employed for sampling severe rainfall events - The calibration curve of the sediment storage tank is a straight line - Usefulness of soil loss data to calibrate and validate models - Accuracy of a soil erosion model can be estimated	- High costs of installation and management - Unsuitable for spatially widespread measurements - Closed system on a hillslope - Difficult extrapolation of soil erosion data from plot to catchment scale - The measurements require the calibration of the sediment storage system or/and the applied sampling system

Table 1. *Cont.*

Process	Variable	Measurement Technique	Limits	Advantages	Disadvantages
		Micro-topographic profiler	- Pins can penetrate into the soil	- Simple - Low cost - Widely applied - Pin disposition can be photographed	- Cumbersome - Time-consuming - The operator is forced to enter the plot - No information on the measurement errors - Measurement accuracy depends on the operator's choices
Rill and gully erosion at field scale	Rill/gully channel geometry	Terrestrial and low-altitude surveys	- Uncertainties of the field operator when determining the correct width of a narrow channel - Accuracy depends on how many measured cross-sections fall in the channel reach - Accuracy of gully measurements is affected by the morphology and size of the channel - Correct and actual definition of the cross-section - Roughness from vegetation or crenulations in the feature can create data voids	- Non-invasive method - Automatic extraction of channel features - No impact of the survey activity - High spatial resolution - Image processing is almost automatic - Accurate DEMS	- Time-consuming work for postprocessing - Incorrect survey of width affects the channel volume measurement
		Terrestrial laser scanner (TLS)	- Roughness from vegetation or crenulations in the feature can create data voids	- Non-invasive method - Proven accuracy - High data density acquisition	- Challenging data processing (large datasets, specialized software)

The rainfall impact energy can be directly measured by some instruments, which are essentially force transducers and use piezoelectric sensors, named impact meters. The measurement principle is based on the conversion, by a pressure transducer, of the rainfall impact to an electric signal whose magnitude is used for determining the kinetic power or momentum using a specific empirical calibration curve, which is the relationship between rainfall variables and the transducer output pulse. The measurement accuracy of impact meters depends on both the design data (sensor dimension, signal/noise ratio) and methodological aspects (signal–rainfall kinetic power calibration curve). A direct measurement method was developed by Jayawardena and Rezaur [98] using a piezoelectric transducer, recording the amplitude of electrical pulses originated by the raindrop impact on the transducer surface. As the calibration curve (electric signal–variable representative of rainfall impact) was a straight line for rainfall momentum, the authors concluded that this direct measurement was less sensitive to errors due to electrical signal acquisition than P_n. This result allows us to state that the measurement of *RM* by the piezoelectric technique has to be preferred to that of P_n.

Recently, a new method to measure rainfall energy and a related device (Figure 4), the subjects of a patent (https://www.uibm.gov.it/bancadati/Number_search/type_url?type=wpn; Patent Number: 102018000010691 (accessed on 3 September 2024)), have been proposed. This method allows rainfall kinetic power and momentum to be obtained by measuring rainfall intensity *I* and the number of raindrops *N*. These variables can be measured by the device proposed in the filed patent (Figure 4) consisting of a pluviograph that measures *I*, equipped with a collar fixed to its upper part that houses four sensors counting the number, *N*, of pulses originating from the collisions of raindrops on each sensor. The main advantages of the patented method and device [97] are (1) the simplicity of the measurement technique; (2) the cheapness of manufacturing the collar; (3) the possibility of modifying the existing pluviographs by applying the collar; and (4) the possibility of integrating the new device in the production and marketing process of the actual pluviographs. Recently, Serio et al. [50,52] suggested that the patented method allows an accurate measurement of P_n in the Mediterranean environment and pointed out the possibility of improving it by detecting a further variable in addition to the rainfall intensity and the number of raindrop impacts [97].

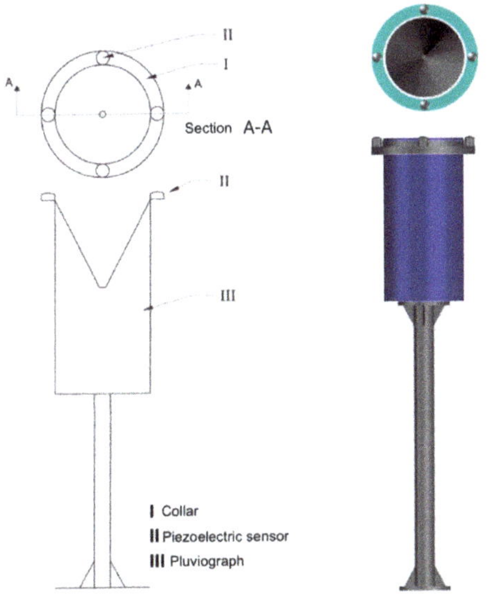

Figure 4. Scheme of the patented device proposed by Carollo et al. [97] (modified from Carollo et al. [97]).

To improve the patented device's measurement technique, Carollo et al. [97] developed a theoretical analysis based on the momentum distribution. Hypothesizing that the momentum distribution is the same as the distribution of the electrical signal detected by a specific piezoelectric sensor on which the raindrops impact, the detection of I, N, and the mean or standard deviation of the diameter deduced by the momentum distribution allowed them to estimate the rainfall kinetic power and momentum. The authors positively tested the new approach using more than 84000 DSDs detected at Palermo (Italy), Sparacia (Italy), and El Teualret (Spain) experimental areas using the same optical disdrometer (model ODM 470 made by Eigenbrodt). Moreover, the analysis suggested that the best estimates of P_n and RM are obtained by the knowledge of I and N, and the mean diameter deduced by the momentum distribution. The practical application of the new suggested approach requires the construction of a prototype of the patented device to test the hypothesis that the mean diameter (or standard deviation) deduced by the momentum distribution is the same as that deduced by the distribution of an electrical signal detected by a specific piezoelectric sensor on which the raindrops impact. The method proposed by Carollo et al. [97], combined with the patented device, represents an important advance for measuring rainfall energy characteristics at a large spatial scale.

3. Measuring Soil Loss at the Plot Scale

3.1. Plot Characteristics

The measurement of soil loss at the plot scale (Figure 5) is frequently carried out to investigate interrill and rill erosion (Table 1) [16,19,21,29,99,100]. When interrill soil erosion processes have to be investigated, the physical variable to be measured is the quantity of soil detached by rainfall impact and transported by overland flow. The plot area has to be small to avoid rill formation while overland flow can occur [17].

Figure 5. Experimental plots for measuring soil loss at the Sparacia experimental area.

Interrill erosion can be measured in controlled laboratory conditions using a small plot and simulated rainfall, or in the field under natural or simulated rainfalls. Laboratory conditions generally refer to sieved soils packed to a given bulk density in pans that are placed with a fixed steepness. Different mesh sizes were used to sieve the soil [101,102], even if the range of 10–20 mm was recommended to simulate field conditions [103]. Soil samples used for laboratory measurements generally have a thickness of about 10 cm [101, 102] and a plot area ranging from 0.20×0.40 m^2 to 1.25×1.78 m^2 [101–107]. For rainfall impact investigations carried out in laboratory conditions, the soil pan is located in the middle of the area wetted by the simulated rainfall. When the soil pan is large, multiple nozzles are used simultaneously to obtain a uniform rainfall distribution [104]. Total runoff (water plus sediments) is collected at the lower end of the small plot using a gutter to convey the suspension to a small storage volume (i.e., a few liters). The weight of the oven-dried sediments is the total soil loss, which can be referred to the sampled area and expressed, for example, in g m^{-2}.

In the field, an appropriate choice of plot area is required as the soil loss data can be determined by the microplot dimensions [103,108]. The microplot area roughly ranges from 0.04 to 10 m^2 [30,99,109]. Bagarello and Ferro [30], comparing soil loss measurements in small (0.04 m^2) and large (0.16 m^2) square microplots, concluded that the mean soil losses in the small microplots were from 1.6 to 3.5 times higher than those obtained in the large ones. The plot size also affects the soil erosion process as, in small microplots, the process is expected to be transport-limited (only a quote of the detached soil particles is transported), while in large microplots (10 m^2, 5 m slope length), the process is detachment-limited (overland flow is not saturated as it could transport more material than detached) [109]. The use of large microplots ensures that only the interrill erosion sub-process is studied.

The measurement of rill plus interrill erosion requires large plots, roughly 2–50 m wide and 10–200 m long [110]. In the US, the standard plot is 22 m long and 1.8 m wide, with a consequent area of 39.6 m^2. A plot was constructed bounding a land portion, with a rectangular shape (Figure 6), with the longest side oriented in the maximum slope direction, and hydraulically disconnected from the neighboring area.

Figure 6. View of the two unit plots at the Sparacia experimental area.

However, the installation of plot borders on three sides and a gutter at the downstream edge creates a closed system, i.e., a soil surface that is not connected with the surrounding hillslope surface. After many years of functioning, closed soil erosion plots can demonstrate the exhaustion of soil material available to be detached due to the selective transport of soil particles, which leads to the armoring of the soil surface, a lack of input of soil particles from outside the plot [21,111,112], and a low soil formation rate [113]. Soil erosion collected in open plots avoids the inconvenience of causing detachment-limited conditions [21] even if this experimental condition can create difficulty in determining the contributing area of the collector [19].

Boix-Fayos et al. [21,100] suggested that bigger plots are more realistic because of their better reproduction of the relationships between the different factors characterizing a natural system. Large plots are more representative of field reality and can avoid some problems occurring in small ones, such as unnatural hydrological disconnection within the investigated system, low energy flows determined by short distances, and induction to a quick hydrological response to runoff. On the contrary, small plots are useful for easy hydrological characterization and investigating the erosion response of different soil surfaces. The direct extrapolation of soil erosion data from plots to catchments is not advisable without considering the connectivity of water and sediment fluxes, the morphological variability (length, slope, aspect) of the hillslopes of a watershed, and the non-linear scale effect [100]. Some studies [114,115] highlighted, for both rill and gully erosion, the relationship between the erosion rate and plot length. Rejman [114] found that, even if some rill characteristics remain similar over the plot length, the rill cross-sectional area (i.e., the total eroded volume) largely varies with slope length. As suggested in the literature [116], this relationship could be due to increasing plot length determining an increased flow rate, characterized by a variation in the sediment transport capacity of the flow moving in the channel. Studying the physical representativeness of plot data is required for the proper interpolation, up-scaling, and use of the obtained measurements.

3.2. The Plot Scale Effect

Recent research has demonstrated that scale effects can be detected on runoff, sediment concentration, and soil loss [108]. In this study [108], a power-scaling relationship was theoretically deduced and calibrated by measurements carried out for 31 rainfall events monitored in bare ploughed plots in Eastern Spain. For 48% of the investigated events, the analysis demonstrated that statistically significant scale effects occurred for all tested variables even if spatial scale effects were less frequent for the composite variable (soil loss, the product of sediment concentration and runoff). Runoff and soil loss always decreased, passing from the reference area (1 m^2) to the largest plot (48 m^2). At the annual temporal scale, the analysis also demonstrated that the scale effect was more temporally stable for soil loss than for both runoff and sediment concentration.

Field erosion research started in the USA for the development of the Universal Soil Loss Equation (USLE) using plots that were 6 ft (1.83 m) wide and 72.6 ft (22.1 m) long, corresponding to a plot area of 1% of an acre (0.00405 ha) [117]. Subsequent investigations varied the plot width w, as an effect of plot width on the measured soil loss cannot be a priori excluded. Bagarello and Ferro [118] and Nearing [119] used w values in the range of 2–8 m, Boix-Fayos et al. [21] used w = 1 and 3 m, and Risse et al. [120] varied the width from 1.83 to 39.0 m. In a narrow plot, the flow direction mainly occurs along the maximum slope, and this circumstance increases the probability of having a concentrated runoff characterized by higher runoff and soil loss. Localized differences in soil erodibility, roughness, and infiltration rate should have a wider effect in a narrow plot than in a large one. Using runoff volume, soil loss, and sediment concentration data collected at the rainfall event scale in two Italian experimental areas to monitor soil erosion (Sparacia in Sicily and Masse in Umbria), Bagarello et al. [121] pointed out that, for low-erosivity events, the effect of plot width on the measured variable can vary from negligible to significant. This variable result can be explained by considering that the initial soil condition before the rainfall event plays a main role in the hydrological response of a plot subjected to a low-erosivity event. This variability of the plot width effect can also be explained by considering that measurement errors increase when the measured variable decreases [30]. On the contrary, for highly erosive events, the relevance of local soil conditions disappears, and the plot width effect tends to be negligible. Considering that the annual soil loss is generally due to a few severe erosive events and that these events affect the effectiveness of soil conservation strategies, a consequence of the analysis by Bagarello et al. [121] is that narrow plots can be usefully employed for sampling severe rainfall events, reducing the required experimental efforts.

The dimensionless factors (topographic factors L and S, crop and management factor C, and support practice factor P) of the USLE allow scaling from the reference condition of the *unit plot* (which is 22.1 m long, 9% steep, bare, and with no soil conservation practice) to any plot with different topographic, cover and management, and support practices. The unit plot is useful for estimating the potential erosion RK, i.e., the product between the rainfall erosivity R and soil erodibility K factors, which is up-scaled or down-scaled by the dimensionless factors. The chosen length and slope (22.1 m, 9%) of the unit plot have no theoretical justification [122–124] and have only the merit of being the most used slope length and the average gradient of the erosion plots supporting the USLE development [125]. According to the mathematical structure of the USLE, the unit plot could be only a mathematical concept useful for defining a reference condition to define the factors of the equation. Laflen and Moldenhauer [117] pointed out that *"the unit plot concept, while very useful, was apparently a myth as far as soil erosion measurements were concerned. A unit plot never existed! Or if it did, data from it was never reported"*. and Laflen and Flanagan [126] confirmed that *"there is little evidence that there ever existed an actual "unit plot", or if one ever existed, data from it has not been found"*. To the best of our knowledge, unit plots are not widespread, and recently some unit plots were used in Iran [127,128] and the Sparacia experimental area, South Italy [129]. The main interest for using unit plot data is the possibility of carrying out field evaluations of the soil erodibility factor $K = A/R$ by using simultaneous measurements of soil loss A and the rainfall erosivity factor R. An analysis of the literature [125] demonstrated that the field evaluation of the K factor was often carried out using plots that were not fully compliant with the unit plot condition. In these cases, soil erodibility factor estimates are affected by the accuracy of the USLE factor relationships for L, S, C, and P, and are not consistent with the K measurement in the unit plot. In other words, using the factor relationships to adjust plot data can determine a different erodibility factor from that which would be measured in the unit plot condition. Considering that the K values from the nomograph by Wischmeier et al. [130] were obtained with this plot data adjustment, the differences detected between the soil erodibility values measured in unit plots in recent investigations [125,127,128] and the nomograph values are potentially explained.

Other doubts on the up-scaling and down-scaling of the potential erosion RK derive from the definition of the factors accounting for slope length, steepness, cover and management, and erosion-control practice. These factors (L, S, C, and P) are all defined as soil loss ratios between the actual soil loss and that from a plot with the unit plot characteristic associated with the factor for other fixed conditions. As a consequence, while K requires an estimate strictly related to the unit plot condition, the same is not necessary for L, S, C, and P. For instance, measurements to estimate the slope effect on soil loss can equally be collected on cropped or bare plots, only varying slope steepness, neglecting the fact that the S factor estimate could be different for the two cover conditions. This assumption is related to the hypothesis of independence of the model factors, which raises doubts for some authors [111].

3.3. *Quality of Soil Loss Data*

According to the out-site measurements of total soil loss reported above, water and sediments intercepted by a gutter, placed at the downstream plot end, are conveyed to a storage system in which the mixture (sediments and runoff) is totally or partially collected [19,109,131]. Plot soil loss is obtained by multiplying the measured runoff volume and the sediment concentration of the water–sediment mixture collected in the storage device. In plots equipped for monitoring soil erosion, a widely applied method for measuring sediment concentration is to store runoff and the eroded sediments in a single tank or a sequence of tanks. If the equipped plots are large and the expected runoff volumes are high, a system of multiple tanks (Figure 3) that can be possibly connected by multislot divisors can be installed [19].

For a tank of known geometry, the suspension volume can be easily calculated by measuring the water depth, while the sediment concentration can be measured by collecting either (1) all sediments stored in the tank after siphoning the supernatant-cleared water or (2) a sample of the mixed suspension [131–133]. The main uncertainties and sources of errors are ascribable to the sampling technique of the stored sediments. The procedure (1) is generally used when the fraction of coarse particles is dominant in the stored sediments and sampling is carried out 2–3 days after the end of the rainfall event to allow the sedimentation of eroded particles at the tank bottom [134]. This procedure, which is cumbersome, allows one to accurately collect, oven-dry, and weigh sediments deposited on the bottom of the tank to determine their mass. The procedure (2) is commonly applied when the stored sediments are mainly composed of fine particles and is based on the practice of stirring the suspension, collecting a sample by immersing a sampling device, like a bottle, to a depth beneath the water surface in the tank, and oven-drying the collected sample [135,136]. This procedure was applied to collect the experimental data used in the United States to empirically derive the USLE [14,15], and it was also adopted worldwide [137–140]. This sampled concentration C_{bt} can be considered representative of the actual sediment concentration C_a in the tank only if a uniform distribution of sediments throughout the depth of water in any vertical can be ensured [135]. In other words, the suspended sediment concentration measured by a single sample C_{bt} becomes representative of the stored water–sediment mixture concentration C_a if a complete mixing condition is reached, as the latter implies that the measured suspended sediment concentration is independent of the measurement point inside the tank and this measured value coincides with the actual one.

Lang [132] tested the technique adopted for the USLE database using a bottle sampler for collecting clay soil–water mixtures and concluded that the actual suspended particle concentration was twice the concentration sampled by the bottle sampler. Consequently, this author raised doubts on the accuracy of plot soil loss data collected by using a single measurement point.

According to Ciesiolka et al. [135], even if the delay time between stirring and sampling is short, the consequent error is relevant, and, in any case, the possible correction of this effect is only effective if stirring allows one to *"achieve a uniform initial distribution of all sediment sizes throughout the depth of mixture in the collection box"*.

Bagarello and Ferro [131] investigated the hypothesis $C_{bt} = C_a$ used to compile the USLE soil loss database. These authors concluded that the calibration curve $C_a - C_{bt}$ is required and can be represented by a straight line, whose slope is dependent on the water depth in the tank and hence on the accumulated suspension volume. As a consequence, the sampling procedure by a single bottle device, adopting the $C_{bt} = C_a$ assumption, produces an increasing soil loss underestimation for increasing soil loss values.

For the case of a storage tank system, Bagarello and Ferro [131] and Bagarello et al. [141] investigated factors that affect the sediment concentration measurements, and they obtained calibration curves of sediment storage tanks (Table 1). In particular, the mean concentration C_m (g L^{-1}) of the water–sediment mixture stored in a tank is measured by the concentration profile, which corresponds to an incomplete mixing condition obtained by sampling given volumes at varying depths. If C_m is calculated by the integration of the concentration profile measured along the axial vertical of a tank (Figure 7), Bagarello and Ferro [131] theoretically deduced the following relationship between the actual concentration C_a (g L^{-1}) and the measured concentration C_m (calibration curve):

$$C_a = bC_m \tag{2}$$

where b is a coefficient to be experimentally determined, and depends on the tank geometry, the soil type, and the water depth inside the tank [142]. For the specific storage system of the Masse experimental station (Central Italy), using the original soil (S1) of the plots and the sediments (S2) obtained in the erosive events, Todisco et al. [143] confirmed the applicability of Equation (2) and estimated the coefficient b. This analysis also demonstrated that the calibration curves corresponding to the soil S1 and the eroded sediment S2 are not statistically different.

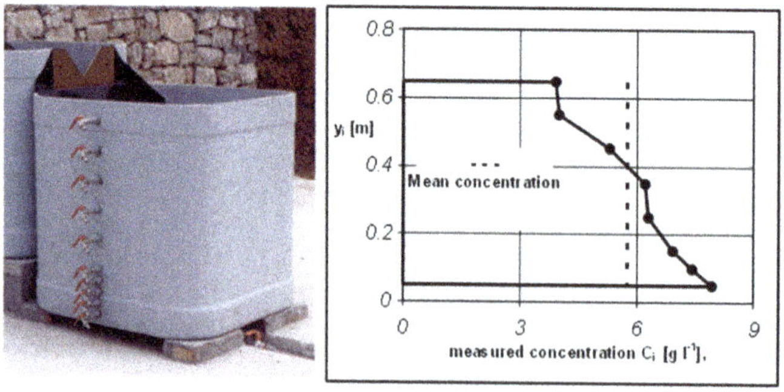

Figure 7. Scheme for integrating the concentration profile measured along the axial vertical of a tank.

Considering that the sediment concentration measurement requires homogeneous mixing, Huang et al. [144] carried out an experimental investigation for determining the calibration curve using four soils and different concentrations in the range 1–1000 g L^{-1}. This investigation confirmed that the measured sediment concentration is less than the actual one and the calibration curve is a straight line.

Nikkami [145] demonstrated that sampling by four bottles installed on a vertical and by a pipette procedure showed an underestimation of the suspension concentration as during the sampling time coarse material settles. Nikkami [145] also proposed a cylindrical sampler that collects a complete column of water–sediment mixture through the depth of the storage tank and tested it for a single low sediment concentration of 1.87 g L^{-1}. This sampler showed the best sampling efficiency in comparison with the bottle sampler and the pipette procedure.

Carollo et al. [146] proposed a new sampler (Figure 8) that is a brass cylinder 120 cm high with an inner diameter of 4.75 cm (Table 1). The sampler was equipped with a closing valve "guillotine" fitted with a sealing gasket. This valve is controlled from above by means of a knob joined to a drive pin. The sampler can be simply introduced into the suspension collected in the tank and, after the closure of the valve, is extracted. The authors carried out some experimental runs using a tank identical to those established at the Sparacia experimental area. These runs demonstrated that the measurement was not affected by the water level in the tank and the sampling vertical, and that the spatial variability of the concentration measured in different verticals became negligible at the highest concentration value. In any case, the average measured concentration was almost equal to the actual concentration, so no calibration curve was required for the proposed sampler.

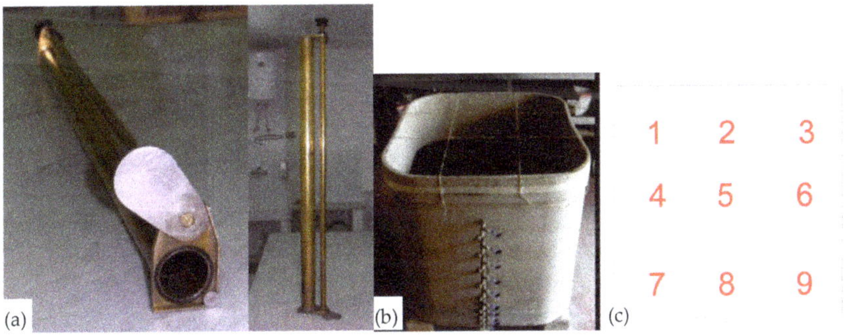

Figure 8. View of the sampler (**a**) and scheme of the partition of the cross-section in nine sampling areas (**b**,**c**) proposed by Carollo et al. [146] (modified from Carollo et al. [146]).

Considering that in many plots the water–sediment mixture is manually sampled, Zobisch et al. [147] also verified the accuracy and the repeatability of the sampling by comparing the procedure results of different field workers. The main results showed a light underestimation of the runoff volume and the very poor accuracy of the soil loss measurements. Zobisch et al. [147] concluded that the results were affected by two different effects: (1) the method applied for stirring the suspended sediments just before sampling and (2) the adopted sampling method (the sample was retrieved by plunging a beaker into the suspension). In fact, the mixing of the suspended sediments affected the mixture homogeneity within the tank, and the timing and plunging depth of the sample beaker affected the sampled sediment concentration.

Finally, considering the variability of soil loss measurements between apparently identical plots [118,148,149], the use of replicated plots is recommended to obtain representative soil erosion measurements for a particular treatment.

Recently, Bagarello and Ferro [150] studied the relationship between the relative performances of the USLE-M [151,152] and USLE-MM [153–156] and soil loss data errors at the event scale. In particular, considering the calibration curve of the storage tank of Equation (2) and the underestimation b_u of the coefficient b ($b_u < b$), the USLE-MM equation assumes the following expression:

$$A_{e,u} = \frac{b_u}{b} a_o (Q_R EI_{30})^{b_1} \quad (3)$$

in which $A_{e,u}$ is the underestimated event soil loss (Mg ha^{-1}); a_o is a coefficient that summarizes the effects of soil erodibility, topographic factors, land use and management, and soil conservation practices; Q_R is the runoff coefficient; EI_{30} (MJ mm h^{-1} ha^{-1}) is the event rainfall erosivity index of Wischmeier and Smith [15]; and b_1 is an exponent equal to 1 for USLE-M and greater than 1 for USLE-MM. The ratio b_u/b is less than 1 and represents the underestimation level. Equation (3) expresses that the underestimation error does not affect the exponent b_1 because the ratio b_u/b only modifies the coefficient a_o. In other words, the USLE-MM model does not rely on a systematic error (underestimation or overestimation) of the measured event soil loss.

3.4. A Plot as a Soil Erosion Model

The performances of a plot soil erosion model need to be evaluated to determine the prediction reliability [157,158]. The determination of the prediction quality needs a criterion to state if the soil loss estimates are acceptable or not. The most obvious method to assess the reliability of the predicted plot soil loss is comparing the estimated and measured values. Generally, this comparison is carried out by using a single or a few replicated measurements for a given treatment. Nevertheless, plots that would be modeled with identical model input parameters give variable data [30,118,149]. Generally, this phenomenon is not considered in the models' evaluation because knowledge of natural variability among plots having the same condition is limited [119]. Thus, it is expected that a rate of any difference between the measured and predicted soil loss depends on model error, but another portion is due to the unexplained variance of the measured sample value from the representative mean value for a particular treatment [148]. The use of many replicated plots for a fixed treatment decreases or removes this type of uncertainty, as it allows the calculation of a more representative mean soil loss to be used as a reference value [149]. However, equipping and operating many plots with a given treatment at a single experimental station has to be performed occasionally, being onerous from an economic point of view and also impractical. Thus, evaluating soil erosion models that are deterministic in nature, like the USLE/RUSLE or the WEPP [14,15,159,160], commonly neglects variability in the measured data [161].

With a view to making the comparison between estimated and measured values more robust, it has to be considered that, as suggested by Nearing [161], the best possible model to predict soil erosion is a physical model of the area with a similar soil type, land use, size, shape, slope, and erosive inputs. In other words, the physical model represented by a replicated plot is the best possible, unbiased, real-world model. Starting from this premise, Nearing [161] used event soil loss data for approximately 3000 pairs of data for individual storms to establish a comparison between the measured soil losses, M, and the corresponding predictions, P_r, obtained on the physical plot model. In other words, P_r was the measured soil loss on a replicate of the plot yielding M. The coefficient of determination, R^2, of the linear regression line of P_r vs. M was equal to 0.77 (Figure 9a). Very similar regression and determination coefficients to those obtained by Nearing [161] were later obtained by Bagarello and Ferro [162], who tested the P_r vs. M relationship with a dataset developed at the Sparacia experimental station in Sicily (Figure 9b), and by Bagarello et al. [142], who used an Italian database that also included plot soil loss data collected in the Umbria region (Figure 9c). With reference to this last database, two alternative definitions of the physical model were compared. The first definition is that proposed by Nearing [161] and used by Bagarello and Ferro [162] of a plot with the same length and steepness but not necessarily the same width as the sampled plot. According to a more restrictive definition, the physical model also has the same width as the sampled plot. Higher R^2 values and a smaller departure of the P_r-M data points from the identity line were detected in the latter case than in the former one.

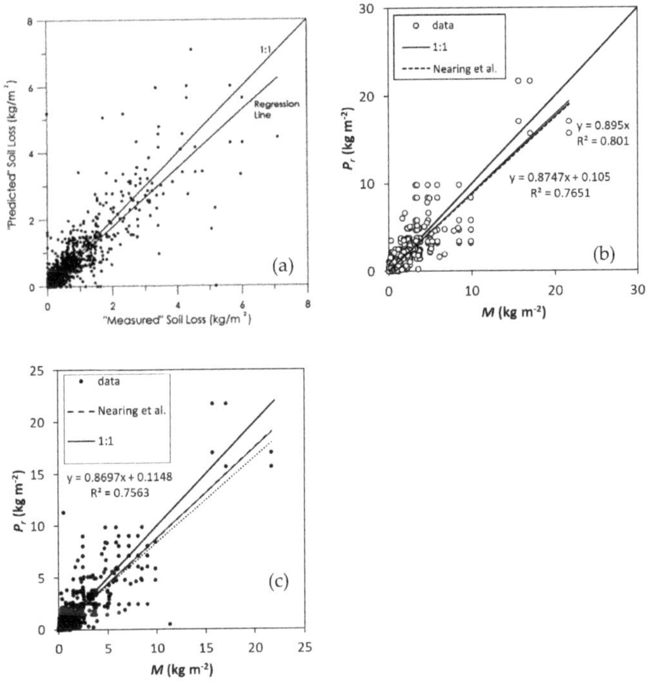

Figure 9. Predicted vs. measured soil losses for the physical model of soil erosion as represented by pairs of replicated plots according to Nearing [161] (**a**), Bagarello and Ferro [162] (**b**), and Bagarello et al. [142] (**c**).

The conclusion by Nearing [161,163], supported by the investigations carried out in Italy, was that it should be unexpected that an uncalibrated erosion model would give a better overall result than an R^2 value equal to 0.77. In other words, this R^2 value has to be considered as a benchmark or best-case prediction scenario in general. Another conclusion from [142] was that the physical model has to be defined in terms of perfect planimetric equivalence. In other terms, the physical model of a plot is another plot with the same steepness, length, and width as that plot.

Taking into account the fact that a replicated plot yields the best prediction of a measured soil loss value, Nearing [119] also suggested that the soil loss prediction can be considered acceptable if the difference between the model prediction and the measured plot data value lies within the population of the differences between pairs of measured values. In other terms, the soil loss prediction is accurate enough if the difference between P_r and M does not exceed the possible differences between pairs of measured values. The problem was therefore finding a method to predict these last differences. The analysis was developed using values of event soil loss from 7 sites in the US, with 2061 replicated storm events in the dataset, and also annual soil loss values from 13 sites, for a total of 797 replicated pairs of plots. The plot widths and slopes ranged from 2 to 8 m and 3 to 16%, respectively, and most of them were 22 m in length. This large database, which was considered to have a general validity, allowed Nearing [119] to (i) recognize that measured data with greater erosion rates have lower relative differences, R_{diff}, between replicates:

$$R_{diff} = \frac{P_r - M}{P_r + M} \quad (4)$$

and (ii) determine the following relationships for the 95% occurrence interval:

$$R_{diff,INF} = 0.236 \times \log(M) - 0.641 \quad (5a)$$

$$R_{diff,SUP} = -0.179 \times \log(M) + 0.416 \quad (5b)$$

where $R_{diff,INF}$ and $R_{diff,SUP}$ are the lower and upper limits of the interval. If a plot soil loss prediction model is applied, the effectiveness coefficient, $e_{(\alpha=0.05)}$, can be determined as the ratio between the number of predictions for which the calculated R_{diff} value falls within the interval ($R_{diff,INF} - R_{diff,SUP}$) and the total number of predictions. According to Nearing [119], an $e_{(\alpha=0.05)}$ value of the order of 0.6 is the most that can be expected from uncalibrated erosion models, while, in the case of a calibrated model, $e_{(\alpha=0.05)}$ can be greater than 0.6.

The applicability of the methodology proposed by Nearing [119] found some support when it was tested at the Sparacia experimental station [162]. In particular, the 95% occurrence interval suggested by Nearing [119] included approximately 88–89% of the data collected at the Sicilian site (Figure 10a). Considering that this discrepancy was overall moderate, the conclusion by Nearing [119] that the proposed 95% occurrence interval should be usable for model validation studies in general was considered acceptable [161].

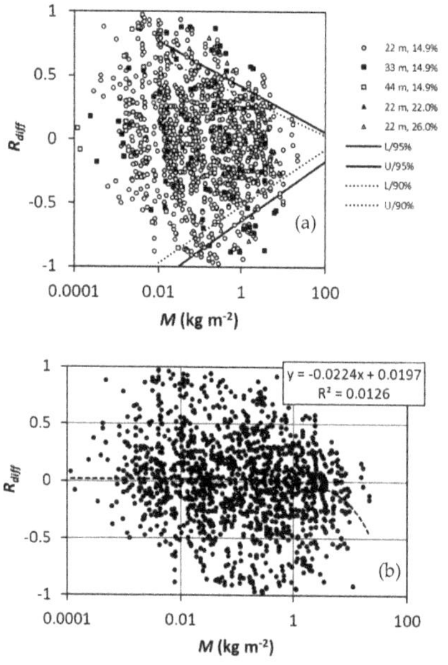

Figure 10. Relative differences in measurements of soil loss between replicated plots, R_{diff}, vs. the measured soil loss value M according to Bagarello and Ferro [162] (**a**) and Bagarello et al. [142] (**b**). The continuous lines in (**a**) define the 95% occurrence intervals for the data calculated by Equations (5a) and (5b).

The same support was not given by an Italian database, which was developed by considering the data collected at both the Masse and Sparacia experimental stations (1468 data pairs) and defining a physical model in terms of perfect equivalence (length, width, and steepness) between the two plots [142]. In particular, three regions were distinguishable on the R_{diff} vs. M plot, with approximate boundaries between adjacent regions at 0.01 and 1 kg m^{-2} (Figure 10b). Relative differences decreasing with an increase in M were only detected for $M > 1$ kg m^{-2}, and, with reference to these data (N = 316), the percentage of R_{diff} values falling within the 95% occurrence interval calculated by Equations (5a) and (5b) was equal to 85.1%. The relative differences also decreased with a reduction in M for $M < 0.01$ kg m^{-2}. This result was considered reasonable since, when the event has a low erosive power, two replicated plots can be expected to yield similar results because the erosion agent detaches and transports only a small amount of soil particles. For $0.01 \leq M \leq 1$ kg m^{-2}, a relationship between R_{diff} and M was not detectable, and the data points practically occupied all the space of the graph, suggesting that plot heterogeneities had a more noticeable impact on soil loss for intermediate levels of the erosion phenomenon. Therefore, the premise of Nearing [119] was only confirmed with reference to the highest erosion rates, but even in this particular case, the correspondence of the predicted occurrence interval with the data was not fully satisfactory.

However, Bagarello et al. [142] also found that $|P_r - M|$ increased with M according to the following relationship (Figure 11):

$$|P_r - M| = 0.356\ M^{0.91} \tag{6}$$

where $R^2 = 0.72$ and there was a 95% confidence interval of the exponent of 0.88–0.94, denoting the non-linearity of the relationship. In relative terms, Equation (6) predicts a departure of the prediction from the measured value that decreases from 80% to 27% as M increases from the minimum (0.00012 kg m^{-2}) to the maximum (21.7 kg m^{-2}) measured value. Equation (6) can be viewed as an alternative approach for applying the physical model concept by Nearing [119] since it predicts, for a given soil loss value (M), the mean absolute difference associated with the sampling of another, identical plot. The soil loss prediction of a model is accurate enough if the absolute difference from the measured value does not exceed the $|P_r - M|$ value calculated by Equation (6).

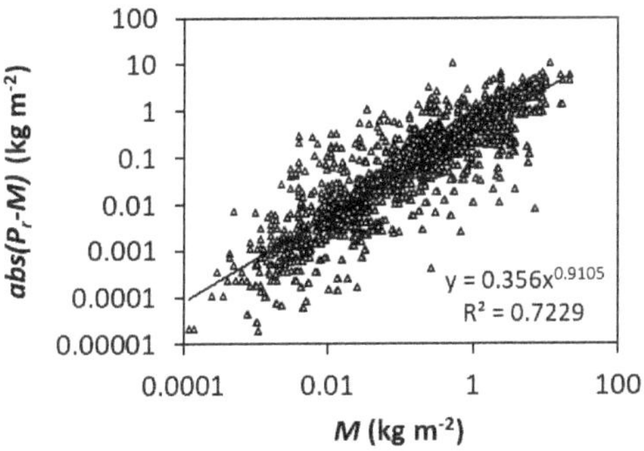

Figure 11. Absolute differences in measured soil loss between replicated plots, *abs* (P_r-M), plotted against the measured values M for the Italian database (modified from Bagarello et al. [142]).

More recently, Bagarello et al. [164] established a direct comparison between the criterion suggested by Nearing [119] and the new criterion developed with the Italian database [142]. In particular, three regions were distinguished on the P_r vs. M plot (Figure 12).

Region 1 shows the zone where the new criterion is certainly more restrictive than the original one. In other words, for the (M, P_r) pairs falling into this region, the expected effectiveness coefficient calculated by the new criterion is certainly lower than the original one. Region 3 refers to the situation in which the original criterion is certainly more restrictive than the new one. An a priori conclusion on the relative selectivity of the two criteria cannot be drawn for the intermediate region 2. Bagarello et al. [164] also demonstrated that most of the P_r vs. M data pairs should fall in region 1. Consequently, the new criterion is more restrictive than the original one in the range of the expected M values.

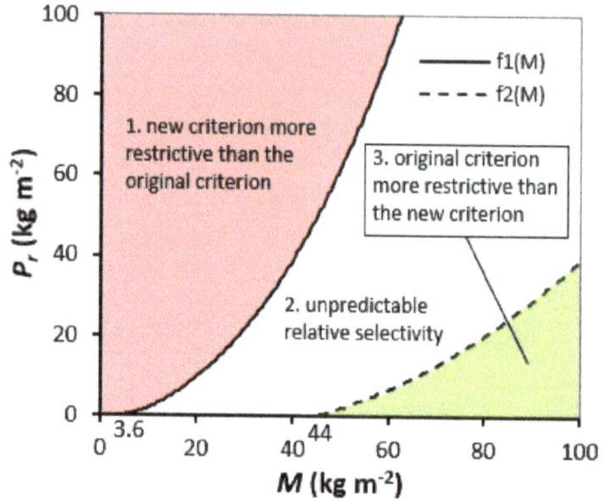

Figure 12. Plot of the three regions of the (M, P_r) reference plane for comparing the selectivity of the original criterion [119] and the new one [142] (modified from Bagarello et al. [164]).

Figure 13 shows the 1468 $(M, |P_r - M|)$ experimental pairs used by Bagarello et al. [142] to develop Equation (6) and the curve representative of this equation, which is expressive, for a given soil loss value, M, of the mean absolute difference associated with the sampling of another, identical plot. Bagarello et al. [164] also tested the least restrictive criterion using a relationship enveloping all data points and an intermediate criterion between Equation (6) and the data-enveloping line. Evidently, the data-enveloping line was not selective, i.e., it did not allow a distinction between acceptable and unacceptable soil loss predictions. The intermediate criterion was developed by carrying out a frequency analysis of the data divided into half-log-cycle intervals. For the data points within each division, the value associated with a given frequency, F, was determined, and a regression curve between these values and the measured ones was fitted (Figure 13). The frequency, F, for which the new and the original [119] criteria gave similar $e_{(\alpha=0.05)}$ was found to be 0.87. When decreasing the values of the frequency from 0.87 to 0.5, which practically yielded Equation (6), the new criterion became more and more restrictive than the original one, while the opposite obviously occurred for frequency values > 0.87.

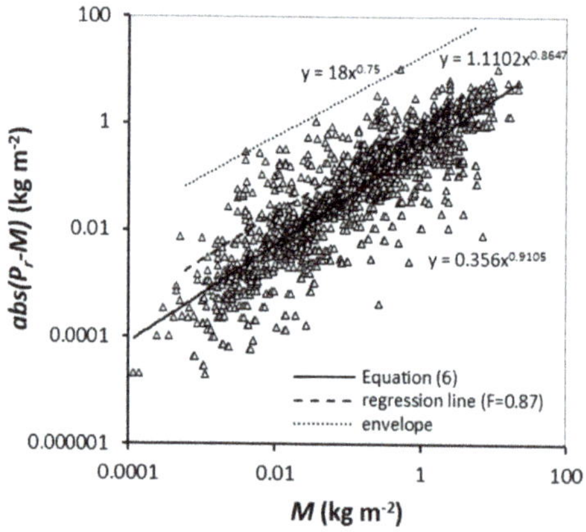

Figure 13. Plot of the absolute difference in measurement of soil loss between replicated plots, *abs* (P_r-M), vs. the measured values, M. The continuous line is the regression line of Equation (6). The dashed line is the regression line associated with a frequency F = 0.87. The dotted line is the data-enveloping line (modified from Bagarello et al. [164]).

4. Measuring Rill and Gully Erosion

When severe rainfall events occur, the concentration of overland flow causes rill development and high erosion rates. The rill component of the erosion process is the result of the channelized transport of the soil particles that are detached from the interrill areas and scoured from the rill wetted perimeter [28,165–167].

Accurate field measurements of rill erosion (Table 1) can be obtained using a ground survey of some cross-sections along the rill channel obtained by a micro-topographic profiler (Figure 14a) or a ruler and a tape. The direct measurements of the rill channel are simple, low-cost, cumbersome, and widely applied [168]. In particular, the rill length, cross-section area, and slope are measured by dividing the channel into segments and measuring each segment length $L_{r,s}$ by a metric ruler [28]. Rill cross-sections are measured in some transversal transects that can have a variable distance interval for considering the spatial variability of rill depth and width and accounting for the occurrence of rill tributaries (Figure 14b).

Figure 14. Example of ten rill spatial variability and formation (**a**) and view of a micro-topographic profiler (**b**) (modified from [28]).

Tape is used for measuring the channel length, while a ruler or rill profiler is used for measuring depths. When these methods are applied, no information on the measurement errors associated with the survey of the cross-section is available [168]. When a micro-topographic profiler (rillmeter) is used, the pin disposition can be photographed, and these images can be used to measure or digitize pin heights. However, the measurement accuracy is mainly dependent on the operator's choice (as an example, the applied technique and the number of performed measurements) rather than on the precision of the specific equipment.

The rill segment volume, $V_{r,s}$, can be obtained using the equation [28,169] that follows:

$$V_{r,s} = \frac{1}{2}\left(A_i + A_{i+1}\right) L_{r,s} \tag{7}$$

in which A_i is the cross-section area at the beginning of the rill segment and A_{i+1} is the cross-section area at the end. The total volume, V, and the total length L of a rill are calculated by adding the volumes $V_{r,s}$ and lengths $L_{r,s}$ of all segments into which the rill is divided, respectively.

After an erosive event, the planned rill network development can be obtained by a photographic survey corrected by using the transverse and longitudinal bands painted on the plot edges (Figure 14).

The direct method, using rillmeters and topographic instruments, is cumbersome and time-consuming, and the operator is forced to enter the plot with the risk of altering the surface to be monitored.

A remote and non-invasive survey of a rilled plot can be carried out by a terrestrial photographic survey (TS) or a low-altitude survey by an unmanned aerial vehicle (UAV) (Table 1). In both cases, these are remote methods and therefore are non-invasive, suitable for the automatic extraction of the characteristic features of rills, and can be performed immediately after an erosive event [31]. Close-range photogrammetry, using terrestrial instrumentation, or low-altitude photogrammetry, using a UAV, are characterized by the minimal impact of the survey activity on ground measurements and their high spatial resolution (centimeters to millimeters). This technique requires both time-consuming work for postprocessing the digital photos and little expertise, even if the image processing is almost automatic [170]. The data accuracy of aerial and terrestrial photogrammetry mainly depends on the instruments and techniques applied [171]. Close-range and low-altitude aerial photogrammetry allow the creation of accurate three-dimensional (3D) and 2.5D models (Digital Elevation Models, DEMs) of the surveyed surface. Images have to be captured, ensuring that any part of the investigated area is included in at least three photographs, considering that the 3D algorithm is designed to work with convergent images. Redundant information from oblique images is used to simultaneously solve the camera model parameters and the scene geometry. It is necessary to scale and georeference the three-dimensional point cloud produced by the workflow using a few ground control points that are situated inside the monitored area.

Measuring rill erosion through an image-based technique (Table 1) requires the use of a 3D model, which represents the three-dimensional topography and considers areas where different z-elevations can be associated with the same couple of plane coordinates x-y. On the other hand, the surface of the rilled area from a vertical perspective is represented by the 2.5D model; in other words, the 2.5D model is not able to represent cross-sections with undercut walls [172] with the x-y coordinates of a ground measurement point that can correspond to different z-elevations. In this case, the eroded rill volume is underestimated by the 2.5D model due to an underestimation of the cross-section area [34].

For a single rill channel, Di Stefano et al. [34] compared the 3D image-based ground measurements of morphological and hydraulic variables with the corresponding measurements performed on a rill gypsum cast (Figure 15). These variables were measured in correspondence with eleven transects of the plot by a profilometer, the 2.5D and 3D models. These authors, using the 3D model as a reference, established that the survey by the profilometer and the 2.5D model had similar accuracy, as the error was generally below 15%. Furthermore, the rill measurements validated the theoretical dimensionless relationship among the morphological variables describing the channelized erosion process as well as the empirical relationship between the rill length and its eroded volume [28,169,173,174].

The analysis, in agreement with a previous study [168], also demonstrated that the accuracy can be affected by the interval distance I between two measured rill cross-sections and the rill length L_r. However, a ratio $i/L_r \leq 0.2$ ensures that the measurement of rill volume can be assumed to be almost independent of the ratio i/L_r.

Low-altitude aerial photogrammetry is conducted by a photo camera installed onboard a radio-controlled microdrone (Figure 16a), which can automatically run in pre-planned flights, programming the number of flight lines and the corresponding flight height by the maps of the studied area. Figure 16b shows the route followed in a flight plane, which follows the classic photogrammetric strips characterized by an overlapping of 80% and a side lapping of 60%.

Orthophotos and DEMs can be used to survey rill features with a "manual" method. The latter is performed by drawing the rill paths from the orthophoto visual interpretation. This manual method cannot be applied for low-depth rills because they are not visually detectable.

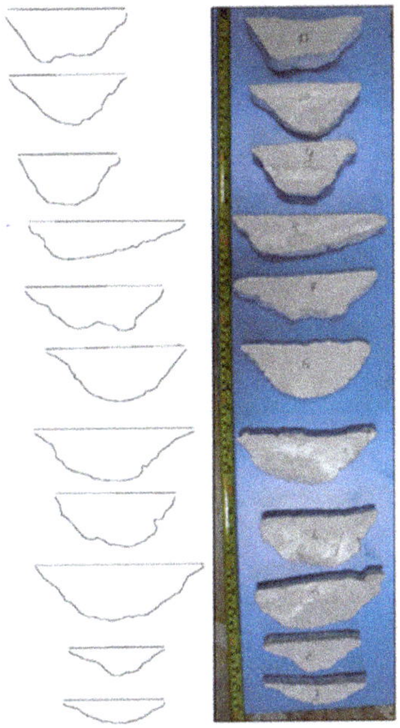

Figure 15. Comparison between the digitized and photographed rill cross-sections obtained from the gypsum cast (modified from Di Stefano et al. [34]).

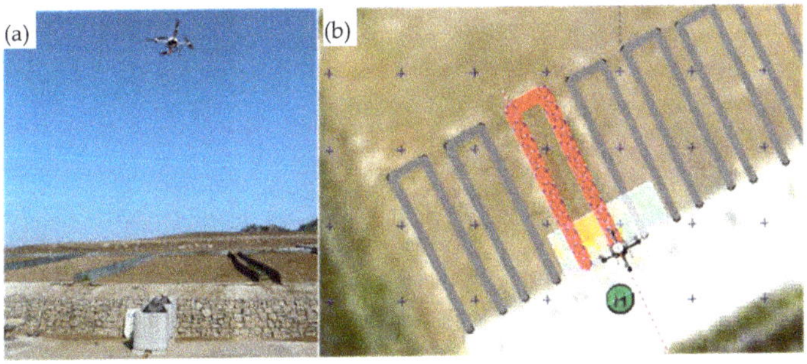

Figure 16. View of a radio-controlled microdrone (a), and an example of a flight plan (b) (modified from Carollo et al. [31]).

Carollo et al. [31] proposed an "automatic" method to extract rills from the DEM, which uses Broscoe's [175] theory of drop analysis. The underlying idea of this theory is that the length of the channels increases and the thalweg slope lowers as the order of the channels within the network increases. The vector coverage of transects orthogonal to the main flow direction—that is, the maximum slope steepness path—was produced by applying the proper calculation routine. For a rilled plot, the intersection of the plot DEM with the transect coverage allowed both the cross-sections and main rill morphological features to be obtained. For a 44 m long plot located in the Sparacia experimental area [31], Figure 17 shows the comparison between the map of rills extracted by the drop analysis (a) and the manual method (b). The developed analysis demonstrated that the number of detected rills of the automatic extraction method was greater than that of the manual one. Moreover, the automatic identified rills were always longer than those determined by the manual method, as the identification of rills by orthophotos was difficult at the downstream end of the plot. For four rills detected in this investigation, Figure 17c shows that the total rill length obtained by the automatic method was higher than those determined by the rillmeter and orthophoto, while the differences between the total rill volumes estimated by the three different methods were negligible (Figure 17d). The comparison in terms of depth and width between the DEM and rillmeter measurements [31] also demonstrated that the DEM yielded underestimations of the depth and overestimations of the width with respect to the rillmeter. Depth underestimation was explained by considering that the rillmeter pins could penetrate into the soil of the rill edge, whereas the overestimation of the surface rill width could be caused by uncertainties of the field operator when determining the correct width of a narrow channel. The obtained cross-sectional area invariance with the survey method was explained by the balance of depth underestimations and width overestimations.

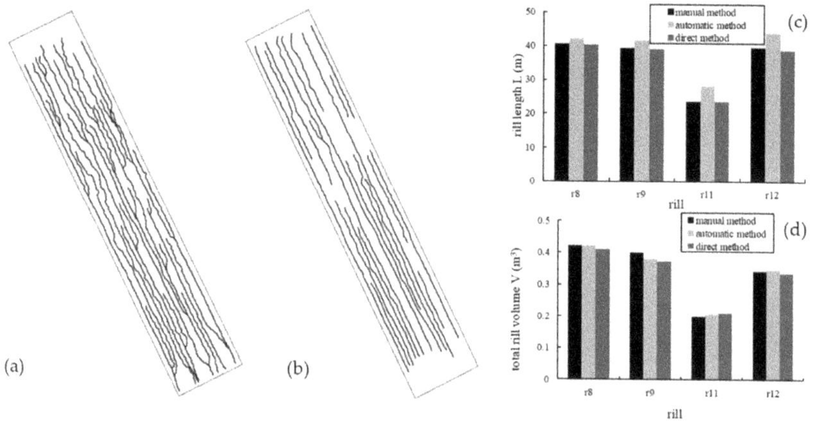

Figure 17. Comparison between the maps of rills extracted by the drop analysis (**a**) and the manual method (**b**), and total rill length (**c**) and volume (**d**) obtained by the different tested methods by Carollo et al. [31] (modified from Carollo et al. [31]).

The measurement of water depth in a rill is generally performed in a cross-section by a point gauge (Figure 18), with this measured value then overlayed on the geometrical cross-section to measure the water surface width, the hydraulic cross-section, and the wetted perimeter. For a rill reach included between two cross-sections, the hydraulic cross-section area, the water depth, the wetted perimeter, and the hydraulic radius can be calculated by averaging all values measured in the reach. The applied measurement technique influences the accuracy of these variables, while their representativeness is affected by how much measuring cross-sections fall in the rill reach. Di Stefano et al. [38] compared the water depth measurements performed by a point gauge (local technique) and those obtained by the survey of the water tracks inside the channel (volumetric technique) obtained by close-range photogrammetry (Figure 19). For each investigated rill reach, the comparison between the local water depth values and those obtained using tracks demonstrates that the data pairs are characterized by wide scattering, which can be explained by the fact that the volumetric technique considers the effects of channel irregularities (geometric variability of the rill channels between two cross-sections with a given distance interval) and the interaction between mobile channel and flow. The analysis also pointed out that the volumetric measurement is affected by the adopted cross-section distance interval applied for calculating the water depth and the hydraulic radius by close-range photogrammetry. Accurate results were also obtained using a distance interval equal to ten times the minimum distance interval (0.00624 m) tested by the authors, having the same order of magnitude as the mean 3D-DTM resolution.

Figure 18. View of point gauges for the measurement of water depth in a rill (modified from Di Stefano et al. [38]).

Figure 19. Survey of the water tracks inside the channel to measure water depth in rills (**a**) and the digital output of the measuring technique (**b**) (modified from Di Stefano et al. [38]).

The influence of different acquisition platforms on the measurement of rill erosion using the SfM technique was investigated by Di Stefano et al. [36]. In particular, these authors measured rill channels produced by a natural intense rainfall event (24 October 2016) that occurred in the Sparacia experimental area using a terrestrial photographic survey (TS) and a low-altitude aerial survey by UAV. Measurements of total and rill erosion were performed on two 6 m $\times$ 22 m plots, named P2 and P3, with steepnesses of 22% and 26%, respectively. For the P2 plot, the point cloud was constructed of all aerial photos and forty terrestrial photographs in order to increase the density of the UAV point cloud and improve its accuracy. The investigation aimed to compare (a) some morphometric parameters of the rill networks surveyed by 3D models by UAV, TS, and UAV + TS photogrammetry and (b) the soil loss measured by these three models with the actual soil loss measured in the sediment storage system of each plot.

The image processing software Agisoft PhotoScan Professional was used to create 3D models, and the total rill channel network was extracted from DEM by combining a drainage algorithm and the convergence index by Thommeret et al. [176]. In particular, the detected rills were automatically divided into contributing rills, which connected to the network reaching the plot outlet and contributed to total soil loss measured at the plot outlet, and non-contributing rills, which were interrupted within the plot. The applied automatic procedure detected rill interruptions where rill deepening was lower than a given threshold, and this condition persisted at least for a rill segment length higher than a threshold value. The three rill networks detected by TS, UAV, and UAV + TS were compared using some morphometric parameters (total length of rills, drainage frequency, and drainage density). The weight of sediment (kg) stored in the tank (rill + interrill erosion), which was assumed as a reference value to evaluate the soil loss measurement error by TS, UAV, and UAV + TS, was compared with the rill component calculated as the product of the total volume V (m^3) of the contributing rills by soil bulk density ρ_s (kg m^{-3}), neglecting the interrill component [169]. The developed analysis of the morphometric parameters pointed out that for both the P3 and P2 plots, the detected length of contributing and non-contributing rill networks was higher in the TS than the UAV measurements, with differences ranging from 4% to 30%. The highest values of drainage frequency (0.74–1.45 m m^{-2}) were also measured by TS, suggesting the better capability of the terrestrial survey for detecting rills, which was consistent with the highest density of the 3D point cloud. The mean length of contributing rills was higher than that of non-contributing rills and did not depend on the data acquisition platform used. The obtained soil loss measurements by TS, UAV, and UAV + TS fell within the 95% confidence interval for the reference soil loss measured by the sampler [31]. Finally, the low measurement errors obtained by all surveys supported the reliability of the methods applied to delineate the contributing rill network and that of the SfM technique to measure rill erosion.

The accuracy of gully measurements is mainly affected by the morphology and size of the channel (Table 1), as an incorrect survey of the width and/or depth of gully cross-sections may cause high errors in the estimate of their volume [177]. In particular, measurements of a gully cross-section can be inaccurate if its shape is narrow and deep [178]. The review of the applications of 3D Photo Reconstruction (3D-PR) in different fields made by Eltner et al. [179] highlighted that some of them have focused specifically on gully/ephemeral gully (EG) erosion. With reference to cost, accuracy, and effectiveness, Castillo et al. [33] supported the superiority of the 3D-PR method in comparison with conventional methods (e.g., terrestrial laser scanner (TLS), profilometer, and total station). These authors also studied how the gully volume error is affected by the ratio between the distance interval of two measurement cross-sections, I, and the gully length, L_c. The variable i/L_c, which is the number of measured cross-sections per unit gully length, is related to the field survey effort, and Castillo et al. [33] obtained a model for calculating this ratio to have a volume error magnitude with a given probability level. Gómez-Gutiérrez et al. [180] verified the applicability of the 3D-PR method for the quantification of headcut gully erosion by TLS as a reference method. Stöcker et al. [181] studied the temporal variability of two gully headcuts by contemporaneously applying UAV and close-range photogrammetry. These last authors obtained an appreciable improvement of the 3D model in comparison with that obtained by the UAV data only. Castillo et al. [33] demonstrated that the 3D-PR is a low-cost and time-saving method when it is applied for surveying a whole gully network. Frankl et al. [172] stated that knowledge of the processes involved in gully erosion needs the use of a 3D model, which is useful for representing the 3D topography, including areas, in which gully cross-sections have undercut walls [172,180], into which multiple elevations can occur for a given couple of plane coordinates x-y. In these cases, the 2.5D model underestimates the area of the gully cross-section and, as a consequence, the eroded soil volume.

Di Stefano et al. [35] investigated the morphological characteristics of an EG, 54 m long and with a drainage basin area of 1.5 ha, formed in the Sparacia experimental area. The EG channel was sectioned in longitudinal segments delimited by two measurement cross-sections, and their lengths I were measured by a metric ruler, while the cross-section was measured by a profilometer (Figure 1). Twenty-four cross-sections with an average distance I = 2.3 m were measured by the profilometer. The 3D model of the EG was created by the SfM technique and Photoscan, using 243 terrestrial photographs, and the DEM with a pixel size of 2.5 mm. The values of morphological (maximum depth, surface width, cross-section area, and cross-section perimeter) and hydraulic (bottom shear stress) variables determined by both the DEM and profilometer were compared using the 3D model as reference. For all the examined variables, the DEM and profilometer gave values close to the reference one as they were characterized by errors falling in the error band of ±10%. For the three different methods (3D, DEM, and profilometer), the relationship between the EG length L_c (measured from the first upstream section to a given measurement section) and the corresponding volume, V, is plotted in Figure 20a. For L_c > 20 m, this figure highlights a light overestimation of the gully volume obtained by the profilometer measurements and a light underestimation of those obtained by the DEM measurements. For DEM, the volume underestimation is probably due to the underestimation of the cross-section area with undercut walls. For the profilometer, this underestimation was probably counterbalanced by the overestimation of the depth in some sections, as the profilometer could have penetrated into the soil of the EG edge. In any case, the profilometer method is characterized by a light overestimation (2.3%) of the total EG volume. Figure 20b shows that the measured L_c-V pairs are always above the curve representing the following relationship:

$$V = 0.0984{L_c}^{1.29} \quad (8)$$

in which V is expressed in m^3 and L_c is expressed in m, obtained by Di Stefano and Ferro [174]. However, the measured pairs L_c-V fall within the data scattering corresponding to the calibration dataset of Equation (8). This result supports the hypothesis that the severity of the EG erosion process can be estimated by the channel length, i.e., neglecting the cross-section area variability. The SfM technique allowed for the measurement of cross-sections by choosing different values of their distance interval, I, and to evaluate the effect of I on the V measurement. Di Stefano et al. [35], using I = 0.25, 0.5, and 1 m, demonstrated that the distance interval did not affect the L_c-V relationship. For the studied cases ($0.25 \le I \le 1$ m, L_c = 54 m), the ratio i/L_c was lower than 0.018. This result, together with those presented by Casalì et al. [168] and Castillo et al. [33], suggested that the influence of the i/L_c ratio on the EG volume error is limited for a large, narrow, and deep gully (average width–depth ratio less than 5 and average cross-section area greater than 0.06 m^2) and $i/L_c \le 0.49$.

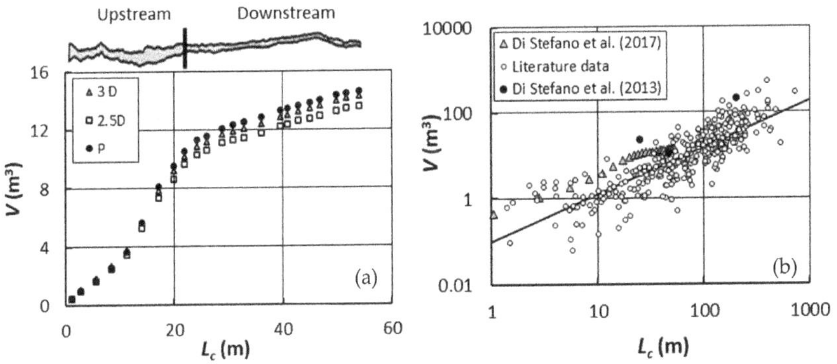

Figure 20. Relationship between the EG length L_c (measured from the first upstream section to a given measurement section) and the corresponding volume V (**a**) and L_c-V pairs obtained by Di Stefano et al. [35] and Di Stefano et al. [169] (**b**) (modified from Di Stefano et al. [35]).

Using estimates of the soil volume bounded by cross-section limits, several studies conducted in recent years have evaluated methods for measuring rill erosion [177,179,182,183]. These methods, which use 2D and 3D approaches, necessitate precisely defining the boundaries of every cross-section of every channel under investigation [184]. Stated differently, there may be a disagreement between the suitability of a volume-based measuring method and the challenges of accurately defining the perimeter of every cross-section and the channel's boundary surface. Currently, two distinct approaches are used: (1) the visual examination of the cross-section profile, and (2) threshold identification for a particular morphological feature. For the second approach, the cross-section limits are localized where there is a sudden change in the slope of the cross-section profile [35,177]. The extraction of the cross-section profile by a DEM is needed to identify a threshold for a given morphological variable, such as slope, plan curvature [178,185,186], or convergence index [176]. Nevertheless, as their applicability is restricted to the investigated rill, the criteria are typically subjective and cannot be used broadly [187]. Castillo et al. [182,183] proposed an automated method based on the calculation of a simple geometrical measurement, which is the bank trapezoid area corresponding to any point of each gully bank. Castillo et al. [182] stated that the point in the bank that corresponds to the largest trapezoid area determines the gully limit for each bank. This automated approach yields results comparable with those obtained by the visual assessment of five scientists experienced in gully erosion assessments. An approach to providing a correct and actual definition of cross-section width was developed by Casalí et al. [177]. This definition can be used to propose a new method for determining the geometry of gully cross-sections, which is based on the availability of both a detailed DEM of the gully whose geometry needs to be determined (DEMpost) and a DEM of the same area (DEMpre) prior to the formation of the investigated gully (not more than 1 year old). This theory relies on the fact that the periodicity (usually one year) of tillage operations influences the formation–obliteration cycle of an ephemeral gully. The gully-eroded area or cross-sections are derived by subtracting the DEMpost from the DEMpre in a vertical plane orthogonal to the channel axis. The width of the gully in a given section is defined by the line that connects the two points where the two DEMs intersect. When the DEM of the eroded area before the gully appeared is not available, the procedure may not work.

In order to identify the actual channel from the hillslope areas, Palmeri et al. [184] suggested a new technique for estimating the actual cross-section depth and determining the cross-section limits in rills and ephemeral gullies. These authors demonstrated that, for the maximum value of the hydraulic radius, the real depth corresponds to a specific bank slope. The ratio of the tested channel's total volume to its boundary wetted surface yielded the hydraulic radius.

5. The Usefulness of Measuring Soil Erosion

Govers et al. [188] suggested that soil erosion is a self-evident process and "*simple visual observations on rills and gullies ....are clear indications that the implementation of conservation measures is necessary*". This consideration could support the idea that a qualitative assessment is sufficient, notwithstanding Govers et al. [188] arguing that "*erosion rates should be reliably quantified in order to assess the necessity for conservation measures*".

Assuming the hypothesis that "*wash erosion*" (i.e., interrill erosion) is negligible as compared to the mass of soil eroded by rills and gullies, Boardman and Evans [189] support the consideration that "*field-scale assessment of erosion is a satisfactory technique*" and it is the best way of assessing erosion. Furthermore, "*monitoring a landscape to assess the severity, frequency and extent of erosion is quick, easy and cheap*". This idea is entirely opposed to using plot experiments to study the processes driving soil erosion and collect data to calibrate and test soil erosion models. Roels [190], discussing the estimation of soil loss at the regional scale based on plot data, correctly stated that "*insufficient attention is given to the sources of bias associated with the use of particular plot sizes or equipment, and to statistics (e.g., sampling procedures, replication, etc.)*". Notwithstanding the listed difficulties and the lack of standardized plots, Roels [190] argued that "*it is possible to collect well-defined and precise measurements"* if the use of statistics in the experimental plot design and in the data treatment is assured. Accurate soil loss estimates at the event scale are required when *"the annual soil loss is dominated by only a few storms per year"*. Farrell and Neff [191] established that when erosion plots are used, reliable estimates of the annual soil loss can be obtained by measurements carried out for 15 or more years. This long measurement period and the consequent costs make the plot approach unattractive for many researchers.

Why is measuring soil erosion important? In the words of Stroosnijder [16], it is important because *"measurements are needed for the design of technology as well as for its evaluation"*. Erosion prediction technology depends on measurements for its development, calibration, and testing. Measurements alone can give empirical evidence of the investigated process and need modeling approaches for its extrapolation in time and space. Beven [192] pointed out that the difficulties in building a hydrological model could be consequent to the unavailability of adequate measuring techniques. Stroosnijder [16] also noticed that data of adequate quality are insufficient as soil erosion experimentation is expensive, new technologies and equipment should be developed to reduce the heaviness of soil loss measurements, and skilled personnel are lacking.

The validity of the measurements depends on the appropriateness of the applied techniques for the research objectives to be achieved [100]. In particular, the spatial scale of measurements has to be related to the hydrologic and geomorphological processes to be investigated. Direct extrapolation from plots to the basin scale must be accurately carried out, as the connectivity of water and sediment fluxes should be considered. The size of the plot is related to the aims of the investigation: large plots are representative of the heterogeneity of the natural system, while wide plots allow for reducing the border effects, and long plots are useful to carry out rill erosion measurements. The size of the plots also affects the number of replicates needed to obtain a representative measurement. Indeed, it should increase as the plot size decreases in order to have similar surface patterns among the different replicates [21].

6. Research Needs and Future Perspectives

Even though several analytical forms of raindrop size distribution have been proposed over the years, the most applied analytical form for meteorological and soil erosion studied is the gamma distribution [47,50,76]. Assouline and Mualem [75], considering the Weibull distribution, suggested a new approach for studying the mechanism of the forces that shape the raindrop size distribution through coalescence and break up processes along their pathway from the cloud to the soil surface. The authors stated that the hypothesis of uniform random raindrop fragmentation led to considering the measured DSD as identical to the Weibull distribution since the probability of a drop breaking up is proportional to its relative volume. Considering the physical meaning of the Weibull distribution, its reliability to reproduce rainfall DSDs should be investigated. Furthermore, a theoretical approach to determining rainfall kinetic power and momentum by Weibull distribution should be developed. Moreover, the drop sizes detected by a disdrometer could be analyzed at the event scale to determine the DSD for the whole rainfall event, to estimate the corresponding parameters of Ulbrich's distribution, and to verify the theoretical expression of P_n and RM at the event scale.

The construction of the prototype of a new device (Figure 4) (https://www.uibm.gov.it/bancadati/Number_search/type_url?type=wpn; Patent Number: 102018000010691 (accessed on 3 September 2024)), could permit the verification of the hypothesis that the rainfall momentum distribution, $M(D)dD$, is the same as the distribution of the electrical signal detected by a specific piezoelectric sensor on which the raindrops impact.

Unit plots are not widespread and this experimental equipment is only rarely available [127–129]. The availability of unit plot data should be increased, as these data allow us to carry out field evaluations of the soil erodibility factor and verify the available mathematical tools such as the nomograph of Wischmeier et al. [130].

Investigations on the ratio of interrill and rill erosion and the weight of the two processes on the plot soil loss should be encouraged. The low measurement errors obtained by terrestrial and aerial surveys encourage the verification of these results in different field conditions. In particular, the reliability of the methods applied to delineate the contributing rill network and that of the SfM technique to measure rill erosion should be verified in rilled surfaces developed in soils with different textures.

Previous results suggested that the severity of the EG erosion process can be estimated by the channel length, i.e., neglecting the cross-section area variability. This result can be used to estimate the contribution of the gully erosion using surveys of areas in which the gullies are localized and their length can be simply measured.

Finally, new field measurements will be beneficial for testing soil erosion models and apply the most accurate ones for developing soil conservation strategies, and for quantifying the soil loss reduction attained with soil conservation practices (i.e., biochar, cover crops, etc.).

7. Conclusions

About twenty years ago, Stroosnijder raised the question of whether the measurement of soil erosion was possible. Although he was certain that "*measurements are needed to develop, calibrate, and validate*" erosion prediction technologies, he believed that the lack of good soil loss data was chronic and attributable to the lack of funds required to realize and maintain field experimental installations, the need for new measurement technologies and equipment, and the lack of skilled personnel.

This question is also relevant to counteract the opinion of other authors that consider soil erosion as a self-evident phenomenon and think that "*simple visual observations on rills and gullies*" are sufficient to assess the need for soil conservation measures. The idea of a field-scale assessment of erosion as a technique sufficient to monitor the phenomenon and establish soil conservation practices strongly contrasts with that of using measurements to calibrate and test soil erosion models useful for supporting soil conservation strategies.

Even if soil erosion experimentation is expensive, in recent years many new measurement technologies and equipment have been developed.

In this paper, a critical review of the most commonly used field measurement techniques and the effect of measurement errors on data quality were presented.

This analysis supported the following main conclusions:

(1) The measurement of rainfall erosivity, which is the climatic forcing of the erosion processes, requires the design of simple devices, with low costs of installation and management, able to detect the raindrop size distribution and the rainfall energy characteristics. The low cost is a requirement that is useful for guaranteeing installation in many sites, which is the prerequisite to spatially distributed information;

(2) The plot scale measurement requires appropriate studies on interpolation, up-scaling, and technologies able to obtain high-quality data with known accuracy;

(3) The effect of systematic errors in soil loss data, due to the assumption that the measured concentration of the suspension collected in the field plot storage system coincides with the actual one, has to be considered in the calibration and testing of soil erosion models;

(4) The natural variability between plots having the same treatment is required for evaluating a model. While a portion of any difference between the measured and predicted soil loss depends on model error, another portion is due to the unexplained deviation of the measured sample value from the representative mean value for a particular treatment; the variability of measured data has to be considered when a deterministic model is tested;

(5) A remote and non-invasive survey of a rilled plot carried out by a terrestrial photographic survey (TS) or a low-altitude aerial survey by UAV are methods suitable for the automatic extraction of the characteristic rill features. Close-range photogrammetry, using terrestrial instrumentation or low-altitude aerial photogrammetry, is characterized by the minimal impact of the survey activity and high spatial resolution;

(6) The rills detected by a remote survey can be separated into contributing rills, which are connected to the network reaching the plot outlet and contribute to total soil loss measured at the plot outlet, and non-contributing rills, which are interrupted within the plot;

(7) The water depth values measured using tracks along the rill are obtained by a volumetric technique that considers the effects of channel irregularities (geometric variability of the rill between two cross-sections) and the interaction between the flow and mobile channel;

(8) The 3D–PR method is a low-cost and time-saving method when it is applied to surveying a whole gully network. The gully erosion needs the use of a 3D model, which can represent the three-dimensional topography, including cross-sections with undercut walls;

(9) The severity of the EG erosion process can be estimated by the channel length.

In conclusion, the main research gaps to overcome are the following:

(1) New devices for the direct measurement of rainfall characteristics;

(2) An increase in the availability of unit plots equipped for the investigation of different soils;

(3) Encouraging the application of new measurement techniques to determine the weight of interrill and rill erosion;

(4) Testing the applicability of the gully channel length as an indicator of erosion process severity;

(5) The use of soil erosion measurement to assess new soil conservation strategies (i.e., biochar, cover crops).

Author Contributions: Conceptualization, A.N., F.G.C., C.D.S., V.P. (Vincenzo Palmeri), V.P. (Vincenzo Pampalone), M.A.S., V.B. and V.F.; writing—original draft preparation, A.N., F.G.C., C.D.S., V. Pal., V.P. (Vincenzo Pampalone), M.A.S., V.B. and V.F.; writing—review and editing, A.N., F.G.C., C.D.S., V.P. (Vincenzo Palmeri), V.P. (Vincenzo Pampalone), M.A.S., V.B. and V.F.; supervision, V.B. and V.F. All authors have read and agreed to the published version of the manuscript.

Funding: This project was funded by the National Recovery and Resilience Plan (NRRP), Mission 4, Component 2, Investment 1.4—Call for tender no. 3138 of 16 December 2021, rectified by Decree no. 3175 of 18 December 2021 of the Italian Ministry of University and Research funded by the European Union—NextGenerationEU Project Code CN_00000033, Concession Decree no. 1034 of 17 June 2022 adopted by the Italian Ministry of University and Research, CUP B73C22000790001, Project Title "National Biodiversity Future Center (NBFC)". This study was funded by the SiciliAn MicronanOTecH Research And Innovation CEnter "SAMOTHRACE" (MUR, PNRR-M4C2, ECS_00000022), spoke 3—Università degli Studi di Palermo "S2-COMMs—Micro and Nanotechnologies for Smart & Sustainable Communities. This study was also carried out within the PRIN project "Soil Conservation for sustainable AgricuLture in the framework of the European green deal" (SCALE) and received funding from the European Union Next-GenerationEU (National Recovery and Resilience Plan—NRPP, M4.C2.1.1., project 2022PB2NSP_002, CUP B53D23017980006).

Data Availability Statement: No new data were created or analyzed in this study. Data sharing is not applicable to this article.

Conflicts of Interest: The authors declare no conflicts of interest.

References

1. European Commission (EC). A Strategy to Keep Europe's Soils Robust and Healthy 2006. Available online: http://ec.europa.eu/environment/soil/index_en.htm (accessed on 3 September 2024).
2. Kendall, H.W.; Pimentel, D. Constraints on the Expansion of the Global Food Supply. *Ambio* **1994**, *23*, 198–205. Available online: http://www.jstor.org/stable/4314199 (accessed on 3 September 2024).
3. Bazzoffi, P. Soil erosion tolerance and water runoff control: Minimum environmental standards. *Reg. Environ. Chang.* **2009**, *9*, 169–179. [CrossRef]
4. Dregne, H.E. Soil erosion: Cause and effect. *Land Use Policy* **1987**, *4*, 412–418. [CrossRef]
5. Lal, R. Soil conservation and ecosystem services. *Int. Soil Water Conserv. Res.* **2014**, *2*, 36–47. [CrossRef]
6. Raclot, D.; Le Bissonnais, Y.; Annabi, M.; Sabir, M.; Smetanova, A. Main issues for preserving Mediterranean soil resources from water erosion under global change. *Land Degrad. Dev.* **2018**, *29*, 789–799. [CrossRef]

7. Carollo, F.G.; Di Stefano, C.; Nicosia, A.; Palmeri, V.; Pampalone, V.; Ferro, V. A new strategy to assure compliance with soil loss tolerance at a regional scale. *Catena* **2023**, *223*, 106945. [CrossRef]
8. Di Stefano, C.; Nicosia, A.; Pampalone, V.; Ferro, V. Soil loss tolerance in the context of the European Green Deal. *Heliyon* **2023**, *9*, e12869. [CrossRef]
9. Kumar, S.; Raj, A.D.; Mariappan, S. Fallout radionuclides (FRNs) for measuring soil erosion in the Himalayan region: A versatile and potent method for steep sloping hilly and mountainous landscapes. *Catena* **2024**, *234*, 107591. [CrossRef]
10. Panagos, P.; Borrelli, P.; Poesen, J.; Ballabio, C.; Lugato, E.; Meusburger, K.; Montanarella, L.; Alewell, C. The new assessment of soil loss by water erosion in Europe. *Environ. Sci. Policy* **2015**, *54*, 438–447. [CrossRef]
11. Borrelli, P.; Robinson, D.A.; Fleischer, L.R.; Lugato, E.; Ballabio, C.; Alewell, C.; Meusburger, K.; Modugno, S.; Schütt, B.; Ferro, V.; et al. An assessment of the global impact of 21st century land use change on soil erosion. *Nat. Commun.* **2017**, *8*, 2013. [CrossRef]
12. Panagos, P.; Standardi, G.; Borrelli, P.; Lugato, E.; Montanarella, L.; Bosello, F. Cost of agricultural productivity loss due to soil erosion in the European Union: From direct cost evaluation approaches to the use of macroeconomic models. *Land Degrad. Dev.* **2018**, *29*, 471–484. [CrossRef]
13. Panagos, P.; Matthews, F.; Patault, E.; De Michele, C.; Quaranta, E.; Bezak, N.; Kaffas, K.; Patro, E.R.; Auel, C.; Schleiss, A.; et al. Understanding the cost of soil erosion: An assessment of the sediment removal costs from the reservoirs of the European Union. *J. Clean. Prod.* **2024**, *434*, 140183. [CrossRef]
14. Wischmeier, W.H.; Smith, D.D. Predicting rainfall-erosion losses from cropland east of the Rocky Mountains: Guide for selection of practices for soil and water conservation. In *Agriculture Handbook*; No. 282; USDA: Washington, DC, USA, 1965.
15. Wischmeier, W.H.; Smith, D.D. Predicting rainfall-erosion losses—A guide to conservation farming. In *Agriculture Handbook*; No. 537; USDA: Washington, DC, USA, 1978.
16. Stroosnijder, L. Measurement of erosion: Is it possible? *Catena* **2005**, *64*, 162–173. [CrossRef]
17. Toy, T.J.; Foster, G.R.; Renard, K.G. *Soil Erosion: Processes, Prediction, Measurement and Control*; Wiley & Sons: Hoboken, NJ, USA, 2002.
18. Bagarello, V.; Ferro, V. Measuring soil loss and subsequent nutrient and organic matter loss on farmland. In *Oxford Research Encyclopedia of Environmental Science*; Oxford University Press: Oxford, UK, 2017. [CrossRef]
19. Bonilla, C.A.; Kroll, D.G.; Norman, J.M.; Yoder, D.C.; Molling, C.C.; Miller, P.S.; Panuska, J.C.; Topel, J.B.; Wakeman, P.L.; Karthikeyan, K.G. Instrumentation for measuring runoff, sediment, and chemical losses from agricultural fields. *J. Environ. Qual.* **2006**, *35*, 216–223. [CrossRef] [PubMed]
20. Raza, A.; Ahrends, H.; Habib-Ur-Rahman, M.; Gaiser, T. Modeling Approaches to Assess Soil Erosion by Water at the Field Scale with Special Emphasis on Heterogeneity of Soils and Crops. *Land* **2021**, *10*, 422. [CrossRef]
21. Boix-Fayos, C.; Martínez-Mena, M.; Calvo-Cases, A.; Arnau-Rosalén, E.; Albaladejo, J.; Castillo, V. Causes and underlying processes of measurement variability in field erosion plots in Mediterranean conditions. *Earth Surf. Proc. Land.* **2007**, *32*, 85–101. [CrossRef]
22. Haigh, M.J. The use of erosion pins in the study of slope evolution. *Tech. Bull.* **1977**, *18*, 31–49.
23. Loughran, R.J. The measurement of soil erosion. *Prog. Phys. Geogr.* **1989**, *13*, 216–233. [CrossRef]
24. Couper, P.; Stott, T.; Maddock, I. Insights into river bank erosion processes derived from analysis of negative erosion-pin recordings: Observations from three recent UK studies. *Earth Surf. Proc. Land.* **2002**, *27*, 59–79. [CrossRef]
25. Hancock, G.R.; Lowry, J.B.C. Hillslope erosion measurement—A simple approach to a complex process. *Hydrol. Process.* **2015**, *29*, 4809–4816. [CrossRef]
26. Hudson, N.W. Field measurement of soil erosion and runoff. In *FAO Soils Bulletin*; Food & Agriculture Organization: Rome, Italy, 1983; p. 139.
27. Lawler, D.M. A new technique for the automatic monitoring of erosion and deposition rates. *Water Resour. Res.* **1991**, *27*, 2125–2128. [CrossRef]
28. Bruno, C.; Di Stefano, C.; Ferro, V. Field investigation on rilling in the experimental Sparacia area, South Italy. *Earth Surf. Proc. Land.* **2008**, *33*, 263–279. [CrossRef]
29. Bagarello, V.; Ferro, V. Chapter 1.4.2. Assessment of soil erosion. In *CIGR Handbook of Agricultural Engineering*; CIGR—The International Commission of Agricultural Engineering, Ed.; American Society of Agricultural Engineers: St. Joseph, MI, USA, 1999; pp. 153–184.
30. Bagarello, V.; Ferro, V. Plot-scale measurement of soil erosion at the experimental area of Sparacia (southern Italy). *Hydrol. Process.* **2004**, *18*, 141–157. [CrossRef]
31. Carollo, F.G.; Di Stefano, C.; Ferro, V.; Pampalone, V. Measuring rill erosion at plot scale by a drone-based technology. *Hydrol. Process.* **2015**, *29*, 3802–3811. [CrossRef]
32. Vinci, A.; Todisco, F.; Brigante, R.; Mannocchi, F.; Radicioni, F. A smartphone camera for the structure from motion reconstruction for measuring soil surface variations and soil loss due to erosion. *Hydrol. Res.* **2017**, *48*, 673–685. [CrossRef]
33. Castillo, C.; Pérez, R.; James, M.R.; Quinton, J.N.; Taguas, E.V.; Gómez, J.A. Comparing the accuracy of several field methods for measuring gully erosion. *Soil Sci. Soc. Am. J.* **2012**, *76*, 1319–1332. [CrossRef]
34. Di Stefano, C.; Ferro, V.; Palmeri, V.; Pampalone, V. Measuring rill erosion using structure from motion: A plot experiment. *Catena* **2017**, *156*, 383–392. [CrossRef]

35. Di Stefano, C.; Ferro, V.; Palmeri, V.; Pampalone, V.; Agnello, F. Testing the use of an image-based technique to measure gully erosion at Sparacia experimental area. *Hydrol. Process.* **2017**, *31*, 573–585. [CrossRef]
36. Di Stefano, C.; Palmeri, V.; Pampalone, V. An automatic approach for rill network extraction to measure rill erosion by terrestrial and low-cost unmanned aerial vehicle photogrammetry. *Hydrol. Process.* **2019**, *33*, 1883–1895. [CrossRef]
37. Ferro, V.; Pampalone, V. Monitoring rill and ephemeral gully erosion by Structure from Motion (SfM) technique. In *Technologies for Integrated River Basin Management*; Edibios: Cosenza, Italy, 2021; pp. 103–110.
38. Di Stefano, C.; Nicosia, A.; Pampalone, V.; Palmeri, V.; Ferro, V. New technique for measuring water depth in rill channels. *Catena* **2019**, *181*, 104090. [CrossRef]
39. Wischmeier, W.H.; Smith, D.D. Rainfall energy and its relationship to soil loss. *Eos Trans. Am. Geophys. Union* **1958**, *39*, 285–291.
40. Nearing, M.A.; Yin, S.; Borrelli, P.; Polyakov, V.O. Rainfall erosivity: A historical review. *Catena* **2017**, *157*, 357–362. [CrossRef]
41. Wang, L.; Li, Y.; Gan, Y.; Zhao, L.; Qin, W.; Ding, L. Rainfall erosivity index for monitoring global soil erosion. *Catena* **2024**, *234*, 107593. [CrossRef]
42. Van Dijk, A.I.J.M.; Bruijnzeel, L.A.; Rosewell, C.J. Rainfall intensity-kinetic energy relationships, a critical literature appraisal-Review. *J. Hydrol.* **2002**, *261*, 1–23. [CrossRef]
43. Carollo, F.; Ferro, V. Modeling rainfall erosivity by measured drop-size distributions. *J. Hydrol. Eng.* **2015**, *20*, C4014006. [CrossRef]
44. Carollo, F.; Ferro, V.; Serio, M.A. Estimating rainfall erosivity by aggregated drop size distributions. *Hydrol. Process.* **2016**, *30*, 2119–2128. [CrossRef]
45. Carollo, F.; Ferro, V.; Serio, M.A. Reliability of rainfall kinetic power-intensity relationships. *Hydrol. Process.* **2017**, *31*, 1293–1300. [CrossRef]
46. Carollo, F.; Ferro, V.; Serio, M.A. Predicting rainfall erosivity by momentum and kinetic energy in Mediterranean environment. *J. Hydrol.* **2018**, *560*, 173–183. [CrossRef]
47. Salles, C.; Poesen, J.; Sempere-Torres, D. Kinetic energy of rain and its functional relationship with intensity. *J. Hydrol.* **2002**, *257*, 256–270. [CrossRef]
48. Parsons, A.J.; Gadian, A.M. Uncertainty in modelling the detachment of soil by rainfall. *Earth Surf. Proc. Land.* **2000**, *25*, 723–728. [CrossRef]
49. Ferro, V. Tecniche di misura e monitoraggio dei processi erosivi. *Quad. Di Idronomia Mont.* **2001**, *21*, 63–128. (In Italian)
50. Serio, M.A.; Carollo, F.G.; Ferro, V. Raindrop size distribution and terminal velocity for rainfall erosivity studies. A review. *J. Hydrol.* **2019**, *576*, 210–228. [CrossRef]
51. Carollo, F.G.; Serio, M.A.; Ferro, V.; Cerdà, A. Characterizing rainfall erosivity by kinetic power-median volume diameter relationship. *Catena* **2018**, *165*, 12–21. [CrossRef]
52. Serio, M.A.; Carollo, F.G.; Ferro, V. A method for evaluating rainfall kinetic power by a characteristic drop diameter. *J. Hydrol.* **2019**, *577*, 123996. [CrossRef]
53. Davies, C.N. Unpublished Ministry of Supply Reports Quoted by Sutton in Air Ministry Report, 1942 M.R.P. 40.
54. Spilhaus, A. Raindrop size and falling speed. *J. Meteorol.* **1948**, *5*, 108–110. [CrossRef]
55. Gunn, R.; Kinzer, G.D. The terminal velocity of fall for water droplets in stagnant air. *J. Meteorol.* **1949**, *6*, 243–248. [CrossRef]
56. Best, A.C. The size distribution of raindrops. *Q. J. R. Meteorol. Soc.* **1950**, *76*, 16–36. [CrossRef]
57. Kessler, E.; Wilks, K.E. Radar measurements of precipitation for hydrologic purposes. Reports on WMO/IHD projects 1968. Report No 5, 46.
58. Lui, J.Y.; Orville, H.D. Numerical modelling of precipitation and shadow effects on mountain-induced cumuli. *J. Atmos. Sci.* **1969**, *26*, 1286.
59. Sekhon, R.S.; Srivastava, R.C. Doppler radar observations of drop-size distributions in a thunderstorm. *J. Atmos. Sci.* **1971**, *28*, 983–994. [CrossRef]
60. Atlas, D.; Ulbrich, C.W. Path- and area-integrated rainfall measurement by microwave attenuation in the 1–3 cm band. *J. Appl. Meteorol.* **1977**, *16*, 1322–1331. [CrossRef]
61. Uplinger, W.G. A new formula for raindrop terminal velocity. In Proceedings of the 20th Conference on Radar Meteorology, Boston, MA, USA, 30 November–3 December 1981; American Meteorological Society: Boston, MA, USA, 1981; pp. 389–391.
62. Leone, A.; Pica, M. Caratteristiche dinamiche e simulazione delle piogge. Parte prima, Fondamenti teorici. *Riv. Di Ing. Agrar.* **1993**, *3*, 167–175.
63. Grosh, R.C. Weighted fall speed parameters. In Proceedings of the 26th Conference on Radar Meteorology, Boston, MA, USA; 1996; pp. 607–610.
64. Beard, K.V. Terminal velocity of cloud and precipitation drops aloft. *J. Atmos. Sci.* **1976**, *33*, 851–864. [CrossRef]
65. Jayawardena, A.W.; Rezaur, R.B. Drop size distribution and kinetic energy load of rainstorms in Hong Kong. *Hydrol. Process.* **2000**, *14*, 1069–1082. [CrossRef]
66. Lenard, P. Uber regen. Meteor. Z. 21, 248–262 for English translation see. *Quart. J. R. Meteorol. Soc.* **1904**, *31*, 62–73.
67. Schmidt, W. Eine unmittelbare Bestimmung der Fallgeschwindigkeit von Regenropfen. *Sitz. Akas. Wiss. Wien. Mathem.-naturw. Klasse.* **1909**, *118*, 71–84.
68. Laws, J.O. Measurements of the fall-velocity of water-drops and raindrops. *Trans. Am. Geophs. Union.* **1941**, *22*, 709–721.
69. Blanchard, D.C. Behavior of water drops at terminal velocity. *Trans. Am. Geophs. Union* **1950**, *31*, 836–842.

70. Epema, G.F.; Riezebos, H.T. Fall velocity of water drops at different heights as a factor influencing erosivity of simulated rain. Rainfall simulation, runoff and soil erosion. *Catena Suppl.* **1983**, *4*, 1–18.
71. Serio, M.A.; Carollo, F.G.; Caruso, R.; Ferro, V. Guidelines for the Energetic Characterization of a Portable Drip-Type Rainfall Simulator for Soil Erosion Research. *Water* **2024**, *16*, 2100. [CrossRef]
72. Assouline, S.; El Idrissi, A.; Persoons, E. Modelling the physical characteristics of simulated rainfall: A comparison with natural rainfall. *J. Hydrol.* **1997**, *196*, 336–347. [CrossRef]
73. Wolfs, J.L. Contribution à L'étude de L'erosion par Simulation de Pluie. Master's Thesis, University Catholique de Louvain-La-Nueve, Louvain-la-Neuve, Belgium, 1981.
74. Serio, M.A. Estimating Rainfall Erosivity by Drop Size Distributions. Ph.D. Thesis, University of Palermo, Palermo, Italy, 2017.
75. Assouline, S.; Mualem, Y. The similarity of regional rainfall: A dimensionless model of drop size distribution. *Trans. ASAE* **1989**, *32*, 1216–1222. [CrossRef]
76. Uijlenhoet, R.; Stricker, J.N.M. A consistent rainfall parameterization based on the exponential raindrop size distribution. *J. Hydrol.* **1999**, *218*, 101–127. [CrossRef]
77. Wiesner, J. Beitragezur Kenntnis des troischenregens (About Contributions to the knowledge of the tropical rain). *Atmos. Electr.* **1895**, *104*, 1397–1434.
78. Bentley, W.A. Studies of raindrops and raindrop phenomena. *Mon. Weather Rev.* **1904**, *32*, 450–456.
79. Laws, J.O.; Parsons, D.A. The relation of raindrop size to intensity. *Trans. Am. Geophys. Union* **1943**, *24*, 452–460.
80. Hudson, N.W. Raindrop size distribution in high intensity rain. *Rhod. J. Agric. Res.* **1963**, *1*, 6–11.
81. Joss, J.; Waldvogel, A. A spectrograph for raindrops with automatic interpretation. *Pure Appl. Geophys.* **1967**, *68*, 240–246. [CrossRef]
82. Carter, C.E.; Greer, J.D.; Braud, H.J.; Floyd, J.M. Raindrop characteristics in South Central United States. *Trans. Am. Soc. Agric. Eng.* **1974**, *17*, 1033–1037. [CrossRef]
83. Eigel, J.D.; Moore, I.D. A simplified technique for measuring raindrop size and distribution. *Trans. Am. Soc. Agric. Eng.* **1983**, *24*, 1079–1083. [CrossRef]
84. Hauser, D.; Amayenc, P.; Nutten, B.; Waldteufe, L.P. A new optical instrument for simultaneous measurement of raindrop diameter and fall speed distributions. *J. Atmos. Ocean. Technol.* **1984**, *1*, 256–269. [CrossRef]
85. Grossklaus, M.; Uhlig, K.; Hasse, L. An optical disdrometer for use in high wind speeds. *J. Atmos. Ocean. Technol.* **1998**, *15*, 1051–1059. [CrossRef]
86. Salles, C.; Poesen, J. Performance of an optical spectro-pluviometer in measuring basic rain erosivity characteristics. *J. Hydrol.* **1999**, *218*, 142–156. [CrossRef]
87. Löffler-Mang, M.; Joss, J. An optical disdrometer for measuring size and velocity of hydrometeors. *J. Atmos. Ocean. Technol.* **2000**, *17*, 130–139. [CrossRef]
88. Herngren, L. Build-Up and Wash-Off Process Kinetics of PAHs and Heavy Metals on Paved Surfaces Using Simulated Rainfall. Ph.D. Thesis, Queensland University of Technology, Brisbane, Australia, 2005.
89. Lanzinger, E.; Theel, M.; Windolph, H. *Rainfall Amount and Intensity Measured by the Thies Laser Precipitation Monitor*; TECO: Geneva, Switzerland, 2006; pp. 4–6.
90. Salvador, R.; Bautista-Capetillo, C.; Burguete, J.; Zapata, N.; Serreta, A.; Playán, E. A photographic method for drop characterization in agricultural sprinklers. *Irrig. Sci.* **2009**, *27*, 307–317. [CrossRef]
91. Parsakhoo, A.; Lotfalian, M.; Kavian, A.; Hoseini, S.A.; Demir, M. Calibration of a portable single nozzle rainfall simulator for soil erodibility study in hyrcanian forests. *Afr. J. Agric. Res.* **2012**, *7*, 3957–3963. [CrossRef]
92. Friedrich, K.; Higgins, S.; Masters, F.J.; Lopez, C.R. Articulating and stationary PARSIVEL disdrometer measurements in conditions with strong winds and heavy rainfall. *J. Atmos. Ocean. Technol.* **2013**, *30*, 2063–2080. [CrossRef]
93. Kathiravelu, G.; Lucke, T.; Nichols, P. Rain drop measurement techniques, a review. *Water* **2016**, *8*, 1–20. [CrossRef]
94. Montero, J.; Tarjuelo, J.M.; Carrión, P. Sprinkler droplet size distribution measured with an optical spectropluviometer. *Irrig. Sci.* **2003**, *22*, 47–56. [CrossRef]
95. Burguete, J.; Playán, E.; Montero, J.; Zapata, N. Improving drop size and velocity estimates of an optical disdrometer: Implications for sprinkler irrigation simulation. *Trans. ASABE* **2007**, *50*, 2103–2116. [CrossRef]
96. Shen, E.; Liu, G.; Abd Elbasit, M.A.; Zhan, X.; Feng, Q.; Dan, C.; Shi, H.; Chen, X.; Zhang, Q.; Guo, Z. Comparison of a laser precipitation monitor, piezoelectric transducer and particle imaging transient visual measurement technology under simulated rainfall in laboratory conditions. *J. Hydrol.* **2022**, *610*, 127978. [CrossRef]
97. Carollo, F.G.; Serio, M.A.; Pampalone, V.; Ferro, V. Advances in a measurement method of rainfall kinetic power and momentum affecting soil erosion processes. *Hydrol. Process.* **2024**, *38*, e15172. [CrossRef]
98. Jayawardena, A.W.; Rezaur, R.B. Measuring drop size distribution and kinetic energy of rainfall using a force transducer. *Hydrol. Process.* **2000**, *14*, 37–49. [CrossRef]
99. Bagarello, V.; Ferro, V. *Erosione e Conservazione del Suolo*; (In Italian). McGraw-Hill: Milano, Italy, 2006; p. 539. ISBN 88-386-6311-4.
100. Boix-Fayos, C.; Martínez-Mena, M.; Arnau-Rosalén, E.; Calvo-Cases, A.; Castillo, V.; Albaladejo, J. Measuring soil erosion by field plots. Understanding the sources of variation. *Earth-Sci. Rev.* **2006**, *78*, 267–285. [CrossRef]
101. Grosh, J.L.; Jarrett, A.R. Interrill erosion and runoff on very steep slopes. *Trans. ASAE* **1994**, *37*, 1127–1133. [CrossRef]

102. Zhang, X.C.; Nearing, M.A.; Miller, W.P.; Norton, L.D.; West, L.T. Modeling interrill sediment delivery. *Soil Sci. Soc. Am. J.* **1998**, *62*, 438–444. [CrossRef]
103. Agassi, M.; Bradford, J.M. Methodologies for interrill erosion studies. *Soil Tillage Res.* **1999**, *49*, 277–287. [CrossRef]
104. Bradford, J.M.; Huang, C. Comparison of interrill soil loss for laboratory and field procedures. *Soil Technol.* **1993**, *6*, 145–156. [CrossRef]
105. Lattanzi, A.R.; Meyer, L.D.; Baumgardner, M.F. Influences of mulch rate and slope steepness on interrill erosion. *Soil Sci. Soc. Am. Proc.* **1974**, *38*, 946–950. [CrossRef]
106. Moldenhauer, W.C.; Long, D.C. Influences of rainfall energy on soil loss and infiltration rates: I. Effect over a range of texture. *Soil Sci. Soc. Am. Proc.* **1964**, *28*, 813–817. [CrossRef]
107. Poesen, J.; Ingelmo-Sanchez, F.; Mucher, H. The hydrological response of soil surfaces to rainfall as affected by cover and position of rock fragments in the top layer. *Earth Surf. Proc. Land.* **1990**, *15*, 653–671. [CrossRef]
108. Bagarello, V.; Ferro, V.; Keestra, S.; Rodrigo Comino, J.; Pulido, M.; Cerdà, A. Testing simple scaling in soil erosion processes at plot scale. *Catena* **2018**, *167*, 171–180. [CrossRef]
109. Chaplot, V.; Le Bissonnais, Y. Field measurement of interrill erosion under different slopes and plot sizes. *Earth Surf. Proc. Land.* **2000**, *25*, 145–153. [CrossRef]
110. Brakensiek, D.O.; Osborn, H.B.; Rawls, W.J. Field manual for research in agricultural hydrology. In *Agriculture Handbook*; No. 224; Department of Agriculture: Washington, DC, USA, 1979.
111. Morgan, R.P.C. *Soil Erosion and Conservation*; Blackwell Publishing: Malden, MA, USA, 2005; p. 299.
112. Ollesch, G.; Vacca, A. Influence of time on measurement results of erosion plot studies. *Soil Tillage Res.* **2002**, *67*, 23–39. [CrossRef]
113. Dunjó, G.; Pardini, G.; Gisbert, M. The role of land use and land cover on runoff generation and sediment yield at a microplot scale, in a small Mediterranean catchment. *J. Arid Environ.* **2004**, *57*, 239–256. [CrossRef]
114. Rejman, J.; Brodowski, R. Rill characteristics and sediment transport as a function of slope length during a storm event on loess soil. *Earth Surf. Process. Landf.* **2005**, *30*, 231–239. [CrossRef]
115. Valentin, C.; Poesen, J.; Li, Y. Gully erosion: Impacts, factors and control. *Catena* **2005**, *63*, 132–153. [CrossRef]
116. Gilley, J.E.; Kottwitz, E.R.; Simanton, J.R. Hydraulic characteristics of rills. *Trans. ASAE* **1990**, *33*, 1900–1906. [CrossRef]
117. Laflen, J.M.; Moldenhauer, W.C. *The USLE Story*; Special Publication: Beijing, China, 2003.
118. Bagarello, V.; Ferro, V. Analysis of soil loss data from plots of differing length for the Sparacia experimental area, Sicily, Italy. *Biosyst. Eng.* **2010**, *105*, 411–422. [CrossRef]
119. Nearing, M.A. Evaluating soil erosion models using measured plot data: Accounting for variability in the data. *Earth Surf. Proc. Land.* **2000**, *25*, 1035–1043. [CrossRef]
120. Risse, L.M.; Nearing, M.A.; Nicks, A.D.; Laflen, J.M. Assessment of error in the universal soil loss equation. *Soil Sci. Soc. Am. J.* **1993**, *57*, 825–833. [CrossRef]
121. Bagarello, V.; Ferro, V.; Giordano, G.; Mannocchi, F.; Pampalone, V.; Todisco, F.; Vergni, L. Effect of plot size on measured soil loss for two Italian experimental sites. *Biosyst. Eng.* **2011**, *108*, 18–27. [CrossRef]
122. Olson, T.C.; Wischmeier, W.H. Soil-erodibility evaluations for soils on the runoff and erosion stations. *Soil Sci. Soc. Am. J.* **1963**, *27*, 590–592. [CrossRef]
123. Meyer, L.D. Evolution of the universal soil loss equation. *J. Soil Water Conserv.* **1984**, *39*, 99–104.
124. Wischmeier, W.H. The USLE: Some reflections. *J. Soil Water Conserv.* **1984**, *39*, 105–107.
125. Carollo, F.G.; Serio, M.A.; Pampalone, V.; Ferro, V. The unit plot of the Universal soil loss equation (USLE): Myth or reality? *J. Hydrol.* **2024**, *632*, 130880. [CrossRef]
126. Laflen, J.M.; Flanagan, D.C. The development of U.S. soil erosion prediction and modeling. *Int. Soil Water Conserv.* **2013**, *1*, 1–11. [CrossRef]
127. Vaezi, A.R.; Sadeghi, S.H.R.; Bahrami, H.A.; Mahdian, M.H. Modeling the USLE K factor for calcareous soils in northwestern Iran. *Geomorphology* **2008**, *97*, 414–423. [CrossRef]
128. Ostovari, Y.; Ghorbani-Dashtaki, S.; Bahrami, H.A.; Naderi, M.; Dematte, J.A.M.; Kerry, R. Modification of the USLE K factor for soil erodibility assessment on calcareous soils in Iran. *Geomorphology* **2016**, *273*, 385–395. [CrossRef]
129. Bagarello, V.; Ferro, V.; Pampalone, V. Measuring the USLE soil erodibility factor in the unit plots of Sparacia (southern Italy) experimental area. In Proceedings of the 2022 IEEE Workshop on Metrology for Agriculture and Forestry (MetroAgriFor), Perugia, Italy, 3–5 November 2022; pp. 271–276. [CrossRef]
130. Wischmeier, W.H.; Johnson, C.B.; Cross, B.V. A soil erodibility nomograph for farmland and construction sites. *J. Soil Water Conserv.* **1971**, *26*, 189–193.
131. Bagarello, V.; Ferro, V. Calibrating storage tanks for soil erosion measurement from plots. *Earth Surf. Proc. Land.* **1998**, *23*, 1151–1170. [CrossRef]
132. Lang, R.D. Accuracy of two sampling methods used to estimate sediment concentration in runoff from soil-loss plots. *Earth Surf. Proc. Land.* **1992**, *17*, 841–844. [CrossRef]
133. Pierson, F.B., Jr.; Van Vactor, S.S.; Blacbum, W.H.; Wood, J.C. Incorporating small scale spatial variability into predictions of hydrologic response on sagebrush rangelands. *Var. Rangel. Water Eros. Process.* **1994**, *38*, 23–34.
134. Porto, P.; Walling, D.E. Validating the use of 137Cs and 210Pbex measurements to estimate rates of soil loss from cultivated land in southern Italy. *J. Environ. Radioact.* **2012**, *106*, 47–57. [CrossRef] [PubMed]

135. Ciesiolka, C.A.A.; Yu, B.; Rose, C.W.; Ghadiri, H.; Lang, D.; Rosewell, C. A simple method for improving soil loss measurement in USLE type experiments. In Proceedings of the ISCO 2004 13th International Soil Conservation Organisation Conference, Brisbane, Australia, 4–8 July 2004.
136. Ciesiolka, C.A.A.; Yu, B.; Rose, C.W.; Ghadiri, H.; Lang, D.; Rosewell, C. Improvement in soil loss estimation in USLE type experiments. *J. Soil Water Conserv.* **2006**, *61*, 223–229.
137. Edwards, K. Runoff and soil loss studies in New South Wales. In *Technical Handbook*; No. 10; Soil Conservation Service of NSW: Sydney, Australia, 1987.
138. Hudson, N. *Soil Conservation*; B.T. Batsford Ltd.: London, UK, 1971.
139. Rosewell, C.J. SOILLOSS: A program to assist in the selection of management practices to reduce soil erosion. In *Technical Handbook*, 2nd ed.; No. 11; Department of Conservation and Land Management: Sydney, Australia, 1993.
140. Argus, I.; Giutings, A.N.; Kurnia, U.; Abdurachman, A.; van der Poel, P. Soil erosion research in Indonesia: Past experience and future directions. In *Soil Erosion in Multiple Scale*; Penning de Vries, E.W.T., Agus, F., Kerry, J., Eds.; CABI Publishing: Wallingfrord, UK, 1998; pp. 255–268.
141. Bagarello, V.; Di Piazza, G.V.; Ferro, V. Manual sampling and tank size effects on the calibration curve of plot sediment storage tanks. *Trans. ASAE* **2004**, *47*, 1105–1112. [CrossRef]
142. Bagarello, V.; Ferro, V.; Giordano, G.; Mannocchi, F.; Pampalone, V.; Todisco, F. A modified applicative criterion of the physical model concept for evaluating plot soil erosion predictions. *Catena* **2015**, *126*, 53–58. [CrossRef]
143. Todisco, F.; Vergni, L.; Mannocchi, F.; Bomba, C. Calibration of the soil loss measurement method at the Masse experimental station. *Catena* **2012**, *91*, 4–9. [CrossRef]
144. Huang, Y.; Zhao, M.; Zhao, H.; Lei, T.; Tang, J.; Wang, W. Correction of measurement errors on sediment concentration sampled by stirring-sampling method from traditional runoff collection tanks. *Int. J. Agric. Biol. Eng.* **2022**, *15*, 134–142. [CrossRef]
145. Nikkami, D. Investigating sampling accuracy to estimate sediment concentrations in erosion plot tanks. *Turk. J. Agric. For.* **2012**, *36*, 583–590. [CrossRef]
146. Carollo, F.G.; Di Stefano, C.; Ferro, V.; Pampalone, V.; Sanzone, F. Testing a new sampler for measuring plot soil loss. *Earth Surf. Proc. Land.* **2016**, *41*, 867–874. [CrossRef]
147. Zobisch, M.A.; Klingspor, P.; Oduor, R.A. The accuracy of manual runoff and sediment sampling from erosion plots. *J. Soil Water Conserv.* **1996**, *51*, 231–233.
148. Nearing, M.A.; Govers, G.; Norton, L.D. Variability in soil erosion data from replicated plots. *Soil Sci. Soc. Am. J.* **1999**, *63*, 1829–1835. [CrossRef]
149. Wendt, R.C.; Alberts, E.E.; Hjelmfelt, A.T., Jr. Variability of runoff and soil loss from fallow experimental plots. *Soil Sci. Soc. Am. J.* **1986**, *50*, 730–736. [CrossRef]
150. Bagarello, V.; Ferro, V. Supporting USLE-MM reliability by analyzing soil loss measurement errors. *Hydrol. Process.* **2017**, *31*, 847–853. [CrossRef]
151. Kinnell, P.I.A. Runoff ratio as a factor in the empirical modeling of soil erosion by individual rainstorms. *Aust. J. Soil Res.* **1997**, *35*, 1–13. [CrossRef]
152. Kinnell, P.I.A.; Risse, L.M. USLE-M: Empirical modeling rainfall erosion through runoff and sediment concentration. *Soil Sci. Soc. Am. J.* **1998**, *62*, 1667–1672. [CrossRef]
153. Bagarello, V.; Di Piazza, G.V.; Ferro, V.; Giordano, G. Predicting unit plot soil loss in Sicily, south Italy. *Hydrol. Process.* **2008**, *22*, 586–595. [CrossRef]
154. Bagarello, V.; Ferro, V.; Giordano, G. Testing alternative erosivity indices to predict event soil loss from bare plots in Southern Italy. *Hydrol. Process.* **2010**, *24*, 789–797. [CrossRef]
155. Bagarello, V.; Ferro, V.; Giordano, G.; Mannocchi, F.; Todisco, F.; Vergni, L. Predicting event soil loss from bare plots at two Italian sites. *Catena* **2013**, *109*, 96–102. [CrossRef]
156. Bagarello, V.; Ferro, V.; Pampalone, V. A new version of the USLE-MM for predicting bare plot soil loss at the Sparacia (South Italy) experimental site. *Hydrol. Process.* **2015**, *29*, 4210–4219. [CrossRef]
157. Foster, G.R.; Lane, L.J. User Requirements: USDA, Water Erosion Prediction Project (WEPP) Draft 6.3. NSERL Report (USA), 1987.
158. Quinton, J.N. The Validation of Physically-Based Erosion Models with Particular Reference to EUROSEM. Ph.D. Thesis, Cranfield University, Bedford, UK, 1994.
159. Renard, K.G.; Foster, G.R.; Weesies, G.A.; McCool, D.K.; Yoder, D.C. Predicting soil erosion by water: A guide to conservation planning with the revised universal soil loss equation (RUSLE). In *Agriculture Handbook*; No. 703; USDA: Washington, DC, USA, 1997.
160. Flanagan, D.C.; Nearing, M.A. *USDA-Water Erosion Prediction Project: Hillslope Profile and Watershed Model Documentation*; NSERL Report No. 10; USDA-ARS National Soil Erosion Research Laboratory: West Lafayette, IN, USA, 1995.
161. Nearing, M.A. Why soil erosion models over-predict small soil losses and under-predict large soil losses. *Catena* **1998**, *32*, 15–22. [CrossRef]
162. Bagarello, V.; Ferro, V. Testing the "physical model concept" by soil loss data measured in Sicily. *Catena* **2012**, *95*, 1–5. [CrossRef]
163. Nearing, M.A. Capabilities and limitations of erosion models and data. In Proceedings of the 13th International Soil Conservation Organization Conference, Brisbane, Australia, 4–8 July 2004; p. 4.

164. Bagarello, V.; Ferro, V.; Pampalone, V. Comparing two applicative criteria of the soil erosion physical model concept. *Vadose Zone J.* **2017**, *16*, 1–10. [CrossRef]
165. Di Stefano, C.; Nicosia, A.; Palmeri, V.; Pampalone, V.; Ferro, V. Rill flow velocity and resistance law: A review. *Earth-Sci. Rev.* **2022**, *231*, 104092. [CrossRef]
166. Govers, G.; Giménez, R.; Van Oost, K. Rill erosion: Exploring the relationship between experiments, modelling and field observations. *Earth-Sci. Rev.* **2007**, *84*, 87–102. [CrossRef]
167. Di Stefano, C.; Nicosia, A.; Palmeri, V.; Pampalone, V.; Ferro, V. Rill flow resistance law under sediment transport. *J. Soil Sediment.* **2022**, *22*, 334–347. [CrossRef]
168. Casalì, J.; Loizu, J.; Campo, M.A.; De Santisteban, L.M.; Àlvarez-Mozos, J. Accuracy of methods for field assessment of rill and ephemeral gully erosion. *Catena* **2006**, *67*, 128–138. [CrossRef]
169. Di Stefano, C.; Ferro, V.; Pampalone, V.; Sanzone, F. Field investigation of rill and ephemeral gully erosion in the Sparacia experimental area, South Italy. *Catena* **2013**, *101*, 226–234. [CrossRef]
170. James, M.R.; Robson, S. Straightforward reconstruction of 3D surfaces and topography with a camera: Accuracy and geosciences applications. *J. Geophys. Res.* **2012**, *117*, 1–17. [CrossRef]
171. Marzolff, I.; Poesen, J. The potential of 3D gully monitoring with GIS using high-resolution aerial photography and a digital photogrammetry system. *Geomorphology* **2009**, *111*, 48–60. [CrossRef]
172. Frankl, A.; Stal, C.; Abraha, A.; Nyssen, J.; Rieke-Zapp, D.; De Wulf, A.; Poesen, J. Detailed recording of gully morphology in 3D through image-based modelling. *Catena* **2015**, *127*, 92–101. [CrossRef]
173. Castillo, C.; Pérez, R.; James, M.R.; Gómez, J.A. SF3M software: 3-D photo-reconstruction for non-expert users and its application to a gully network. *Soil* **2015**, *1*, 583–594. [CrossRef]
174. Di Stefano, C.; Ferro, V. Measurements of rill and gully erosion in Sicily. *Hydrol. Process.* **2011**, *25*, 2221–2227. [CrossRef]
175. Broscoe, A.J. Quantitative Analysis of Longitudinal Stream Profiles of Small Watersheds; Technical Report 1959; Columbia University, Department of Geology, 1959; No. 18.
176. Thommeret, N.; Bailly, J.S.; Puech, C. Extraction of thalweg networks from DTMs: Application to badlands. *Hydrol. Earth Syst. Sci.* **2010**, *14*, 1527–1536. [CrossRef]
177. Casalí, J.; Giménez, R.; Campo-Bescós, M.A. Gully geometry: What are we measuring? *Soil* **2015**, *1*, 509–513. [CrossRef]
178. Daba, S.; Rieger, W.; Strauss, P. Assessment of gully erosion in eastern Ethiopia using photogrammetric technique. *Catena* **2003**, *50*, 273–291. [CrossRef]
179. Eltner, A.; Kaiser, A.; Castillo, C.; Rock, G.; Neugirg, F.; Abellán, A. Image-based surface reconstruction in geomorphometry–merits, limits and developments. *Earth Surf. Dyn.* **2016**, *4*, 359–389. [CrossRef]
180. Gómez-Gutiérrez, A.; Schnabel, S.; Berenguer-Sempere, F.; Lavado-Contador, F.; Rubio-Delgado, J. Using 3D photo-reconstruction methods to estimate gully headcut erosion. *Catena* **2014**, *120*, 91–101. [CrossRef]
181. Stöcker, C.; Eltner, A.; Karrasch, P. Measuring gullies by synergetic application of UAV and close range photogrammetry—A case study from Andalusia, Spain. *Catena* **2015**, *132*, 1–11. [CrossRef]
182. Castillo, C.; Ángel Campo-Bescós, M.; Casalí, J.; Giménez, R. A new automated method for the determination of cross-section limits in ephemeral gullies. In Proceedings of the EGU General Assembly Conference, Vienna, Austria, 23–28 April 2017; p. 5340.
183. Castillo, C.; Marín-Moreno, V.J.; Pérez, R.; Muñoz-Salinas, R.; Taguas, E.V. Accurate automated assessment of gully cross-section geometry using the photogrammetric interface FreeXSapp. *Earth Surf. Proc. Land.* **2018**, *43*, 1726–1736. [CrossRef]
184. Palmeri, V.; Di Stefano, C.; Ferro, V. A Maximizing Hydraulic Radius (MHR) method for defining cross-section limits in rills and ephemeral gullies. *Catena* **2021**, *203*, 105347. [CrossRef]
185. Evans, M.; Lindsay, J. High resolution quantification of gully erosion in upland peatlands at the landscape scale. *Earth Surf. Proc. Land.* **2010**, *35*, 876–886. [CrossRef]
186. Perroy, R.L.; Bookhagen, B.; Asner, G.P.; Chadwick, O.A. Comparison of gully erosion estimates using airborne and ground-based LiDAR on Santa Cruz Island, California. *Geomorphology* **2010**, *118*, 288–300. [CrossRef]
187. Castillo, C.; Campo-Bescós, M.A.; Giménez, R.; Pérez, R.; Casalí, J. The Optimal Lid Method for the objective definition of cross-section limits in ephemeral gullies. *Catena* **2019**, *176*, 381–393. [CrossRef]
188. Govers, G.; Merckx, R.; van Wesemael, B.; Van Oost, K. Soil conservation in the 21st century: Why we need smart agricultural intensification. *Soil* **2017**, *3*, 45–59. [CrossRef]
189. Boardman, J.; Evans, R. The measurement, estimation and monitoring of soil erosion by runoff at the field scale: Challenges and possibilities with particular reference to Britain. *Prog. Phys. Geogr.* **2020**, *44*, 31–49. [CrossRef]
190. Roels, J.M. Estimation of soil loss at a regional scale based on plot measurements—Some critical considerations. *Earth Surf. Proc. Land.* **1985**, *10*, 587–595. [CrossRef]
191. Farrell, D.A.; Neff, E.L. Resolutions and recommendations. In Proceedings of the Workshop on Estimation Erosion and Sediment Yield on Rangelands Tucson, Arizona, 7–9 March 1981; pp. 209–212.
192. Beven, K. How far can we go in distributed hydrological modelling? *Hydrol. Earth Syst. Sci.* **2001**, *5*, 1–12. [CrossRef]

water

MDPI

Article

Integrating UAV-LiDAR and Field Experiments to Survey Soil Erosion Drivers in Citrus Orchards Using an Exploratory Machine Learning Approach

Jesús Rodrigo-Comino [1,2,*], Laura Cambronero-Ruiz [1,2], Lucía Moreno-Cuenca [1,2], Jesús González-Vivar [1], María Teresa González-Moreno [1,2,3] and Víctor Rodríguez-Galiano [4]

[1] Departamento de Análisis Geográfico Regional y Geografía Física, Facultad de Filosofía y Letras, Campus Universitario de Cartuja, University of Granada, 18071 Granada, Spain; lauracambro@ugr.es (L.C.-R.); luciamc1012@ugr.es (L.M.-C.); jesusvivar@ugr.es (J.G.-V.); mteresagm@ugr.es (M.T.G.-M.)
[2] Andalusian Research Institute in Data Science and Computational Intelligence, University of Granada, 18016 Granada, Spain
[3] Medialab UGR—Research Laboratory for Digital Culture and Society, University of Granada, 18016 Granada, Spain
[4] Departamento de Geografía Física y Análisis Geográfico Regional, Universidad de Sevilla, 41004 Seville, Spain; vrgaliano@us.es
* Correspondence: jesusrc@ugr.es

Abstract: Citrus orchards are especially vulnerable owing to low inter-row vegetation cover, and frequent tillage. Here, we combine controlled field experiments with proximal remote sensing–derived geomorphometric variables and machine learning (ML) to identify key factors of erosion in a Mediterranean climate citrus plantation located close to Seville and the National Park of Doñana (Southern Spain) on Gleyic Regosols (clayic, arenic). We conducted rainfall simulations with 30 s sampling, measured infiltration (mini-disc infiltrometer), saturated hydraulic conductivity (Kfs; Guelph permeameter), compaction (penetrologger), and soil respiration (gas analyzer) at multiple points, and derived high resolution morphometric indices from proximal sensing (UAV-LiDAR). Linear models and Random Forests were trained to explain three responses: soil loss, sediment concentration (SC), and runoff. Results show that soil loss is most strongly associated with maximum compaction and Kfs (multiple regression: $R^2 = 0.68$; adjusted $R^2 = 0.52$; $p = 0.063$), while SC increases with surface compaction and exhibits weak relationships with topographic metrics. Runoff decreases with average infiltration, which is related to compaction ($\beta = -4.83 \pm 2.38$; $R^2 = 0.34$; $p = 0.077$). Diagnostic checks indicate centered residuals with mild heteroscedasticity and a few high leverage observations. Random Forests captured part of the variance for soil loss (≈29%) but performed poorly for runoff, consistent with limited sample size and modest nonlinear signal. Morphometric analysis revealed gentle relief but pronounced convergent–divergent patterns that modulate hydrological connectivity. There were strong differences in the experiments conducted close to the trees and in the tractor trails. We conclude that compaction and near surface hydraulic properties are the most influential and measurable controls of erosion at plot scale and the UAV-LiDAR could not give us extra-insights. We highlight that integrating standardized field protocols with proximal morphometrics and ML can be the best method to prioritize a small set of explanatory variables, helping to reduce experimental effort while maintaining explanatory power.

Keywords: soil erosion proximal sensing; geomorphometry; machine learning; citrus orchards

1. Introduction

Soil erosion and land degradation are among the most pressing environmental challenges of the 21st century, particularly in the context of accelerating climate change, intensive land use, and resource depletion [1,2]. Climate projections suggest that the frequency and intensity of extreme rainfall events will increase in many regions, exacerbating soil erosion and the associated loss of soil fertility, water storage capacity, and ecosystem services (IPCC, 2022; https://www.ipcc.ch). Unsustainable agricultural practices, especially those characterized by intensive soil management, further aggravate these processes as demonstrated many years ago [3,4]. However, this has resulted in widespread topsoil depletion, declining crop productivity, and deterioration of water quality [5–7]. Tackling soil erosion is therefore critical to achieving both global food security and sustainable land management, as the degradation of arable land directly threatens the resilience of agricultural systems [8,9].

In perennial cropping systems such as citrus orchards, soil erosion rates are particularly severe due to sloping terrain, frequent tillage, and low ground cover between rows [10–12]. This issue has severe economic and environmental consequences, especially in major citrus-producing regions [13]. For example, studies in China in citrus have reported soil loss rates exceeding tolerable thresholds by up to several times [14,15], while research in Spain has highlighted high sediment yields and off-site impacts in Mediterranean orchards under conventional management [16]. Similarly, in South America, citrus plantations in countries such as Brazil show significant rates of soil and nutrient loss, which compromise soil fertility and water quality downstream [17,18]. These examples illustrate that soil erosion in perennial orchards is a global problem with direct implications for agricultural sustainability and ecosystem services.

Addressing this challenge effectively requires robust and reliable field data. High-quality datasets with sufficient replication, standardized equipment, and reproducible protocols are essential for quantifying soil erosion processes and for calibrating and validating predictive models [19–21]. Experimental approaches such as rainfall simulation experiments, infiltration and permeability measurements, and soil compaction assessments provide crucial insights into the physical mechanisms driving erosion [22,23]. However, these techniques are labor-intensive, time-consuming, and often restricted to small spatial scales, which limits their broader applicability, especially as a nature-based solutions [24,25].

Recent advances in machine learning (ML) offer an opportunity to overcome some of these limitations by detecting patterns and predicting soil erosion risks from complex datasets [26]. While ML applications have become increasingly common in soil science, the integration of proximal remote sensing data such as drones—which allows the derivation of high-resolution geomorphological and morphometric information [27,28]—with field experiments remains rare. Traditionally, methodologies such as rainfall simulation experiments, infiltration and compaction measurements, or soil respiration analysis have been applied independently, without linking them to spatially explicit datasets. The main novelty of our article is the integration of a specific approach at the same time, which enables the identification of the key explanatory factors behind soil erosion from multiple perspectives, thereby reducing the number of experiments and variables required and improving the scalability and cost-effectiveness of monitoring programs. The new results would be related to the identification of the most influential explanatory factors controlling soil loss and sediment dynamics in woody crop systems, thereby improving the design of soil erosion monitoring strategies and reducing the need for extensive experimental campaigns.

In this study, our main novelty and primary objective is to combine controlled field experiments with proximal remote sensing-derived morphometric variables to better understand soil erosion dynamics in citrus orchards. We conducted a series of rainfall simulation experiments, infiltration, permeability, compaction, and soil respiration measurements at multiple sampling points (10), which were then integrated with detailed geomorphometric information derived from proximal sensing (UAV-LiDAR) using machine learning algorithms to detect patterns and relationships.

Our working hypothesis is that integrating standardized field experiments with proximal remote sensing-derived geomorphometric data through machine learning would allow the detection of a small subset of explanatory variables that account for most of the observed variability in soil erosion. If confirmed, this approach could substantially optimize the monitoring process by minimizing the number of field experiments required, improving the spatial extrapolation of results to larger areas, and providing a cost-effective framework for sustainable soil management in citrus orchards worldwide.

2. Materials and Methods

2.1. Study Area Description

The study was conducted in a commercial citrus orchard located in the province of Seville (southern Spain; Figure 1a), in proximity to the Doñana National Park, a region characterized by its Mediterranean agroecosystem and flat topography. The selected site lies on tow inter-row lines with minimal slope (<5%; Figure 1b) and an elevation of approximately 5.4 m above sea level (Figure 1c). The orchard is cultivated with orange trees under irrigated conditions for most of the year due to very poor natural drainage and twice tilled per year using heavy machinery. Soil sampling and field experiments were carried out near the coordinates 209599.5 E, 4119656.5 N (EPSG:25830; UTM Zone 30N). The experimental orchards were planted in April 2010 in the Lower Guadalquivir Valley and comprise four citrus varieties: Navel Powell and Barnfield (both grafted onto Carrizo citrange), Nova (Citrumelo rootstock), and Valencia Midknight (Macrophylla rootstock). All plots share the same planting frame (7 m $\times$ 4 m), tree age (15.7 years), and similar management practices, including drip irrigation, annual tillage, and chemical weed control. This homogeneity ensured consistency across the experimental conditions while allowing for comparison across scion–rootstock combinations.

The soil at the site is classified as a Gleyic Regosol (clayic, arenic) according to the IUSS-WRB [29] classification system (Figure 1d). It developed from a heterogeneous mix of silt, sand, and clay, with surface compaction observable in tractor traffic lanes and lower horizons, though no surface sealing was detected. The surface horizon (A) shows a silty clay loam texture, moderate blocky and prismatic structure, and ferric mottling indicating hydromorphic processes. Subsurface horizons (B and B/C) are compact, with low biological activity and poor aggregation stability (0.18–0.25), high pH values (7.89–7.99), and low to moderate electrical conductivity (421–462 μS/cm). The orchard exhibits signs of surface erosion, particularly between the rows on slight mounds, and features wide desiccation cracks (2–5 cm width, up to 20 cm deep). Rock outcrops are absent, and coarse fragments are scarce (<10%), which, together with the weak aggregation and high compaction, suggest vulnerability to both waterlogging and erosion. The site's accessibility and contrasting microtopographic features made it suitable for the installation of rainfall simulation plots, infiltration tests, and soil physical and biological monitoring. The Bajo Guadalquivir region, located in southwestern Andalusia (Spain), experiences a Mediterranean sub-humid to semi-arid climate, characterized by hot, dry summers and mild, wet winters. According to the Junta de Andalucía's climate classification (https://www.juntadeandalucia.es/medioambiente/portal/areas-tematicas/cambio-climatico-y-clima/clima-en-andalucia/regiones-climaticas-de-andalucia, accessed on 10 December 2025), the area is characterized by mean annual temperatures between 17 °C and 19 °C, with maximum summer temperatures often exceeding 40 °C, especially in July and August. Annual precipitation ranges from 500 to 700 mm, distributed irregularly across the year and concentrated in autumn and winter months (October–March). The prolonged dry season, combined with intense rainfall episodes, increases the vulnerability of the soil to erosion processes. This climatic regime exerts a strong influence on hydrological dynamics and vegetation cover, with important implications for runoff generation, infiltration patterns, and the intensity of erosive processes in cultivated lands such as citrus orchards.

Figure 1. Study area. (**a**): localization in Europe of the study site; (**b**): exact localization of the sampled points (1–10) in the citrus orchard plantation; (**c**): general overview using a UAV; (**d**): soil profile in the study area (Gleyic regosol; clayic, arenic; IUSS-WRB, 2022 [29]).

2.2. *UAV-Based Photogrammetry, Field Experiments and Laboratory Analysis*

Due to the logistical complexity of conducting rainfall simulations and in-situ soil measurements across a working commercial citrus plantation, we opted for a high-replication, low-spatial-extent sampling design. Each sampling point was selected to represent a distinct morphometric unit derived from proximal sensing data, and field measurements were repeated per point using standardized protocols. The average values of these replicates were used to characterize each point (Supplementary Material S1). While this approach limits the number of spatial units (n = 10), it enhances internal consistency and minimizes intra-plot variability because they were carried out the same days. This trade-off is not common in soil erosion studies under real-field conditions because the necessity of joining various trained researchers and specific calibrated devices the same days at the same moment, also where physical constraints restrict the number of simulated rainfall events that can be realistically deployed.

2.2.1. UAV Survey and LiDAR Data Acquisition Workflow

Ground Control Points (GCPs) were strategically placed throughout the site using a high-precision GNSS receiver (Leica Zeno 20; Leica Geosystems, Heerbrug, Switzerland), with centimeter-level accuracy). The UAV platform employed in this study was a DJI Matrice 350 RTK (Shenzhen, China) equipped with a LiDAR sensor. This quadcopter features a Real-Time Kinematic (RTK) positioning system, enabling real-time, high-accuracy georeferencing. The onboard sensor, a DJI Zenmuse L2 (Shenzhen, China), integrates LiDAR technology with a high-resolution RGB camera. The onboard RGB camera delivered 20 MP resolution, enabling point cloud colorization. The mission comprised three flights conducted at mean altitudes of 51.9 m, 56.5 m, and 77.8 m, with average flight speeds between 4.1 and 4.2 m/s. Processing in DJI Terra, set to high point density (100%), produced point clouds with an average density of 1702 points/m^2 across the study area (0.378 km^2), from which a digital elevation model (DEM) was generated at 0.1 m/pixel resolution.

This setup allows for the generation of high-density 3D point clouds, which are crucial for detailed topographic reconstruction, detection of vegetation cover changes, and the production of digital terrain and surface models. The acquired data were processed using DJI Terra software (V.5) and subsequently analyzed and visualized through Geographic Information System (GIS) tools (ArcGIS Pro (https://www.esri.com/en-us/arcgis/products/arcgis-pro/overview, accessed on 10 December 2025), ESRI, Louisville, CO, USA). In the case of LiDAR data, the processed outputs included 3D point clouds and both Digital Surface Models (DSM) and Digital Terrain Models (DTM). For DSM and DTM products derived from LiDAR, a Canopy Height Model (CHM) was computed by subtracting the DTM from the DSM. This allowed estimation of vegetation height at each sampling location. However, since the sampling points were deliberately placed in bare inter-row areas free of canopy cover, the CHM was only used as a verification layer and not included in the predictive analysis. Vegetation density was not computed from LiDAR data; field observations and imagery were used to confirm the absence of vegetative obstruction. Raster values for all morphometric indices were extracted using a 3 × 3 pixel window (equivalent to ~1.8 m^2), centered on each sampling point. This window size balances spatial precision with the need to reduce noise from microtopographic variability and positional uncertainty between UAV-derived rasters and GNSS-located field samples.

A total of 12 morphometric and topographic variables were derived from the high-resolution LiDAR-based Digital Elevation Model (DEM) using SAGA GIS (SAGA User Group/Hamburg University of Technology; Germany). The parameters include both average values and their corresponding standard deviations to capture the spatial heterogeneity within the sampling window. Specifically, slope characteristics were represented by Slope_mean and Slope_std, while curvature metrics included both Plan_Curvature and Profile_Curvature, providing insights into flow direction and profile convexity, respectively. Hydrological and catchment properties were described using Channel_Network_Base_Level, Channel_Network_Distance, Total_Catchment_Area, and the Topographic Wetness Index (TWI), all expressed as both mean and standard deviation values. Additional descriptors such as Aspect, Hillshade, LS Factor (from the RUSLE model), CIT (connectivity Index; adapted from [30]), and Convergence Index were also extracted to reflect solar radiation, erosive potential, flow accumulation, and network initiation thresholds. The inclusion of standard deviation values enhances the dataset's capability to represent microtopographic heterogeneity, a critical factor in runoff generation and erosion processes. Finally, to extract local terrain attributes around each sampled point, a Python-based workflow was implemented using GeoPandas, Rasterio, and NumPy. A shapefile containing GPS-located citrus sampling points was reprojected to match the coordinate reference system (EPSG:25829) of the SAGA-derived raster layers. For each raster, mean and standard deviation values were extracted within a 3×3 pixel window centered on each point, effectively capturing microtopographic variability. The resulting values were compiled into a structured table and exported as a CSV file for subsequent statistical and machine learning analyses.

2.2.2. Rainfall Simulation Experiments

A total of ten rainfall simulation experiments were conducted to evaluate the response of the soil to controlled rainfall conditions, maintaining consistent parameters such as intensity, droplet size, water volume applied per unit area, and duration [31]. For this purpose, we used the mini-portable rainfall simulator developed by Eijkelkamp (The Netherlands), which applies artificial rainfall over a confined plot of 0.0625 m^2 (Figure 2a). This device simulates rainfall through a pressurized sprinkler system, enabling the collection of quantitative data on runoff production, infiltration, and sediment detachment [32]. Although limited in spatial scale, the simulator is highly practical for field use due to its low water consumption, short setup time, and minimal personnel requirements [33]. It is particularly well-suited for estimating soil erodibility under standardized conditions, following methodologies such as those proposed by [34]. During each test, an identical volume of water was applied to each plot, and runoff and eroded sediment were collected through an outlet positioned at the base of the simulator. The simulated rainfall had a drop size of approximately 5.9 mm and was applied for 360 s at a steady rate of 6 $mm \cdot min^{-1}$. The equipment consists of three main components: a calibrated sprinkler head containing 49 capillary tubes to ensure uniform rainfall distribution; an adjustable aluminum stand to control drop height (maintained at 35–40 cm in our case); and a base frame to stabilize the system and direct water flow toward the outlet without lateral losses. Prior to each simulation, site-specific surface characteristics were recorded to assess their influence on runoff activation and erodibility [35,36]. These included: (i) vegetation cover (%), which was <5%, estimated through grid analysis of overhead photographs following (ii) rock fragment cover (%), also <5%, assessed using the same method; (iii) surface roughness, measured in both directions using the chain method described by [37]; and (iv) slope angle, determined with the LiDAR drone. During the simulation, the same operator—ensuring consistency and avoiding observer bias—recorded key hydrological response times: (i) Time to Ponding (Tp), indicating when the entire plot surface became visibly saturated; (ii) Time to Outlet (To), marking the appearance of the first drop at the outlet; and (iii) Time to Runoff Initiation (Tr), corresponding to the start of continuous surface flow [38]. After each experiment, the volume of runoff (liters) and sediment yield (grams) were determined in the laboratory. Runoff samples were evaporated at 110 °C to isolate solid content, and sediment concentration ($g \cdot L^{-1}$) was calculated by dividing the mass of solids by the volume of water collected.

2.2.3. Infiltration and Permeability Experiments

To evaluate soil permeability and infiltration capacity at the pedon scale (parallelly), two devices were employed at selected locations within the catchment area: a mini-disc infiltrometer (Figure 2b) and Guelph Permeameter (GP; Figure 2c). Simultaneously with the rainfall simulation experiments, infiltration measurements were conducted using a mini-disc infiltrometer [19,38]. A total of 20 measurements were performed, with 10 repetitions carried out on each side of the simulator to capture the spatial variability of infiltration near the runoff plots. These measurements provided valuable data on infiltration rates and near-surface hydraulic conductivity under field conditions. The mini-disc infiltrometer consists of a transparent acrylic reservoir connected to a semi-permeable disk at its base and a vertical suction regulation tube. For each measurement, the reservoir was filled with water, the suction was set to 4.5 cm, and the decrease in water volume was recorded using a stopwatch at fixed time intervals (10 s, 30 s, 1 min, 2 min, and 5 min). This approach allowed for a consistent assessment of infiltration behavior across the experimental area. A total of 10 permeability tests were conducted following the methodology outlined by [39]. Each test lasted approximately 30 min, with water level readings taken every 2 min to calculate infiltration dynamics over time. The final infiltration rate (cm·min^{-1}) was determined by averaging the values recorded once a quasi-steady flow had been reached [40,41]. Based on the GP measurements, both the infiltration flux and the saturated hydraulic conductivity (Kfs) were computed using the equations proposed by [42], which have been widely applied in field hydrology studies.

Figure 2. Methods employed in the field experiments and measurements. (**a**): Rainfall simulations; (**b**): Infiltration (mini-disc); (**c**): Guelph permeameter; (**d**): Soil compaction; (**e**): CO_2 analysis; (**f**): soil profile description.

2.2.4. Soil Compaction and CO_2

In parallel, soil respiration was measured using an EGM-5 infrared gas analyzer (PP Systems, Amesbury, MA, USA; Figure 2d), which detects CO_2 emissions, a proxy for microbial activity and soil biological function [43,44]. The system uses a non-dispersive infrared (NDIR) sensor operating at 4.26 μm to selectively quantify CO_2 concentrations, reducing cross-sensitivity to other gases. Measurements were conducted with a closed-loop system connecting the analyzer to a soil respiration chamber (SRC, 1171 mL) that covered an area of 78 cm^2. CO_2 fluxes were measured over 60 s, generating both linear and quadratic models of gas accumulation. Each measurement was replicated three times at different positions within the same plot, resulting in 20 individual readings, all recorded in ppm and stored digitally via USB for subsequent processing and analysis.

Soil penetration resistance was assessed using a Penetrologger (Eijkelkamp, Zevenaar, The Netherlands) equipped with a medium-sized cone featuring a 60° apex angle, following standard field protocols [45–47]. Measurements were conducted within a 2.5 m^2 plot established adjacent to the rainfall simulation area to ensure consistency in surface conditions. Penetration depth and resistance were recorded systematically at 50 cm intervals along a regular grid (Figure 2e), allowing for high-resolution spatial analysis of compaction patterns and subsoil structure. The device automatically logged depth-resistance profiles, facilitating rapid data acquisition and minimizing operator-induced variability.

The dataset was designed from the monitored citrus field with variables derived from UAV-based photogrammetry, field surveys, and laboratory analysis. The dataset includes hydrological indicators (runoff, soil loss, sediment concentration), soil properties (e.g., compaction, infiltration rate), and morphometric terrain attributes (e.g., slope, aspect, curvature, LS factor). All preprocessing and analyses were performed in 4.4.0 (R Core Team, 2024) and RStudio 2025.05.1 ("Mariposa Orchid"). All data manipulation steps were conducted using the tidyverse (data manipulation and visualization), janitor (data cleaning), corrplot (correlation matrix visualization), ggplot2 (visualization), randomForest (machine learning) and caret (model training and validation support). Variables were standardized and renamed using clean_names() for consistency. Only numeric variables were retained for correlation and modeling purposes.

2.3. *Dataset Description, Preprocessing, Analysis and Model Construction*

2.3.1. Correlation Analysis and Exploratory Visualization

To reduce dimensionality and assess collinearity, Pearson correlation coefficients were calculated [48,49] using the cor() function with pairwise complete observations. A threshold of $|r| > 0.7$ was used to select explanatory variables relevant to the target variable (e.g., soil_loss, runoff, or sc). The results were visualized with corrplot. Although Pearson correlation was initially used to filter highly collinear variables ($|r| > 0.7$), the primary objective was not to exclude non-linear associations but to simplify the model due to the low sample size. We recognize that alternative methods like VIF or PCA could also be employed in future work to retain variables with non-linear explanatory power. Scatterplot matrices (pairwise plots) were created to explore linear and non-linear trends using the base R function pairs(). Additionally, simple linear relationships were visualized using ggplot2 with geom_point() and geom_smooth (method = "lm").

2.3.2. Model Construction

Random Forest models were implemented using the randomForest package in R with 500 trees (ntree = 500). The number of variables randomly sampled at each split (mtry) was set to 1, a conservative choice aimed at reducing the risk of overfitting given the limited dataset (n = 10). We did not constrain tree depth or minimum node size, as the primary objective was to explore variable importance rather than optimize predictive accuracy. Variable importance was assessed using two standard metrics: the percentage increase in mean squared error (%IncMSE) and the increase in node purity (IncNodePurity). To validate model performance, we randomly split the dataset into a training set (80%) and a hold-out test set (20%) using the caret::createDataPartition() function. All training, feature selection, and internal validation (via 5-fold cross-validation) were conducted on the training subset. Model performance was then evaluated on the independent test set using Root Mean Square Error (RMSE). Given the small sample size, results are interpreted heuristically to identify potentially relevant explanatory variables rather than for predictive generalization.

Complementary to Random Forest models, multiple linear regression (MLR) models were fitted using the lm() function in R. Diagnostic plots (e.g., residuals vs. fitted, Q-Q plots, scale-location, and leverage) were used to evaluate model assumptions. Both MLR and RF approaches were applied to the three erosion-related response variables (soil loss, sediment concentration, and runoff), and their performance was compared using R^2 and RMSE values.

The methodological workflow implemented in this study is summarized in the following flowchart (Figure 3), which integrates UAV-based data acquisition, field experiments, laboratory analyses, and statistical modeling.

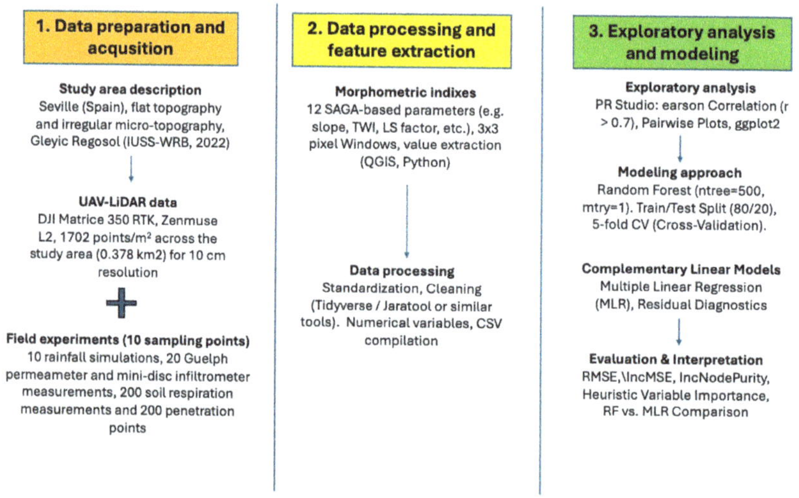

Figure 3. Methodological workflow followed in this study [29].

3. Results and Discussion

3.1. Descriptive Statistics of the In-Situ Experiments and Measurements, and UAV-Survey

The analysis of infiltration dynamics using the mini-disc infiltrometer revealed relatively homogeneous surface conditions across the citrus orchard, which coincided with the study conducted by [50] in the conventional citrus plantation. Average infiltration rates, based on ten replicated measurements, ranged from approximately 1.4 to 3.1 mm·min^{-1}, values typical of moderately compacted soils with low to medium permeability [51,52]. The boxplot shows low dispersion and a nearly symmetric distribution, supporting the presence of uniform infiltration patterns. Although a few individual points were recorded above the whiskers, these likely represent minor field-scale variability rather than systemic heterogeneity (Figure 4a).

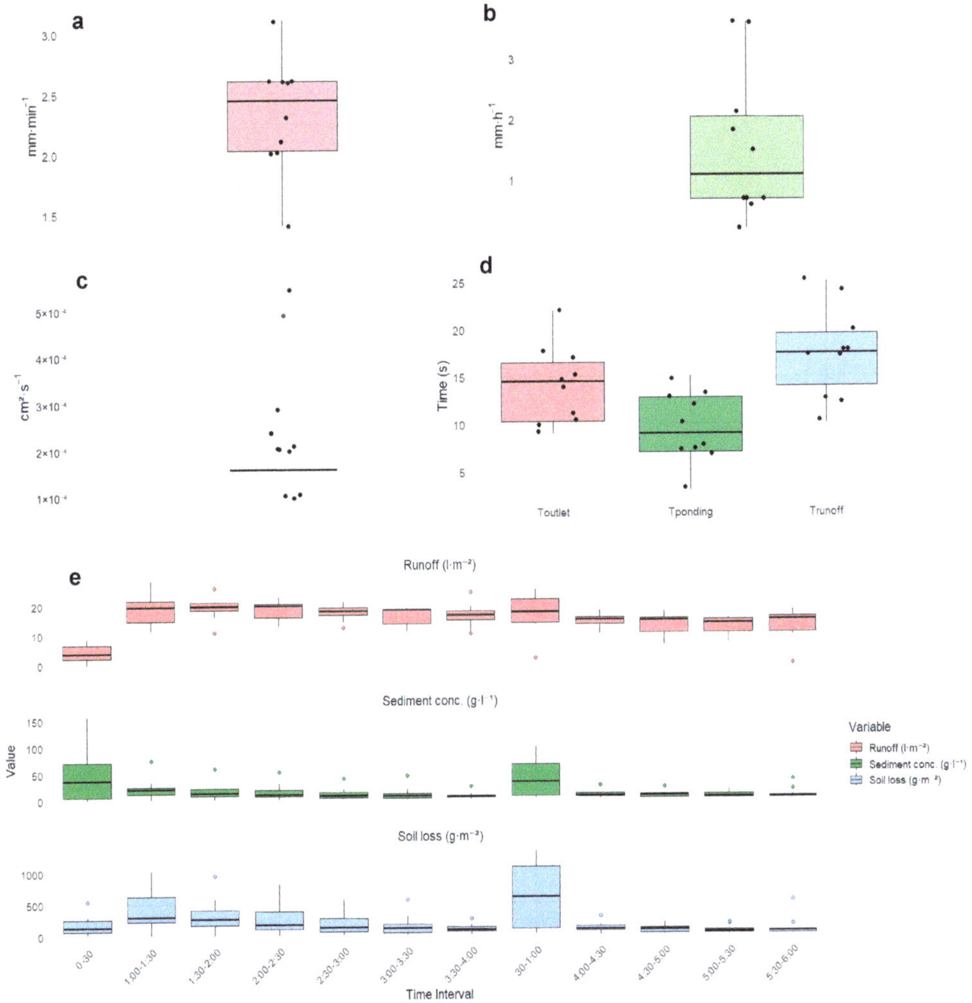

Figure 4. Hydrological response of the experiments and measurements. (**a**): average soil infiltration rates (mini-disc infiltrometer); (**b**): average soil saturated hydraulic conductivity (Guelph permeameter); (**c**): average soil flux matric potential; (**d**): Toutlet (time to outlet), Tponding (time to ponding) and Trunoff (time to runoff generation); (**e**): runoff, sediment concentration and soil loss per interval using a rainfall simulator.

In contrast, the saturated hydraulic conductivity (Kfs) values, obtained with the Guelph permeameter (Figure 4b), exhibited substantial variability among samples, spanning from <2.0 mm·h^{-1} to >3.5 mm·h^{-1}. The distribution was strongly skewed by several high outliers, pointing to possible structural differences in the subsoil and heterogeneous pore connectivity, likely influenced by localized macropore flow, coinciding with other crops such as olive orchards and vineyards [53,54]. This variability underscores the complexity of water movement in structured agricultural soils, where preferential pathways can strongly alter infiltration rates [55,56]. Similarly, matric flux potential (MFP) values remained generally stable across the orchard, clustering tightly around 1.59×10^{-4} $cm^2 \cdot min^{-1}$, except for a single elevated observation (Figure 4c). This consistency suggests relatively uniform matric suction and water release potential under unsaturated conditions.

The rainfall simulation experiments provided further insight into the initial hydrological response of the soils. The temporal parameters—Time to Ponding (Tp), Time to Outlet Flow (To), and Time to Runoff (Tr); Figure 4d—captured distinct phases in water movement. Ponding occurred rapidly (3–15 s) and with minimal variability, reflecting limited storage capacity at the surface. In contrast, outlet and runoff initiation times showed greater dispersion (9–22 s and 10–25 s, respectively), suggesting microtopographic differences and variability in flow connectivity across the plots. The narrow ranges observed for these parameters are consistent with the uniform surface texture inferred from infiltration measurements [57,58]. The temporal evolution of runoff ($L \cdot m^{-2}$), sediment yield ($g \cdot m^{-2}$), and sediment concentration ($g \cdot L^{-1}$) revealed a consistent erosive pattern across replicates (Figure 4e). Runoff increased sharply in the first minutes, peaking between 1:00 and 2:30 min, before gradually stabilizing. Sediment yield followed a similar pattern, with maximum detachment and transport during early stages, when soil cohesion was lowest and surface sealing most active. Sediment concentration peaked in the initial interval (0–30 s), consistent with splash erosion dominance, before dilution effects reduced concentrations [36,59]. These findings collectively highlight the effectiveness of the rainfall simulator in reproducing early-stage erosive dynamics [60,61].

In Figure 5, soil respiration and compaction results are depicted in a different heatmaps. CO_2 measurements peak in plots 6 and 9, suggesting hotspots of microbial activity likely driven by favorable moisture or organic matter conditions [62,63]. However, the temperature at plot 6 appears anomalously high or incorrectly recorded (gray bar), which warrants further investigation. Soil compaction is highest in the center column, particularly from plots 2 to 10, indicating consistent mechanical pressure likely from repeated machinery traffic [64,65]. Interestingly, despite elevated compaction, high CO_2 levels still appear, implying that microbial activity may persist even under physical soil stress, or that surface conditions remain aerated [66]. These observations support the need for integrated monitoring, as respiration hotspots may not always align with assumptions based solely on compaction or temperature.

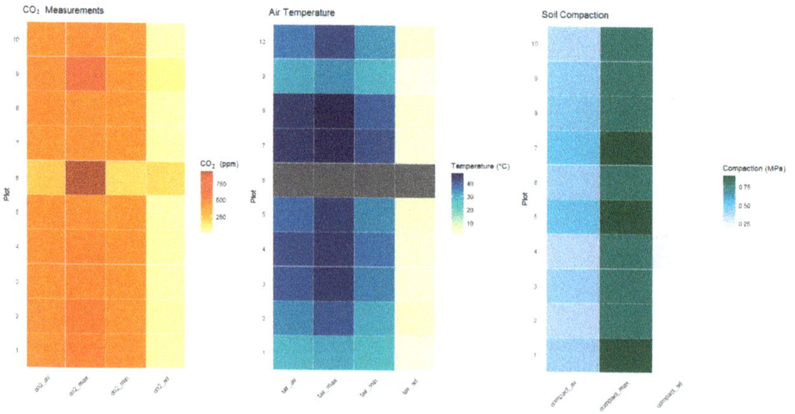

Figure 5. Gas analysis and air temperature, and soil compaction. av: average; max: maximum; min: minimum; sd: standard deviation.

Finally, the morphometric characterization using SAGA GIS (Table 1) depicts a landscape of gentle relief but marked spatial heterogeneity. While the orchard's topography was predominantly flat, high-resolution LiDAR enabled the detection of microtopographic features such as mounds, furrows, and subtle drainage paths [16,67]. These features, though minor in gradient, significantly influenced local hydrological behavior. Morphometric indices were derived from the DTM after vegetation filtering using standard ground classification algorithms to eliminate canopy influence. Slopes were generally low (0.1–0.2) and hillshade values consistent (~0.8), yet several indices indicate pronounced contrasts. Channel Network Base Level (1.8–4.5) and Distance (0.1–0.6) varied little, pointing to stable relative incision. By contrast, the Convergence Index (−13.3 to +13.7) and CIT Index (0.0–5.5) emphasize the alternation between convergent hollows and divergent ridges, which directly influences runoff concentration and sediment redistribution. LS Factor (0.6–3.3) and TWI (2.6–6.0) variability further suggest differences in erosion potential and wetness conditions, while Total Catchment Area showed the highest variability (1.4–427.7), reflecting sharp contrasts in upslope contributing area. The combination of these metrics indicates a terrain with alternating convergent–divergent patterns, variable hydrological accumulation, and localized zones more susceptible to erosion and moisture accumulation [68–70].

Table 1. Basic morphometric statistics derived from LiDAR-based terrain analysis in SAGA GIS. Values correspond to the mean and standard deviation of a 3 × 3 pixel window centered on each sampling point. The indices include: (a) CNBL—Channel Network Base Level (m); (b) CND—Channel Network Distance (m); (c) Convergence Index (unitless); (d) Catchment Area (m^2); and (e) TWI—Topographic Wetness Index (unitless, log-transformed ratio of upslope area and slope).

Aspect	CNBL	CND	CIT Index	Conv. Index	Hillshade	LSFactor	Plan Curv.	Prof. Curv.	Slope	Catch. Area	TWI (Mean)
2.1 ± 1.4	4.5	0.3 ± 0.1	0.1 ± 0	−2.6 ± 14.6	0.8	0.8 ± 0.4	0.0	0.0	0.1	6.6 ± 5.9	3.4 ± 1.5
2.2 ± 1.3	3.4	0.5 ± 0.1	1.7 ± 2.3	−3.3 ± 22.3	0.8	2.4 ± 2.2	0.0	0.0	0.1	129.8 ± 156.6	4.8 ± 2.6
3.0 ± 2.0	2.6	0.6 ± 0.1	3.6 ± 6.7	−13.3 ± 12.7	0.8	3.1 ± 3.9	0.0	0.0	0.1	246.7 ± 345.1	5.0 ± 2.8
2.6 ± 1.8	2.2	0.3 ± 0.1	3.4 ± 5.9	−8.3 ± 6.6	0.8	2.5 ± 3.0	0.0	0.0	0.1	344.5 ± 483.9	5.0 ± 3.0
2.1 ± 1.3	1.9	0.1 ± 0.1	5.5 ± 9.9	−6.1 ± 8.6	0.8	3.3 ± 4.1	0.0	0.0	0.1	427.7 ± 601.9	4.7 ± 3.2
2.5 ± 1.7	4.4	0.3 ± 0.1	0.2 ± 0.2	−11.7 ± 13.7	0.8	1.5 ± 1.0	0.0	0.0	0.1	8.4 ± 6.9	3.8 ± 1.5
2.2 ± 1.8	3.4	0.5 ± 0.1	1.6 ± 2.7	−9.6 ± 10.1	0.8	2.6 ± 2.2	0.0	0.0	0.2	108.5 ± 150.7	4.0 ± 2.8
3.6±	2.7	0.6 ± 0.1	0.4 ± 1.0	−7.4 ± 27.7	0.8	0.9 ± 1.1	0.0	0.0	0.1	183.6 ± 260.6	6.0 ± 3.1
2.6±	2.1	0.6 ± 0	0.0 ± 0	13.7 ± 9.0	0.8	0.6 ± 0.5	0.0	0.0	0.1	1.4 ± 0.3	2.6 ± 0.8
2.2±	1.8	0.2 ± 0.1	3.7 ± 9.1	1.8 ± 7.6	0.8	2.5 ± 3.5	0.0	0.0	0.1	290.5 ± 539.6	4.0 ± 3.0

3.2. Correlation Matrix to Define Key Related Hydro-Pedological Variables

Figure 6 shows the Pearson correlation coefficients among all numerical predictors and the target variable (soil loss), allowing visual identification of multicollinearity and key pairwise relationships between measured variables and the target parameters of soil loss and runoff. The results indicated that strong correlations ($r > 0.6$) were found primarily with soil loss, while no substantial associations were observed with runoff. Maximum soil compaction also exhibited a strong positive correlation ($r = 0.71$), suggesting that surface consolidation contributes significantly to increased erosion potential [71]. Similarly, matric flux potential ($r = 0.69$) and saturated hydraulic conductivity (Kfs) ($r = 0.65$) were positively associated with soil loss, indicating that hydraulic properties play a critical role in modulating infiltration-excess runoff and sediment yield [54]. Interestingly, a negative correlation was found between the standard deviation of air temperature (Tair_SD) and soil loss ($r = -0.67$), potentially reflecting the influence of temperature variability on soil moisture dynamics and vegetation cover, although this is an idea to be explored in the future. These findings highlight the combined effects of surface conditions, hydraulic behavior, and microclimatic variability on erosion responses in citrus orchards, reinforcing the importance of integrating multi-scale field and terrain data in soil degradation assessments [72–74].

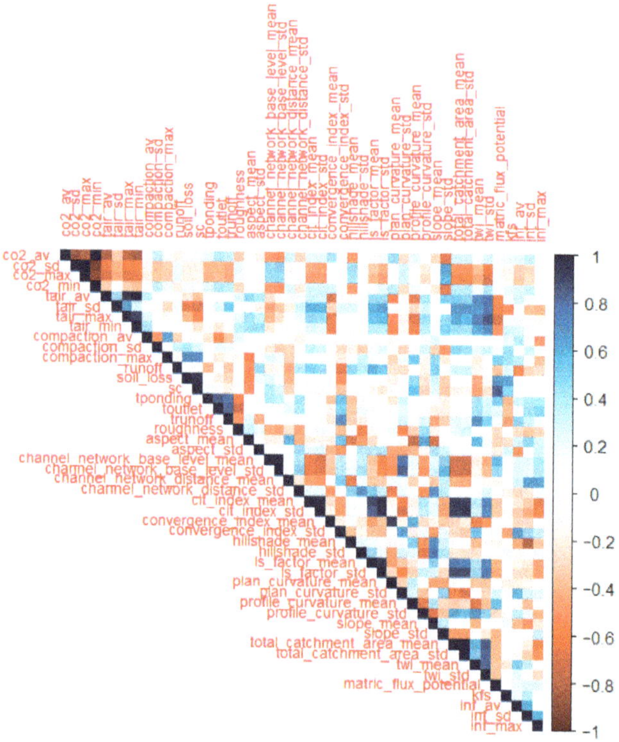

Figure 6. Correlation matrix using Pearson linear relationships. av: average; max: maximum; min: minimum; sd: standard deviation; co2: gas analysis of CO_2; tair: air temperature; sc: sediment concentration; Toutlet (time to outlet), Tponding (time to ponding) and Trunoff (time to runoff generation); cit_index: connectivity index; TWI: Topographic Wetness Index; ls_factor: LS factor (USLE); kfs: soil saturated hydraulic conductivity.

3.3. Prediction of Key Hydro-Pedological Variables and Processes

As a starting point for the results related to the modeling approach (Figure 7), we jointly address the three hydrological responses: soil loss, sediment concentration (SC), and runoff.

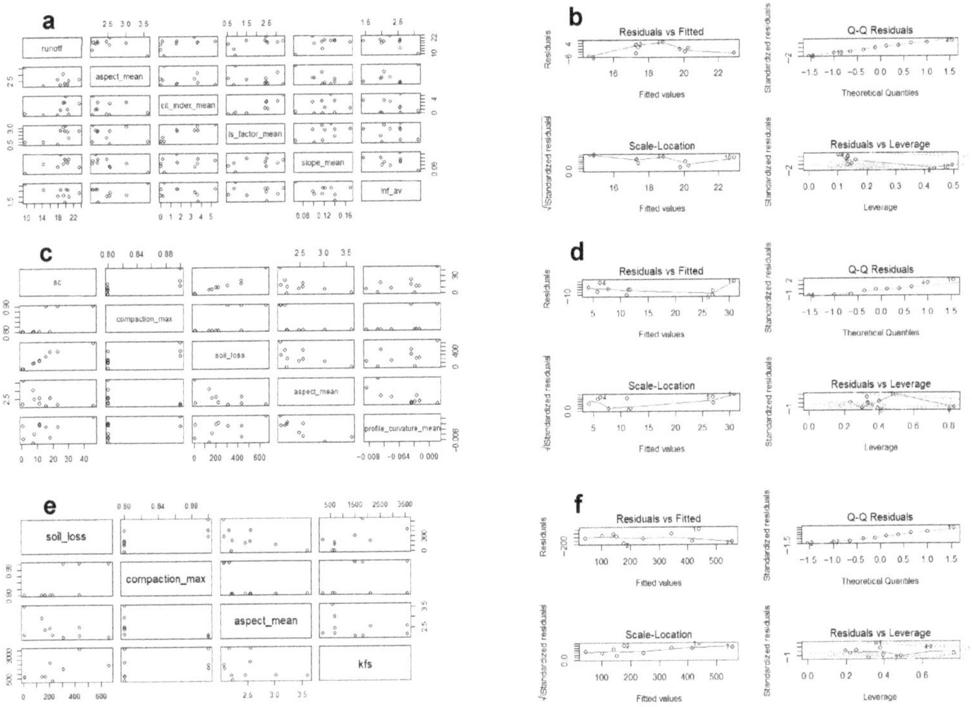

Figure 7. Pairwise scatterplot matrix of runoff, soil loss, and key terrain variables. Each subplot shows the relationship between two variables with fitted trendlines and dispersion patterns. Values are standardized. This matrix helps assess multicollinearity and co-variability between predictors and erosion responses. Pairwise (**a**,**c**,**e**) and residual (**b**,**d**,**f**) graphs show the main results obtained after applying the random forest machine learning technique.

Table 2 summarizes the model performance metrics obtained using the 20% test data. The performance of the soil loss and SC models were moderate (R^2 = 0.28–0.29), while the runoff model performed poorly. Moreover, to assess model performance beyond internal fit, we implemented a hold-out validation approach by splitting the dataset into a training set (80%) and a test set (20%), repeated for each target variable (soil loss, runoff, and sediment concentration). In the training phase, Random Forest models achieved moderate to strong fits, with R^2 values ranging from 0.68 to 0.75 depending on the variable. However, performance on the test set was noticeably lower (R^2 between 0.12 and 0.30), reflecting the expected instability due to the limited sample size (n = 10). Root Mean Square Error (RMSE) values on the test data confirmed this discrepancy, showing increased prediction error relative to the training phase. These results underscore the exploratory nature of our analysis: the machine learning models were not intended to produce robust predictive outputs but to help identify the most influential explanatory variables under the specific conditions of a citrus orchard. The limited generalization is thus an inherent feature of the dataset's structure and sampling scope.

Table 2. Summary of the prediction performance for each model (including RMSE and correlation coefficients between predicted and observed values in the test folds).

Model	Dataset	R^2	RMSE	MAE	Notes
RF (soil_loss)	Test (20%)	0.29	141.9	113.6	Train/test split, ntree = 500
RF (SC)	Test (20%)	0.28	124.2	95.4	Low signal, short range
RF (Runoff)	Test (20%)	−0.46	NaN	NaN	Poor model fit

3.3.1. Runoff

For runoff, a simple linear regression with infiltration average (*inf_av*) indicates a clear inverse trend: each 1-unit increase in *inf_av* is associated with a −4.83 unit change in runoff ($\beta = -4.83 \pm 2.38$), explaining 33.9% of the variance (adjusted $R^2 = 0.26$; $F_{1,8} = 4.10$; $p = 0.077$). The scatterplot shows a steady decline in runoff as infiltration increases, although the 95% CI remains wide, reflecting small sample size. Diagnostic plots reveal centered residuals, mild heteroscedasticity and tail departures, and a few high-leverage points, none exceeding usual Cook's distance thresholds. Pairwise displays suggest weak or no linear associations of runoff with *aspect_mean*, *cit_index_mean*, *ls_factor_mean*, or *slope_mean*. A random-forest model performed poorly (% variance explained = −46%, MSE ≈ 20.34), with unstable importance scores—consistent with limited data and weak nonlinear signal. Overall, the evidence supports infiltration capacity as a primary control on runoff, but results should be considered tentative, warranting larger samples and, potentially, transformations or mixed/regularized models.

3.3.2. Sediment Concentration

Exploratory analyses indicate a clear positive association between SC and maximum compaction: SC increases steadily from low values at ~0.80 compaction to markedly higher values near 0.90, although the 95% confidence band is wide, reflecting limited sample size and dispersion. Pairwise plots show weak or no linear relationships between SC and *aspect_mean* or *profile_curvature_mean*, and no consistent monotonic pattern with *soil_loss*. Diagnostic plots for the SC-compaction fit show centered residuals with mild heteroscedasticity and tail departures from normality, plus a few high-leverage observations. Overall, these results suggest that surface compaction is the main proximal control on SC in our dataset, while topographic metrics contribute little; however, inference should be cautious given leverage points and the small n.

3.3.3. Soil Loss

The multiple linear regression model for soil loss (*soil_loss*) using maximum values of compaction (*compaction_max*) *aspect_mean*, and *kfs* explained 67.8% of the variability (R^2), with an adjusted R^2 of 0.52 and a global test close to significance ($F = 4.22$; $p = 0.063$). Maximum compaction showed the largest positive effect ($\beta = 1.74 \times 10^3$), followed by kfs ($\beta = 0.069$), while *aspect_mean* had a negative effect ($\beta = -89.5$); none of these were statistically significant ($p > 0.15$–0.47). Diagnostic plots indicated centered residuals, slight heteroscedasticity, and tail deviations, with some high-leverage points. The random forest model explained 28.8% of the variance (MSE $\approx 2.67 \times 10^4$) and confirmed the relevance of compaction_max and kfs (highest %IncMSE and IncNodePurity), while *aspect_mean* contributed less information. Overall, the results suggest that compaction and hydraulic conductivity are key factors influencing soil loss; however, the small sample size and observed variability highlight the need for expanding the dataset and exploring transformations or regularization.

3.4. Challenges and Main Weaknesses

One important limitation of this study is the reduced number of spatial sampling points (n = 10), which constrains the generalizability and statistical power of both conventional and machine learning analyses. While each point was thoroughly characterized with repeated field measurements, and averaged to obtain representative values, this aggregation inevitably reduces the ability to model uncertainty within sampling units. Future work should consider hierarchical or mixed-effects modeling frameworks to incorporate both intra-point variance (via more replicates od the rainfall simulations) and spatial heterogeneity (upper, backslope and footslope). Similarly, expanding the number of plot units would allow for more robust machine learning applications and external validation. Nevertheless, the current approach serves as a methodological proof-of-concept for integrating proximal sensing and field experimentation in erosion diagnostics under real-world constraints. We acknowledge that Random Forest is generally robust to multicollinearity and capable of handling a large number of predictors, including the detection of non-linear patterns. Thus, in principle, a preliminary filtering of variables based on Pearson correlation is not strictly necessary. However, in our initial experiments the performance of the models was relatively weak. For this reason, and following recommendations in the literature (e.g., [48,49]), we applied a correlation-based variable reduction ($|r| > 0.7$) to minimize redundancy and to facilitate model interpretation. This approach was also combined with scatterplot matrices and pairwise plots to examine both linear and non-linear relationships. Nevertheless, we agree that it would be valuable to also test Random Forest without prior filtering in order to verify whether model performance improves when all predictors are retained. In addition, it will be worthy to explore other feature selection strategies embedded within the model, as well as dimensionality reduction through PCA, to compare the outcomes. This would allow us to assess more comprehensively whether the correlation-based filtering might have constrained the predictive capacity of the models. We acknowledge that the aggregation of repeated field measurements into averaged values at each sampling point may have reduced the capacity to model intra-point variability. Future studies should explore hierarchical modeling strategies (e.g., linear mixed-effects models) to integrate replication structure and better represent uncertainty components at multiple scales.

Future studies may consider dimension-reduction methods such as Partial Least Squares (PLS) regression, which are well suited for small sample sizes with high-dimensional predictors and have been extensively applied in environmental and chemometric contexts.

The field experiments conducted in a citrus orchard near Doñana National Park (Seville, southern Spain) revealed crucial insights into erosion processes in flat Mediterranean agroecosystems. Through the integration of UAV-LiDAR–derived topographic variables and detailed field measurements, we found that compaction and infiltration dynamics are the dominant controls on hydrological responses [15,50]. Soil loss was most strongly associated with maximum soil compaction and near-surface saturated hydraulic conductivity (Kfs), with *aspect_mean* showing a negative relationship—likely reflecting the influence of solar exposure and biological soil cover, particularly moss. Indeed, LiDAR-derived aspect maps showed shaded microsites beneath the citrus canopy (north- and east-facing orientations) with visible moss growth, which appeared to modulate erosion by increasing surface stability (Figure 8a,b). These subtle biophysical differences significantly impacted model outputs [61]. For sediment concentration (SC), surface compaction was the strongest predictor, while topographic variables contributed little explanatory power although they demonstrated the relevance of the sediment mobilisation close to the trees and the compaction in the inter-row areas (Figure 8c; [64,75]). Infiltration emerged as the primary driver for runoff generation, with an inverse relationship between infiltration and runoff clearly established. However, all models were constrained by limited sample size, moderate heteroscedasticity, and the presence of high-leverage observations. Random forest models confirmed the central role of compaction and Kfs but struggled to capture variance in runoff and SC, underscoring the challenge of applying machine learning with small, noisy datasets. Overall, our results highlight the feasibility and value of integrating proximal sensing with field experiments to identify erosion hotspots and optimize monitoring protocols in perennial cropping systems. Yet, future studies should prioritize larger datasets, additional microclimatic and biological indicators which affect aggregate stability (Figure 8a,d), and regularized modeling frameworks to improve predictive reliability.

(a) (b) (c) (d)

Figure 8. Some key factors affecting soil erosion in citrus orchards detected after finishing the study without possible measurements. (**a**). Biocrusts and catch crops; (**b**). inter-row areas with bare soils; (**c**). concave micro-topography where runoff activates soil erosion; (**d**). example of compacted soil aggregate.

The relatively weak contribution of UAV-LiDAR to the erosion models in our study is primarily explained by the very gentle relief and limited topographic variability of the citrus orchard as other researchers confirmed [76]. In such flat agroecosystems, morphometric indices derived from high-resolution LiDAR data can provide little differentiation among sampling points, thereby constraining their explanatory capacity. This result should not be interpreted as a limitation of the technology itself but rather as a reflection of the site-specific conditions, which reduce the sensitivity of terrain attributes to capture erosion drivers using visual inspections [77]. Nevertheless, UAV-LiDAR offered important indirect benefits by ensuring centimeter-level georeferencing accuracy, maintaining microtopographic consistency, and providing a replicable framework for future monitoring [78]. These strengths highlight that while UAV-LiDAR may add limited explanatory power in low-relief environments, it remains a valuable tool for scaling up erosion assessments, especially when applied in more heterogeneous landscapes.

4. Conclusions

This study demonstrates that integrating UAV-LiDAR–derived morphometric variables with standardized field experiments provides valuable insights into the drivers of soil erosion in citrus orchards under Mediterranean conditions. Despite the limited sample size, considering the difficulties to perform such number of parallel experiments with four different devices at the same time during summer (dry soils) with enough trained researchers, consistent physical controls emerged across multiple models: soil loss was primarily influenced by maximum soil compaction and saturated hydraulic conductivity (Kfs); sediment concentration was associated with surface compaction; and runoff was inversely related to average infiltration. The aspect orientation derived from LiDAR data revealed the presence of moss and more stable surface conditions in shaded areas under the citrus canopy, emphasizing the role of microclimatic and biological factors in erosion dynamics. Although machine learning approaches such as random forests were useful in identifying relevant predictors, their performance was constrained by data variability and sample limitations. Overall, this framework highlights a cost-effective and scalable strategy for monitoring erosion risk, prioritizing key variables, and guiding sustainable land management in perennial cropping systems.

Supplementary Materials: The following supporting information can be downloaded at: https://www.mdpi.com/article/10.3390/w17243541/s1.

Author Contributions: J.R.-C.: Conceptualization, Methodology, Formal Analysis, Data Collection, Investigation, Resources, Writing—Original Draft, Writing—Review & Editing, Visualization, Supervision; L.C.-R.: Data Collection, Laboratory work, Writing—Original Draft; L.M.-C.: Data Collection, Laboratory work, Writing—Original Draft; J.G.-V.: Data Collection, Writing—Original Draft; M.T.G.-M.: Data Collection, Writing—Original Draft; V.R.-G.: Conceptualization, Methodology, Data Collection, Investigation, Resources, Writing—Review & Editing, Supervision, and Project Administration. All authors have read and agreed to the published version of the manuscript.

Funding: This research was carried out within the framework of the projects: (i) "Monitorización hiper-espectral para el manejo de los cítricos en el contexto del cambio climático (CITRUSMART)", Junta de Andalucía (GOPG-SE-23-0024, 2023–2025); (ii) "Desarrollo de productos basados en los nuevos sensores satelitales hiperespectrales europeos e IA para la caracterización de estresores en tierras de cultivo (HIPROESTRES)" (grant number PID2023-152656OB-I00), within the Programa Estatal de Investigación Científica, Técnica y de Innovación (2021–2023) funded by the Ministerio de Ciencia, Innovación y Universidades.

Data Availability Statement: The original contributions presented in this study are included in the article. Further inquiries can be directed to the corresponding author.

Acknowledgments: During the preparation of this work the authors used Chat GPT in order to review the Flow and grammar of the text. After using this tool/service, the authors reviewed and edited the content as needed and take full responsibility for the content of the published article.

Conflicts of Interest: The authors declare no conflicts of interest.

References

1. Seeger, M. (Ed.) *Understanding and Preventing Soil Erosion*; Burleigh Dodds Science Publishing: London, UK, 2024; ISBN 978-1-003-58788-0.
2. Chen, Y.; De Geeter, S.; Poesen, J.; Matthews, F.; Campforts, B.; Borrelli, P.; Panagos, P.; Vanmaercke, M. Global Patterns of Gully Occurrence and Their Sensitivity to Environmental Changes. *Int. Soil Water Conserv. Res.* **2025**. [CrossRef]
3. Visser, S.; Keesstra, S.; Maas, G.; de Cleen, M.; Molenaar, C. Soil as a Basis to Create Enabling Conditions for Transitions Towards Sustainable Land Management as a Key to Achieve the SDGs by 2030. *Sustainability* **2019**, *11*, 6792. [CrossRef]
4. Sanaullah, M.; Usman, M.; Wakeel, A.; Cheema, S.A.; Ashraf, I.; Farooq, M. Terrestrial Ecosystem Functioning Affected by Agricultural Management Systems: A Review. *Soil Tillage Res.* **2020**, *196*, 104464. [CrossRef]
5. Alliaume, F.; Rossing, W.A.H.; Tittonell, P.; Jorge, G.; Dogliotti, S. Reduced Tillage and Cover Crops Improve Water Capture and Reduce Erosion of Fine Textured Soils in Raised Bed Tomato Systems. *Agric. Ecosyst. Environ.* **2014**, *183*, 127–137. [CrossRef]
6. Casalí, J.; Gastesi, R.; Álvarez-Mozos, J.; De Santisteban, L.M.; de Lersundi, J.D.V.; Giménez, R.; Larrañaga, A.; Goñi, M.; Agirre, U.; Campo, M.A.; et al. Runoff, Erosion, and Water Quality of Agricultural Watersheds in Central Navarre (Spain). *Agric. Water Manag.* **2008**, *95*, 1111–1128. [CrossRef]
7. Fu, B.; Merritt, W.S.; Croke, B.F.W.; Weber, T.R.; Jakeman, A.J. A Review of Catchment-Scale Water Quality and Erosion Models and a Synthesis of Future Prospects. *Environ. Model. Softw.* **2019**, *114*, 75–97. [CrossRef]
8. Keesstra, S.; Nunes, J.; Novara, A.; Finger, D.; Avelar, D.; Kalantari, Z.; Cerdà, A. The Superior Effect of Nature Based Solutions in Land Management for Enhancing Ecosystem Services. *Sci. Total Environ.* **2018**, *610–611*, 997–1009. [CrossRef]

9. Albaladejo, J.; Díaz-Pereira, E.; de Vente, J. Eco-Holistic Soil Conservation to Support Land Degradation Neutrality and the Sustainable Development Goals. *Catena* **2021**, *196*, 104823. [CrossRef]
10. Cerdan, O.; Govers, G.; Le Bissonnais, Y.; Van Oost, K.; Poesen, J.; Saby, N.; Gobin, A.; Vacca, A.; Quinton, J.; Auerswald, K.; et al. Rates and Spatial Variations of Soil Erosion in Europe: A Study Based on Erosion Plot Data. *Geomorphology* **2010**, *122*, 167–177. [CrossRef]
11. García-Ruiz, J.M.; Beguería, S.; Nadal-Romero, E.; González-Hidalgo, J.C.; Lana-Renault, N.; Sanjuán, Y. A Meta-Analysis of Soil Erosion Rates across the World. *Geomorphology* **2015**, *239*, 160–173. [CrossRef]
12. Panagos, P.; Borrelli, P.; Meusburger, K.; Alewell, C.; Lugato, E.; Montanarella, L. Estimating the Soil Erosion Cover-Management Factor at the European Scale. *Land Use Policy* **2015**, *48*, 38–50. [CrossRef]
13. Cerdà, A.; Rodrigo-Comino, J.; Giménez-Morera, A.; Novara, A.; Pulido, M.; Kapović-Solomun, M.; Keesstra, S.D. Policies Can Help to Apply Successful Strategies to Control Soil and Water Losses. The Case of Chipped Pruned Branches (CPB) in Mediterranean Citrus Plantations. *Land Use Policy* **2018**, *75*, 734–745. [CrossRef]
14. Duan, J.; Liu, Y.-J.; Yang, J.; Tang, C.-J.; Shi, Z.-H. Role of Groundcover Management in Controlling Soil Erosion under Extreme Rainfall in Citrus Orchards of Southern China. *J. Hydrol.* **2020**, *582*, 124290. [CrossRef]
15. Jianjun, W.; Quansheng, L.; Lijiao, Y. Effect of intercropping on soil erosion in young citrus plantation—A simulation study. *Chin. J. Appl. Ecol.* **2017**, *8*, 143–146.
16. Bayat, F.; Monfared, A.B.; Jahansooz, M.R.; Esparza, E.T.; Keshavarzi, A.; Morera, A.G.; Fernández, M.P.; Cerdà, A. Analyzing Long-Term Soil Erosion in a Ridge-Shaped Persimmon Plantation in Eastern Spain by Means of ISUM Measurements. *Catena* **2019**, *183*, 104176. [CrossRef]
17. Anache, J.A.A.; Wendland, E.C.; Oliveira, P.T.S.; Flanagan, D.C.; Nearing, M.A. Runoff and Soil Erosion Plot-Scale Studies under Natural Rainfall: A Meta-Analysis of the Brazilian Experience. *Catena* **2017**, *152*, 29–39. [CrossRef]
18. Bernoux, M.; Cerri, C.C.; Cerri, C.E.P.; Neto, M.S.; Metay, A.; Perrin, A.-S.; Scopel, E.; Tantely, R.; Blavet, D.; de Piccolo, M.C.; et al. Cropping Systems, Carbon Sequestration and Erosion in Brazil: A Review. In *Sustainable Agriculture*; Lichtfouse, E., Navarrete, M., Debaeke, P., Véronique, S., Alberola, C., Eds.; Springer: Dordrecht, The Netherlands, 2009; pp. 75–85. ISBN 978-90-481-2665-1.
19. Di Prima, S.; Bagarello, V.; Lassabatere, L.; Angulo-Jaramillo, R.; Bautista, I.; Burguet, M.; Cerdà, A.; Iovino, M.; Prosdocimi, M. Comparing Beerkan Infiltration Tests with Rainfall Simulation Experiments for Hydraulic Characterization of a Sandy-Loam Soil. *Hydrol. Process.* **2017**, *31*, 3520–3532. [CrossRef]
20. Iserloh, T.; Ries, J.B.; Arnáez, J.; Boix-Fayos, C.; Butzen, V.; Cerdà, A.; Echeverría, M.T.; Fernández-Gálvez, J.; Fister, W.; Geißler, C.; et al. European Small Portable Rainfall Simulators: A Comparison of Rainfall Characteristics. *Catena* **2013**, *110*, 100–112. [CrossRef]
21. Vishwakarma, D.K.; Yadav, D.; Kumar, R.; Kumar, R.; Bhat, S.A.; Mirzania, E.; Kuriqi, A. Assessing the Performance of Various Infiltration Models to Improve Water Management Practices. *Paddy Water Environ.* **2025**, *23*, 77–93. [CrossRef]
22. Bogunovic, I.; Telak, L.J.; Pereira, P. Experimental Comparison of Runoff Generation and Initial Soil Erosion Between Vineyards and Croplands of Eastern Croatia: A Case Study. *Air Soil Water Res.* **2020**, *13*, 1178622120928323. [CrossRef]
23. Telak, L.J.; Dugan, I.; Bogunovic, I. Soil Management and Slope Impacts on Soil Properties, Hydrological Response, and Erosion in Hazelnut Orchard. *Soil Syst.* **2021**, *5*, 5. [CrossRef]
24. Fini, A.; Frangi, P.; Mori, J.; Donzelli, D.; Ferrini, F. Nature Based Solutions to Mitigate Soil Sealing in Urban Areas: Results from a 4-Year Study Comparing Permeable, Porous, and Impermeable Pavements. *Environ. Res.* **2017**, *156*, 443–454. [CrossRef]
25. Khan, M.O.; Keesstra, S.D.; Słowik-Opoka, E.; Klamerus-Iwan, A.; Liaqat, W. Determining the Role of Urban Greenery in Soil Hydrology: A Bibliometric Analysis of Nature-Based Solutions in Urban Ecosystem. *Water* **2025**, *17*, 322. [CrossRef]
26. Gayen, A.; Pourghasemi, H.R.; Saha, S.; Keesstra, S.; Bai, S. Gully Erosion Susceptibility Assessment and Management of Hazard-Prone Areas in India Using Different Machine Learning Algorithms. *Sci. Total Environ.* **2019**, *668*, 124–138. [CrossRef]
27. Alexiou, S.; Efthimiou, N.; Karamesouti, M.; Papanikolaou, I.; Psomiadis, E.; Charizopoulos, N. Measuring Annual Sedimentation through High Accuracy UAV-Photogrammetry Data and Comparison with RUSLE and PESERA Erosion Models. *Remote Sens.* **2023**, *15*, 1339. [CrossRef]
28. Gómez Gutiérrez, Á.; Schnabel, S.; Lavado Contador, F.; De Sanjosé, J.J.; Atkinson Gordo, A.; Pulido Fernández, M.; Sánchez Fernández, M. Studying the Influence of Livestock Pressure on Gully Erosion in Rangelands of SW Spain by Means of the Uav+sfm Workflow. *Boletín Asoc. Geógrafos Españoles* **2018**, *78*, 66–88. [CrossRef]
29. IUSS-WRB. *World Reference Base for Soil Resources. International Soil Classification System for Naming Soils and Creating Legends for Soil Maps*, 4th ed.; International Union of Soil Sciences (IUSS): Vienna, Austria, 2022.
30. Crema, S.; Cavalli, M. SedInConnect: A Stand-Alone, Free and Open Source Tool for the Assessment of Sediment Connectivity. *Comput. Geosci.* **2018**, *111*, 39–45. [CrossRef]

31. Ries, J.B.; Iserloh, T.; Seeger, M.; Gabriels, D. Rainfall Simulations—Constraints, Needs and Challenges for a Future Use in Soil Erosion Research. *Z. Geomorphol.* **2013**, *57*, 1–10. [CrossRef]
32. Carollo, F.G.; Caruso, R.; Ferro, V.; Serio, M.A. Characterizing the Kamphorst Rainfall Simulator for Soil Erosion Investigations. *J. Hydrol.* **2024**, *643*, 132025. [CrossRef]
33. Hlavčová, K.; Danáčová, M.; Kohnová, S.; Szolgay, J.; Valent, P.; Výleta, R. Estimating the Effectiveness of Crop Management on Reducing Flood Risk and Sediment Transport on Hilly Agricultural Land—A Myjava Case Study, Slovakia. *Catena* **2019**, *172*, 678–690. [CrossRef]
34. Iserloh, T.; Fister, W.; Seeger, M.; Willger, H.; Ries, J.B. A Small Portable Rainfall Simulator for Reproducible Experiments on Soil Erosion. *Soil Tillage Res.* **2012**, *124*, 131–137. [CrossRef]
35. Cerdà, A. The Influence of Aspect and Vegetation on Seasonal Changes in Erosion under Rainfall Simulation on a Clay Soil in Spain. *Can. J. Soil Sci.* **1998**, *78*, 321–330. [CrossRef]
36. Cerdà, A.; Jurgensen, M.F. Ant Mounds as a Source of Sediment on Citrus Orchard Plantations in Eastern Spain. A Three-Scale Rainfall Simulation Approach. *Catena* **2011**, *85*, 231–236. [CrossRef]
37. Saleh, A. Soil Roughness Measurement: Chain Method. *J. Soil Water Conserv.* **1993**, *48*, 527–529. [CrossRef]
38. Ortega, J.P.; Cambronero, L.; Alcarria Salas, M.; Rodríguez, J.L.; Durán-Zuazo, V.H.; Keesstra, S.D.; Rodrigo-Comino, J. Conducting an in Situ Evaluation of Erodibility in a Mediterranean Semi-Arid and Conventional Vineyard in Granada Province (Southern Spain) through Rainfall Simulation Experiments. *Euro-Mediterr. J. Environ. Integr.* **2024**, *9*, 797–808. [CrossRef]
39. Rodrigo-Comino, J.; Bogunovic, I.; Mohajerani, H.; Pereira, P.; Cerdà, A.; Ruiz-Sinoga, J.; Ries, J. The Impact of Vineyard Abandonment on Soil Properties and Hydrological Processes. *Vadose Zone J.* **2017**, *16*, 1–7. [CrossRef]
40. Jobbágy, J.; Krištof, K.; Angelovič, M.; Zsembeli, J. Evaluation of Soil Infiltration Variability in Compacted and Uncompacted Soil Using Two Devices. *Water* **2023**, *15*, 1918. [CrossRef]
41. Naik, A.P.; Pekkat, S. Time Dependence of Hydraulic Parameters Estimation from Transient Analysis of Mini Disc Infiltrometer Measurements. *Eur. J. Soil Sci.* **2022**, *73*, e13228. [CrossRef]
42. Reynolds, W.D. The Guelph Permeameter Method for in Situ Measurement of Field-Saturated Hydraulic Conductivity and Matric Flux Potential. Ph.D. Thesis, University of Guelph, Guelph, ON, Canada, 1986.
43. Arredondo, A.M.; Mirzaei, M.; Wu, G.-L.; Keshavarzi, A.; Durán-Zuazo, V.H.; Bernardo, F.S.; Rodrigo-Comino, J. Evaluating Soil Respiration and Water Infiltration in Esparto Grasslands: The Effects of Hillslope Position and Soil Management in Arid, Human-Affected Mediterranean Environments. *Discov. Sustain.* **2025**, *6*, 421. [CrossRef]
44. Madalina, V.; Valentina, C.; Natalia, E.; Lucian, L.; Monica, M.; Anda, R.; Norbert, B.; Madalina, B.; Silvius, S.; György, D. Experimental Determination of Carbon Dioxide Flux in Soil and Correlation with Dependent Parameters. *IOP Conf. Ser. Earth Environ. Sci.* **2020**, *616*, 012010. [CrossRef]
45. Keller, T.; Colombi, T.; Ruiz, S.; Schymanski, S.J.; Weisskopf, P.; Koestel, J.; Sommer, M.; Stadelmann, V.; Breitenstein, D.; Kirchgessner, N.; et al. Soil Structure Recovery Following Compaction: Short-Term Evolution of Soil Physical Properties in a Loamy Soil. *Soil Sci. Soc. Am. J.* **2021**, *85*, 1002–1020. [CrossRef]
46. Khemis, C.; Abrougui, K.; Mohammadi, A.; Gabsi, K.; Dorbolo, S.; Mercatoris, B.; Mutuku, E.; Cornelis, W.; Chehaibi, S. Development of Artificial Neural Networks to Predict the Effect of Tractor Speed on Soil Compaction Using Penetrologger Test Results. *Processes* **2022**, *10*, 1109. [CrossRef]
47. Pulido, M.; Schnabel, S.; Lavado Contador, J.F.; Lozano-Parra, J.; Gómez-Gutiérrez, Á.; Brevik, E.C.; Cerdà, A. Reduction of the Frequency of Herbaceous Roots as an Effect of Soil Compaction Induced by Heavy Grazing in Rangelands of SW Spain. *Catena* **2017**, *158*, 381–389. [CrossRef]
48. Dormann, C.F.; Elith, J.; Bacher, S.; Buchmann, C.; Carl, G.; Carré, G.; Marquéz, J.R.G.; Gruber, B.; Lafourcade, B.; Leitão, P.J.; et al. Collinearity: A Review of Methods to Deal with It and a Simulation Study Evaluating Their Performance. *Ecography* **2013**, *36*, 27–46. [CrossRef]
49. Kyriazos, T.; Poga, M. Dealing with Multicollinearity in Factor Analysis: The Problem, Detections, and Solutions. *Open J. Stat.* **2023**, *13*, 404–424. [CrossRef]
50. Di Prima, S.; Rodrigo-Comino, J.; Novara, A.; Iovino, M.; Pirastru, M.; Keesstra, S.D.; Cerdà, A. Soil Physical Quality of Citrus Orchards Under Tillage, Herbicide, and Organic Managements. *Pedosphere* **2018**, *28*, 463–477. [CrossRef]
51. Basche, A.D.; DeLonge, M.S. Comparing Infiltration Rates in Soils Managed with Conventional and Alternative Farming Methods: A Meta-Analysis. *PLoS ONE* **2019**, *14*, e0215702. [CrossRef] [PubMed]
52. Jagdale, S.; Nimbalkar, P. Infiltration Studies of Different Soils under Different Soil Conditions and Comparison of Infiltration Models with Field Data. *Int. J. Eng. Technol. Sci.* **2012**, *3*, 154–157.
53. Rodrigo-Comino, J.; Taguas, E.; Seeger, M.; Ries, J.B. Quantification of Soil and Water Losses in an Extensive Olive Orchard Catchment in Southern Spain. *J. Hydrol.* **2018**, *556*, 749–758. [CrossRef]

54. Rodrigo-Comino, J.; Ruiz Sinoga, J.D.; Senciales González, J.M.; Guerra-Merchán, A.; Seeger, M.; Ries, J.B. High Variability of Soil Erosion and Hydrological Processes in Mediterranean Hillslope Vineyards (Montes de Málaga, Spain). *Catena* **2016**, *145*, 274–284. [CrossRef]
55. Amini, M.; Rasoulzadeh, A.; Mobaser, J.A.; Ahmadabad, Y.H.; Pollacco, J.A.P.; Fernández-Gálvez, J. Hydro-Pedotransfer Functions to Estimate Evaporation from Bare Soil Surface in Semi-Arid Regions. *Air Soil Water Res.* **2025**, *18*, 11786221251349047. [CrossRef]
56. Angulo-Jaramillo, R.; Bagarello, V.; Iovino, M.; Lassabatere, L. *Infiltration Measurements for Soil Hydraulic Characterization*; Springer: Cham, Switzerland, 2016; ISBN 978-3-319-31788-5.
57. Cerdà, A.; González-Pelayo, Ó.; Giménez-Morera, A.; Jordán, A.; Pereira, P.; Novara, A.; Brevik, E.C.; Prosdocimi, M.; Mahmoodabadi, M.; Keesstra, S. Use of Barley Straw Residues to Avoid High Erosion and Runoff Rates on Persimmon Plantations in Eastern Spain under Low Frequency–High Magnitude Simulated Rainfall Events. *Soil Res.* **2016**, *54*, 154–165. [CrossRef]
58. Keesstra, S.; Pereira, P.; Novara, A.; Brevik, E.C.; Azorin-Molina, C.; Parras-Alcántara, L.; Jordán, A.; Cerdà, A. Effects of Soil Management Techniques on Soil Water Erosion in Apricot Orchards. *Sci. Total Environ.* **2016**, *551–552*, 357–366. [CrossRef]
59. Cerdà, A.; Ackermann, O.; Terol, E.; Rodrigo-Comino, J. Impact of Farmland Abandonment on Water Resources and Soil Conservation in Citrus Plantations in Eastern Spain. *Water* **2019**, *11*, 824. [CrossRef]
60. de Lima, J.L.M.P.; Carvalho, S.C.P.; de Lima, M.I.P. Rainfall Simulator Experiments on the Importance o of When Rainfall Burst Occurs during Storm Events on Runoff and Soil Loss. *Z. Geomorphol.* **2013**, *57*, 91–109. [CrossRef]
61. Seeger, M. Uncertainty of Factors Determining Runoff and Erosion Processes as Quantified by Rainfall Simulations. *Catena* **2007**, *71*, 56–67. [CrossRef]
62. Fiener, P.; Dlugoß, V.; Korres, W.; Schneider, K. Spatial Variability of Soil Respiration in a Small Agricultural Watershed—Are Patterns of Soil Redistribution Important? *Catena* **2012**, *94*, 3–16. [CrossRef]
63. Parkin, T.B.; Doran, J.W. Field and Laboratory Tests of Soil Respiration. In *Methods for Assessing Soil Quality*; Doran, J.W., Jones, A.J., Eds.; Soil Science Society of America Incorporated: Madison, WI, USA, 1996; Volume SSSA Special Publication 49, pp. 231–246.
64. Bogunović, I.; Kisić, I.; Maletić, E.; Perčin, A.; Matošić, S.; Roškar, L. Soil Compaction in Vineyards of Different Ages in Pannonian Croatia. Part I. Influence of Machinery Traffic and Soil Management on Compaction of Individual Horizons. *J. Cent. Eur. Agric.* **2016**, *17*, 533–544. [CrossRef]
65. Duttmann, R.; Schwanebeck, M.; Nolde, M.; Horn, R. Predicting Soil Compaction Risks Related to Field Traffic during Silage Maize Harvest. *Soil Sci. Soc. Am. J.* **2014**, *78*, 408–421. [CrossRef]
66. Cleveland, C.C.; Nemergut, D.R.; Schmidt, S.K.; Townsend, A.R. Increases in Soil Respiration Following Labile Carbon Additions Linked to Rapid Shifts in Soil Microbial Community Composition. *Biogeochemistry* **2007**, *82*, 229–240. [CrossRef]
67. Moradi, E.; Rodrigo-Comino, J.; Terol, E.; Mora-Navarro, G.; Marco da Silva, A.; N. Daliakopoulos, I.; Khosravi, H.; Pulido Fernández, M.; Cerdà, A. Quantifying Soil Compaction in Persimmon Orchards Using ISUM (Improved Stock Unearthing Method) and Core Sampling Methods. *Agriculture* **2020**, *10*, 266. [CrossRef]
68. Arabameri, A.; Cerda, A.; Pradhan, B.; Tiefenbacher, J.P.; Lombardo, L.; Bui, D.T. A Methodological Comparison of Head-Cut Based Gully Erosion Susceptibility Models: Combined Use of Statistical and Artificial Intelligence. *Geomorphology* **2020**, *359*, 107136. [CrossRef]
69. Keshavarzi, A.; Omran, E.-S.E.; Bateni, S.M.; Pradhan, B.; Vasu, D.; Bagherzadeh, A. Modeling of Available Soil Phosphorus (ASP) Using Multi-Objective Group Method of Data Handling. *Model. Earth Syst. Environ.* **2016**, *2*, 157. [CrossRef]
70. Vermeulen, D.; Van Niekerk, A. Machine Learning Performance for Predicting Soil Salinity Using Different Combinations of Geomorphometric Covariates. *Geoderma* **2017**, *299*, 1–12. [CrossRef]
71. Capello, G.; Biddoccu, M.; Ferraris, S.; Cavallo, E. Effects of Tractor Passes on Hydrological and Soil Erosion Processes in Tilled and Grassed Vineyards. *Water* **2019**, *11*, 2118. [CrossRef]
72. Esteves, M.; Lapetite, J.M. A Multi-Scale Approach of Runoff Generation in a Sahelian Gully Catchment: A Case Study in Niger. *Catena* **2003**, *50*, 255–271. [CrossRef]
73. Martínez-Carreras, N.; Krein, A.; Gallart, F.; Iffly, J.F.; Pfister, L.; Hoffmann, L.; Owens, P.N. Assessment of Different Colour Parameters for Discriminating Potential Suspended Sediment Sources and Provenance: A Multi-Scale Study in Luxembourg. *Geomorphology* **2010**, *118*, 118–129. [CrossRef]
74. Nunes, J.P.; Seixas, J.; Keizer, J.J. Modeling the Response of Within-Storm Runoff and Erosion Dynamics to Climate Change in Two Mediterranean Watersheds: A Multi-Model, Multi-Scale Approach to Scenario Design and Analysis. *Catena* **2013**, *102*, 27–39. [CrossRef]
75. Alameda, D.; Villar, R.; Iriondo, J.M. Spatial Pattern of Soil Compaction: Trees' Footprint on Soil Physical Properties. *For. Ecol. Manag.* **2012**, *283*, 128–137. [CrossRef]
76. Niu, Y.H.; Wang, L.; Wan, X.G.; Peng, Q.Z.; Huang, Q.; Shi, Z.H. A Systematic Review of Soil Erosion in Citrus Orchards Worldwide. *Catena* **2021**, *206*, 105558. [CrossRef]

77. Đomlija, P.; Bernat Gazibara, S.; Arbanas, Ž.; Mihalić Arbanas, S. Identification and Mapping of Soil Erosion Processes Using the Visual Interpretation of LiDAR Imagery. *ISPRS Int. J. Geo-Inf.* **2019**, *8*, 438. [CrossRef]
78. Donmez, C.; Villi, O.; Berberoglu, S.; Cilek, A. Computer Vision-Based Citrus Tree Detection in a Cultivated Environment Using UAV Imagery. *Comput. Electron. Agric.* **2021**, *187*, 106273. [CrossRef]

water

MDPI

Article

Integrated Approach to Assess Simulated Rainfall Uniformity and Energy-Related Parameters for Erosion Studies

Roberto Caruso [1], Maria Angela Serio [1,*], Gabriel Búrdalo-Salcedo [2], Francesco Giuseppe Carollo [1], Almudena Ortiz-Marqués [3], Vito Ferro [1,4] and María Fernández-Raga [5,*]

[1] Department of Agricultural, Food, and Forest Sciences, University of Palermo, 90128 Palermo, Italy; roberto.caruso01@unipa.it (R.C.); vito.ferro@unipa.it (V.F.)
[2] Department of Mechanical, Computer, and Aerospace Engineering, I4 Institute, University of León, 24071 León, Spain; gabriel.burdalo@unileon.es
[3] Department of Mining Technology, Topography, and Structures, University of León, 24071 León, Spain; almudena.ortiz@unileon.es
[4] NBFC—National Biodiversity Future Center, 90133 Palermo, Italy
[5] Department of Chemistry and Applied Physics, I4 Institute, University of León, 24071 León, Spain
* Correspondence: mariaangela.serio@unipa.it (M.A.S.); mferr@unileon.es (M.F.-R.)

Abstract: Rainfall simulators are crucial devices in erosion research, enabling the controlled reproduction of precipitation characteristics for both laboratory and field investigations. This study presents a comprehensive characterization of a rainfall simulator originally designed to assess the erosive effects of precipitation on heritage surfaces. The simulator, installed at the University of León, was evaluated using volumetric methods and disdrometric techniques, employing a Parsivel^2 optical disdrometer. Simulations were conducted with a falling height of 10 m and high-intensity rainfalls. Spatial uniformity was assessed through thematic mapping and the Christiansen Uniformity (CU) coefficient, revealing limited uniformity across the full wetted area, but an improved performance within the central zone (CU up to 80%). Disdrometric data provided detailed insights into drop size and velocity distributions, enabling the estimation of rainfall intensity, kinetic energy, and momentum, as well as the spatial uniformity of the energetic parameters. Empirical models to estimate the raindrop's fall velocity were tested against disdrometric measurements, confirming the simulator's ability to generate rainfall with velocity characteristics comparable to those of natural precipitation. Moreover, the findings underscore the importance of integrating multiple measurement approaches to enhance the reliability and accuracy of rainfall simulator characterization.

Keywords: rainfall simulation; rain erosivity; disdrometric measurements; fall velocity; raindrop size distribution; heritage erosion

1. Introduction

The importance of rainfall simulators in erosion research lies in their capacity to reproduce key precipitation characteristics, such as rainfall intensity, kinetic energy, and momentum, in a rapid and controlled manner while eliminating the variability inherent to natural rainfall [1–4]. Due to these advantages, rainfall simulators have been used as a systematic approach for both laboratory and field studies [5–7].

Over time, a wide range of simulators has been designed [8], each adapted to specific research topics, including soil erosion [9–12], hydrology [13,14], water infiltration [15,16], runoff and sediment transport [17–19], heritage surfaces erosion [20,21], soil texture [22,23], the roles of runoff and splash erosion in soil degradation and landform development [24–27], or the assessment of environmental impacts [28–30].

Nevertheless, the development of high-fidelity rainfall simulators remains a considerable challenge [31]. Moreover, despite their widespread use, the lack of standardized protocols and the heterogeneity of research goals represent additional limitations [32–36], constraining the comparison of research results [8,37]. Indeed, devices range from simple drip systems with capillary tubes [38] to pressurized nozzles/sprinklers [31] and aspersion mechanisms [20]; some operate under highly controlled laboratory conditions, while others are deployed in full-scale field plots; rainfalls can be applied from microplots [39] to large field plots [40].

Few recognized protocols provide standardized procedures for rainfall simulator calibration, testing configuration, and data collection and analysis. As an example, the American Society for Testing and Materials (ASTM) developed the ASTM D6459 and the ASTM D7101 protocols for rolled erosion control [41–45]; both aim to ensure the reproducibility and comparability of erosion-related experiments, promoting consistent evaluation of hydraulic performance under simulated rainfall conditions.

Common approaches for characterizing rainfall simulators typically include the use of rain collectors to quantify rainfall intensity, photographic techniques to determine raindrop fall velocity, and disdrometers to evaluate the size and velocity distribution of simulated raindrops [46–48]. However, the choice of a particular characterization method can significantly influence the measured rainfall parameters [37,49–51], potentially introducing variability into the results and highlighting the need for standardized protocols to ensure comparability across studies [52–54].

Green and Pattison [55] stated that the number, spatial arrangement, and physical characteristics of rainfall collectors, such as their size and shape, directly affect the representation of rainfall distribution over a plot surface. According to the authors, these factors influence the assessment of uniformity, which is usually quantified using the Christiansen Uniformity (*CU*) coefficient:

$$CU = 100\left(1 - \frac{\sum_{i=1}^{n_c}|I_i - I_S|}{I_S n_c}\right) \tag{1}$$

in which I_i is the rainfall intensity of a single rain collector, I_S is the mean rainfall intensity, and n_c (−) is the number of rainfall collectors. Thus, as the *CU* value approaches 100%, the rainfall pattern becomes increasingly uniform.

In other words, the *CU* method is susceptible to the sampling resolution and the layout of collectors, resulting in differences in a rainfall simulator characterization when different experimental setups are employed. As a consequence, the *CU* reflects both the performance of the applied testing procedure and the simulator's actual performance. However, the CU coefficient remains one of the most widely applied methods due to its simplicity, widespread acceptance, and long-standing use as a standard metric for comparing rainfall simulator performance across different studies [55].

In previous literature, various thresholds of the *CU* coefficient have been proposed to evaluate the spatial uniformity of rainfall simulators. A classification, originally developed by the American Society of Agricultural and Biological Engineers (ASABE) [56] for irrigation systems, also adopted in rainfall simulation research as a benchmark for experimental reliability [12,20], defines drippers and nozzles as follows: unacceptable ($CU < 60\%$), poor ($60\% \leq CU < 70\%$), fair ($70\% \leq CU < 80\%$), good ($80\% \leq CU < 90\%$), and excellent ($CU \geq 90\%$). Other authors [10,11,57,58] consider rainfall to be uniform when *CU* exceeds 80%, especially in small-scale experiments. For large plot areas, a threshold of 70% or less has been accepted in some studies [26,59].

Regarding the use of photographic methods for estimating the fall velocity of raindrops, although these techniques have proven reliable, particularly in drip-type rainfall simulators [49,50], their application in pressurized or other complex devices faces several practical limitations. Manual procedures are time-consuming and labor-intensive; in addition, results depend on the operator, and errors can occur in the estimated kinetic energy of up to ±28% compared with the results obtained using automated methods [60]. Additional limitations arise from drop deformation during free fall, which tends to flatten drops, affecting velocity estimation [61]. Furthermore, drops located too close or too far from the camera calibration plane appear enlarged or reduced, respectively, requiring the filtering of out-of-focus data [61]. Finally, in multi-nozzle simulators with high drop densities, drop overlap in the image plane can lead to identification and sizing errors [61].

Currently, disdrometers represent the last advancement in the field, enabling a comprehensive characterization of precipitation. Specifically, these devices provide detailed measurements of the rainfall Particle Size and Velocity Distribution (PSVD), which is crucial for a more reliable estimation of the rainfall kinetic energy, P_n ($J/m^2/s$), and momentum, M (N/m^2). These are two key parameters to define rainfall erosivity [62] and play a central role in soil detachment and splash erosion processes [8,63], infiltration and runoff dynamics [64], and rainfall-induced deterioration of cultural heritage surfaces [20].

Most disdrometers use a laser optical technology, such as the Eigenbrodt ODM 470 [65,66], the Laser Precipitation Monitor (LPM) [67], the Multi-Angle Snowflake Camera [68], the Particle Size Velocity (Parsivel) [69] and its integrated version (Parsivel^2) [70], and the two-dimensional video disdrometer [71,72].

Among the various disdrometers available, the Parsivel^2 (OTT Hydromet GmbH, Kempten, Germany) and the LPM (Thies Clima, Göttingen, Germany) are particularly prevalent in the field of rainfall simulator characterization [8,73–78].

Research on natural precipitation using Parsivel^2 and LPM disdrometers [79] has demonstrated that both instruments systematically tend to underestimate raindrop fall velocity and exhibit substantial discrepancies in the measured PSVD, with these differences becoming more pronounced at higher rainfall intensities. According to the authors [79], such differences and underestimations may arise from both hardware- and software-related factors, including geometric variations in the laser configuration.

Further potential biases intrinsic to optical disdrometers include the nonuniform power distribution across the laser beam, the misinterpretation of multiple particles as a single drop, and the occurrence of margin fallers, i.e., drops that only partially intersect the beam and are therefore not fully detected [80]. Although the manufacturers state that these devices incorporate corrections for margin fallers and multiple particles, the specific methodologies employed to address these sources of error remain undisclosed [79].

Therefore, due to the widespread tendency of optical disdrometers to underestimate fall velocity, several studies [72,79,81–88] have avoided the use of measured velocity distributions, opting instead for empirical relationships [88–91] calibrated on natural rainfall data, which relate terminal velocity, defined as the equilibrium speed at which gravitational and aerodynamic drag forces are balanced [92], to raindrop diameter.

Based on the experimental data of Gunn and Kinzer [88], Atlas et al. [89] and Atlas and Ulbrich [90] adopted the following empirical relationship to estimate the terminal velocity of raindrops, $V_{(D)}$ (m/s), as a function of their diameter D (mm):

$$V_{(D)} = 9.65 - 10.3e^{-0.6D} \tag{2}$$

Beard [91] did not propose a closed equation for estimating fall velocity but developed a physically based model in which environmental variables (such as altitude, latitude, temperature, relative humidity, air pressure) are considered. Therefore, this parametric model uses several fit curves to represent the terminal velocity over specific diameter ranges and atmospheric conditions.

Another formulation used in literature [93,94], also to estimate raindrop terminal velocity in disdrometers that do not provide direct velocity measurement [92,95,96], is that of Ferro [97], who used measurements from several studies [88,91,98–101] to propose the following relationship:

$$V_{(D)} = V_h[1 - \exp(-a_h D)] \quad (3)$$

in which D is expressed in cm, while the coefficients V_h (m/s) and a_h (cm) vary with the falling height, assuming values, respectively, equal to 9.5 m/s and 6 cm for natural precipitations.

However, previous empirical formulations are not commonly applied in simulated rainfall experiments, since the limited drop falling height under laboratory conditions often prevents raindrops from reaching their terminal velocity.

An alternative approach is to apply empirical relationships validated under laboratory conditions to estimate the fall velocity of raindrops, in which $V_{(D)}$ is explicitly expressed as a function not only of the raindrop diameter but also of the drop falling height set during the experiment. Recently, after having analyzed fall velocity measurements in literature in the falling height range of 0.5 m to 20 m [98], Serio et al. [49] proposed the following relationship:

$$V_{(D)} = \frac{11.50}{1 + \frac{1.65}{h^{0.8}}}\left[1 - e^{-11.43h^{-0.24}D}\right] \quad (4)$$

in which D is expressed in cm and h (m) is the falling height, i.e., the distance from the sprinkler's outlet of the rainfall simulator to the disdrometer measuring surface. The reliability of Equation (4) was positively verified using both experimental fall velocity measurements, up to 1.3 m, obtained with a drip-type rainfall simulator [49] and literature data up to 13 m [99].

In this context, the primary contribution of this study is the comprehensive characterization of a laboratory rainfall simulator, which is currently being employed in studies on heritage erosion. This characterization is performed by comparing different approaches, i.e., a volumetric method using rain collectors and a disdrometric one employing a Parsivel^2 disdrometer, and highlighting the discrepancies that can arise depending on the method chosen.

This is critical because erosion studies require the quantification of not only rainfall intensity but also crucial erosivity parameters such as rainfall kinetic energy and momentum, and their spatial uniformity across the tested surface.

The results presented here will deepen our understanding of the simulated rainfall's characteristics, enabling us to adapt the current configuration to different events, thereby enhancing the simulator's applicability to a variety of different situations focused on soil and heritage erosion.

In addition, disdrometric data and raindrop velocity distribution collected with an optical disdrometer are analyzed to evaluate the simulator's ability to reproduce raindrop velocities comparable to those observed in natural rainfall.

2. Materials and Methods

2.1. Experimental Setup

In the present study, the rainfall simulator (RS) recently proposed by Fernández-Raga et al. [20] is used; this RS was originally designed for studies on stone-based heritage erosion [20,102]. The device is located at the Department of Chemistry and Applied Physics, Industrial Engineering School, University of León. The characterization of the RS was performed using rain collectors for the volumetric calibration and an optical disdrometer (OTT Hydromet GmbH, model Parsivel^2, Kempten, Germany) for the disdrometric one. The falling height, defined as the distance from the RS to the optical disdrometer sampling surface, was set to 10 m.

2.1.1. Rainfall Simulator Design

The RS is suspended and anchored by tensioned steel cables to the walls of an internal courtyard (Figure 1), whose vertical walls reduce the wind influence and prevent droplet drift.

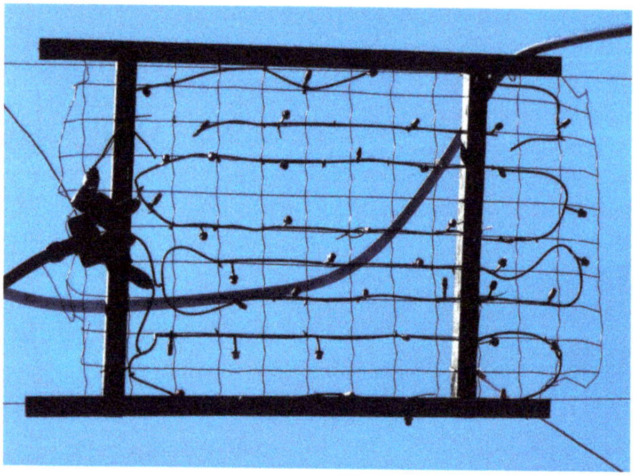

Figure 1. Rainfall simulator (RS) structure.

Tap water was avoided to supply the RS as its mineral content could, over time, compromise the hydraulic efficiency of the system by depositing calcium, especially in the smallest parts of the system (microsprinklers). Moreover, deionized water better represents natural rain from a physical/chemical point of view since the presence of salts modifies the water surface tension [103]. Therefore, using a purification system (Wasserlab, model DE7003, Barbatáin, Navarra, Spain), deionized water was continuously supplied to a 100 L tank equipped with a submersible pump (LISTA, Ehlis S.A., model Hidrosub AS-216, Illescas, Toledo, Spain).

Water was then conveyed to the main supply line using a surface pump (Hidrobex model EH-125, Parets del Vallés, Barcelona, Spain), which delivered it to the top of the internal courtyard. A constant pressure was selected for the present work. In particular, since weathering and erosion processes on stone-based cultural heritage occur over large temporal scales, an appropriate methodology already established for these scenarios [20,102,104] was followed, resulting in rainfall intensity values higher than those observed under natural precipitation conditions.

Still, since the surface pump loses power over time [20], its functioning was automated with an Arduino program. Therefore, after a continuous and constant run of 55 s, the program stops the pump for 5 s, thereby allowing for a constant mean flow rate over time. The cycle can be programmed for the desired time. For this investigation, a total rainfall cycle of 5 min was chosen.

Once water reaches the upper part of the hydraulic system, the main pipeline feeds the RS structure (Figure 1), which consists of a metal frame supporting four independent pipes equipped with 46 emitters with a flow rate that can range from 1 to 70 L/h (Figure 2). Specifically, two emitter models commonly employed in gardening applications were selected: 26 drippers releasing droplets in eight peripheral directions (Figure 2a), and 20 minisprinklers with a single central outlet for drop release, which can generate a spray radius of 0–0.5 m (Figure 2b).

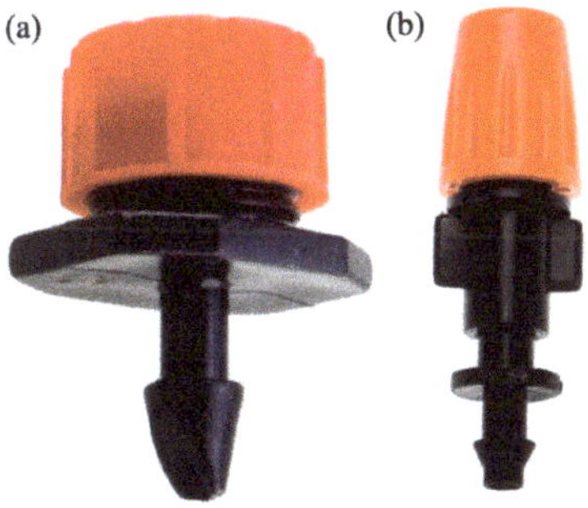

Figure 2. Emitters employed in the RS: drippers (**a**) and minisprinklers (**b**).

Both emitter models allow a fine regulation of the outlet pressure by screwing/unscrewing, thus impacting the size of simulated raindrops. Consequently, to generate simulated rainfalls with the widest possible drop size variability, both the positioning and the opening of the drippers and the minisprinklers were set randomly.

2.1.2. Optical Disdrometer

The disdrometric calibration was performed using a Parsivel^2 disdrometer (Figure 3).

Figure 3. Parsivel^2 optical disdrometer (OTT Hydromet GmbH, Kempten, Germany).

This optical device is widely employed in both simulated and natural rainfall studies. It measures the diameter and fall velocity of individual raindrops, classifying them into 1024 possible size–velocity combinations: 32 diameter classes ranging from 0.2 mm to 25 mm, with class widths progressively increasing from 0.125 mm to 3 mm, and 32 velocity classes ranging from 0.2 m/s to 20 m/s, with class widths increasing from 0.1 m/s to 3.2 m/s. The assignment of each drop to a specific combination is governed by an undisclosed internal algorithm [84].

2.2. *Laboratory Protocol for the Rainfall Simulator Characterization*

2.2.1. Volumetric Characterization

The volumetric calibration was focused on a rectangular sampling area of 1.8 m × 1.8 m (Figure 4).

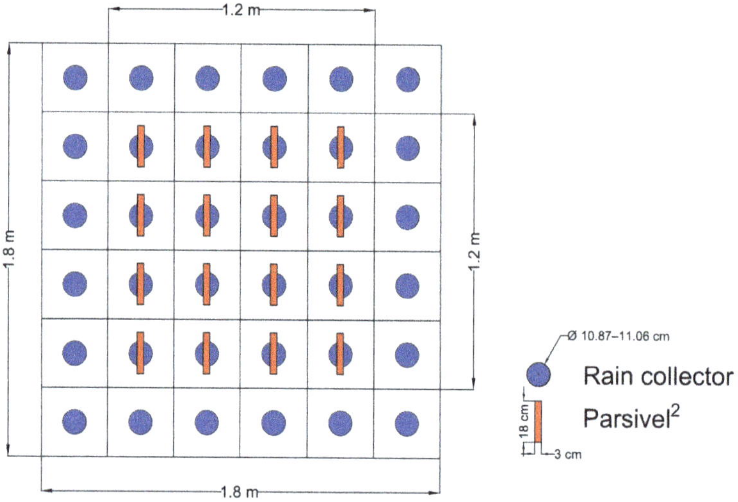

Figure 4. Schematic representation of the volumetric and disdrometric characterization method. All 36 rain collectors were contemporarily employed in the 1.8 m × 1.8 m wetted area, while a single Parsivel^2 disdrometer was alternatively positioned at the center of each of the 16 measuring surfaces of the 1.2 m × 1.2 m wetted area.

Since the floor is composed of regular square tiles measuring 0.3 × 0.3 m, the characterization was performed by placing rain collectors at the center of each of the 36 tiles. Moreover, given the simulator's configuration, characterized by a great falling height and high simulated rainfall intensities, concerns arose around the droplet impact potentially inducing splash phenomena. This could lead to water losses from within the collectors or, conversely, to additional inflow caused by droplets rebounding off the floor, thereby compromising the accuracy of volumetric measurements.

To address this issue, short plastic vessels (p_v), with a capacity of 0.5 L and a sampling surface of 9.607 × 10^{-3} m^2 (Figure 5a), and tall plastic cups (p_c), with a capacity of 1 L and a sampling surface of 9.276 × 10^{-3} m^2 (Figure 5b), were tested. For both rainfall collector types, the larger upper diameter was used to determine the effective sampling surface.

The two distinct collector designs were employed to experimentally evaluate and quantify the potential error introduced by droplet splash (splash-in/splash-out), ensuring the robustness of the volumetric calibration.

In addition, a third calibration dataset was obtained using p_cs across two successive rainfall simulation cycles to verify the temporal consistency of the simulator performance.

In other words, three different treatments (T) were employed: p_v for 5 min (T1), p_c for 5 min (T2), and p_c for 10 min (T3). Due to their limited capacity, p_vs could not be used for the 10 min simulation.

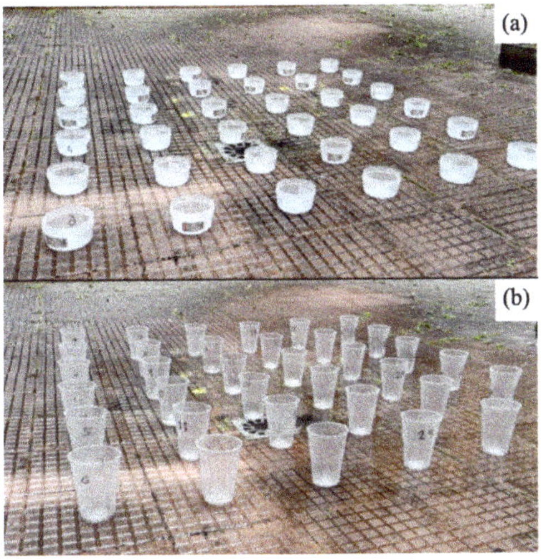

Figure 5. Rain collectors used for the volumetric calibration: plastic vessels (**a**) and plastic cups (**b**).

Before each rainfall simulation, p_v and p_c were calibrated using a digital scale. After the simulation, their outer surfaces were dried, and the collected water volume (*Vol*, L) was weighed. Knowing the volume collected by each collector, their sampling area (A, m^2), and the simulation time (t_S, h), the simulated rainfall intensity (I_s, mm/h) was calculated as:

$$I_S = \frac{Vol}{A\, t_S} \quad (5)$$

For each treatment, three repetitions were performed, leading to a total of 324 rainfall intensity measurements. To evaluate the rainfall spatial distribution of T1, T2, and T3, the mean value of the three repetitions was considered to determine the *CU* coefficient according to Equation (1).

Moreover, the I_S spatial distribution was also graphically evaluated using a geographic information system (GIS) software (QGIS, version 3.34.14-Prizren) by creating thematic maps, one for each treatment. To allow direct visual comparison among treatments, a fixed graduated color scale was applied, ranging from light yellow to dark red, to represent increasing rainfall intensity.

To determine whether the observed differences in terms of simulated rainfall intensity were statistically significant, a two-way analysis of variance (ANOVA) with replication was performed. The analysis considered two factors: treatment (T1, T2, T3), which reflects variations in both collector type and simulated rainfall duration, and collector position within the simulator. That is, the ANOVA test determines whether the considered factors, either individually or through their interaction, are responsible for any statistically significant differences in terms of simulated rainfall intensity. Specifically, F-values and corresponding p-values were calculated for each main effect and for the interaction term, and these were compared against the critical F-values to assess significance. The analysis was conducted using a significance threshold of $\alpha = 0.05$.

2.2.2. Disdrometric Characterization

Regarding the disdrometric characterization, a 1.2 m × 1.2 m wetted area was considered (Figure 4). This focus was selected because the previous volumetric calibration suggested that the central 1.2 m × 1.2 m plot provided the most consistent and uniform rainfall distribution. Thus, the Parsivel² disdrometer was placed at the center of each of the resulting 16 tiles (Figure 4).

For each position, three 5 min rainfall cycle repetitions were performed, resulting in a total of 240 1 min raw PSVD.

For each minute, rain intensity was determined using the following relationship:

$$I_P = \frac{\pi}{6}\frac{t}{3600}\sum_{i=1}^{32}\frac{D_i^3\, n_i}{\sigma_{eff},} \quad (6)$$

in which t (s) is the sampling time, equal to 60 s for the present investigation, D_i (mm) is the mean diameter of the i-th diameter class, n_i (−) is the number of drops recorded in the i-th class, and σ_{eff} (mm²) is the effective sampling area that accounts for the margin faller effect. This means that, to account for the undercounting of drops falling near the laser-beam edges, a correction was applied to the nominal Parsivel² sampling area (5400 mm²), according to the following relationship [79,84,105,106]:

$$\sigma_{eff} = L\left(W - \frac{n_i}{N}\frac{D_i}{2}\right) \quad (7)$$

where L (mm) and W (mm) are the Parsivel² laser beam length and width, respectively, which are equal to 180 mm and 30 mm, and N (−) is the total number of drops recorded during the 1 min interval.

Regarding the rainfall kinetic power and momentum, each raw PSVD was processed as follows for their determination:

$$P_n = \frac{1}{2}\frac{\rho\pi}{6t}\sum_{i=1}^{32}\sum_{j=i}^{32}\frac{\left(\frac{D_i}{1000}\right)^3 v_j^2\, n_{ij}}{\sigma_{eff}} \quad (8)$$

$$M = \frac{\rho\pi}{6t}\sum_{i=1}^{32}\sum_{j=i}^{32}\frac{\left(\frac{D_i}{1000}\right)^3 v_j\, n_{ij}}{\sigma_{eff}} \quad (9)$$

in which ρ (kg/m³) is the water density, equal to 1000 kg/m³, v_j (m/s) is the mean fall velocity of the j-th velocity class, and n_{ij} is the number of drops recorded in the i-th and j-th class.

The mean I_P, P_n, and M values of each position were employed to determine the spatial uniformity of the measured variables, according to Equation (1). Also, similarly to what was described in Section 2.2.1, the same procedure using a GIS software was followed to obtain thematic maps of the I_P, P_n, and M spatial distributions.

The rainfall intensity values obtained from the volumetric characterization using rain collectors were compared with those derived from the disdrometric calibration. As the previous two-way ANOVA with replication (Section 2.2.1) revealed no significant differences among the volumetric treatments (T1, T2, and T3), their mean value, denoted as I_T (mm/h), was used to represent the volumetric rainfall intensity.

To evaluate whether the RS was capable of generating raindrops resembling the terminal velocity of natural ones, an analysis of the measured fall velocity using the Parsivel^2 disdrometer was conducted.

Specifically, at first, for each i-th diameter class of the Parsivel^2 disdrometer, the weighted mean raindrop fall velocity measured with the Parsivel^2 disdrometer, v_P (m/s), relative to the entire dataset and thus independently of the sampled positions was obtained as follows:

$$v_P = \frac{\sum_{j=i}^{32} v_j \, n_i}{n_i} \quad (10)$$

Then, the resulting v_P values were compared with the calculated fall velocities, $V_{(D)}$, derived from Equations (2)–(4).

3. Results and Discussion

3.1. Rainfall Intensity Spatial Distribution: Volumetric and Disdrometric Differences

Figure 6 shows the spatial distribution of the rainfall intensity (I_S) obtained from the three volumetric characterization treatments: T1 (Figure 6a), T2 (Figure 6b), and T3 (Figure 6c). Each circle represents the average rainfall intensity recorded at a specific sampling position, calculated from three repetitions per treatment.

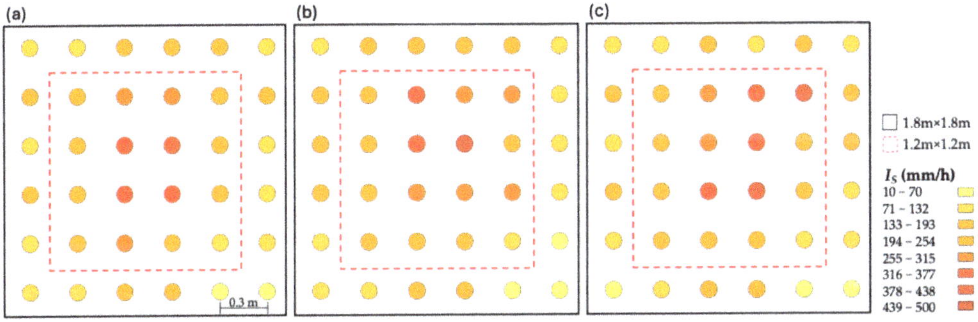

Figure 6. Spatial distribution of rainfall intensity (I_S) for the three characterization treatments (T) employed in this investigation: T1 (**a**), T2 (**b**), and T3 (**c**).

The thematic maps reveal a certain degree of variability in rainfall intensity, with lower values observed in the peripheral zones and higher values concentrated in the central area. When comparing the full sampling area (1.8 m × 1.8 m) with a slightly reduced central portion (1.2 m × 1.2 m), the latter, despite exhibiting higher intensity values, shows improved spatial homogeneity.

This central concentration may be attributed to the simulator's configuration: the 46 emitters, randomly oriented and alternately screwed in to reproduce a complex rainfall spectrum, may contribute more significantly to the central area, where drops from multiple emitters overlap, while the peripheral zones receive input from fewer emitters. As a result, the volumetric distribution is not perfectly uniform in the full sampling area.

These observations are supported by the data presented in Table 1, which reports, for each treatment, the mean I_S value, standard deviation, and CU coefficient calculated according to Equation (1), for both the full and reduced sampling areas.

The analysis reveals that the simulator delivers, on average, a nearly constant rainfall intensity across treatments (Table 1), considering both the full and reduced sampling areas.

However, the RS shows noticeable spatial differences within the full sampling area (*CU* from 56% to 62%), but improved uniformity across the 1.2 m × 1.2 m one (*CU* from 68% to 80%).

These results indicate that, within the full wetted area, the RS exhibits poor uniformity according to ASABE standards [56]. However, when considering only the central portion (Figure 6), the simulator reaches a 70% threshold across all treatments (mean *CU* equal to 75.16%) and even approaches 80% in T1 and T2, a threshold commonly adopted as a benchmark for uniformity in small-scale experiments [10,11,57,58] (Table 1).

Table 1. Mean rainfall intensity (I_S), standard deviation, and Christiansen uniformity (CU) coefficient (*CU*) for the three characterization treatments (T) employed in this investigation, evaluated in both the 1.8 m × 1.8 m and the 1.2 m × 1.2 m sampling surfaces.

Treatment	Mean I_S (mm/h)	Standard Deviation	*CU* (%)	Mean I_S (mm/h)	Standard Deviation	*CU* (%)
	1.8 m × 1.8 m Wetted Area			1.2 m × 1.2 m Wetted Area		
T1: p_v, 5 min	183.1	87.2	61.7	256.9	66.6	79.5
T2: p_c, 5 min	181.8	91.0	59.9	252.9	71.0	77.6
T3: p_c, 10 min	182.9	108.3	55.9	257.6	98.1	68.4

Although measuring two rainfall cycles might intuitively seem to reduce variability, T3 shows the highest variability (Figure 6c), the lowest *CU* values, and the highest standard deviations (Table 1), indicating the presence of more extreme intensity values. This is likely not due to manual interventions (the system runs continuously under Arduino control), but rather to nonstationary effects inherent to the simulator during its operation.

To further investigate whether significant statistical differences exist between the employed treatments (effect 1) and the position of the rain collectors (effect 2, accounting for the spatial distribution), a two-factor ANOVA was performed. The results in terms of sum of squares (*SQ*), degrees of freedom (*df*), mean squares (*MQ*), *F*-value, *p*-value, and *F*-critical are presented in Table 2.

Table 2. Two-factor ANOVA investigating whether significant statistical differences exist between the three employed characterization treatments and the rain collector positions.

Source of Variation	*SQ*	*df*	*MQ*	*F-Value*	*p-Value*	*F-Critical*
Rain collector position	2,496,505.58	35	71,328.73	59.14	6.40×10^{-92}	1.48
Treatments	107.17	2	53.59	0.04	0.96	3.04
Interaction	197,798.09	70	2825.69	2.34	1.43×10^{-6}	1.36
Intragroup	260,497.13	216	1206.01	—	—	—
Total	2,954,907.96	323	—	—	—	—

The results of the two-factor ANOVA show highly significant differences due to rain collector position (*F-value* > *F-critical*; $p < 0.001$), demonstrating spatial heterogeneity to represent the primary source of variability in rainfall intensity. In contrast, treatment effect, assessing differences due to collector type and simulation duration, is not statistically significant (*F-value* < *F-critical*; $p = 0.96$), indicating that rainfall intensity does not change appreciably among treatments.

Considering the previous results, rainfall intensity at each position can be expressed as a single representative value, I_T, obtained by averaging the results from the three different treatments.

Furthermore, to exclude positions less relevant from an erosive point of view, thus characterized by lower rainfall intensity values, and to characterize a more uniform pattern of the simulator from a disdrometric standpoint, the outer positions of the 1.8 m × 1.8 m pattern were excluded from subsequent analyses. Thus, only the central 1.2 m × 1.2 m wetted area was considered (Figure 6). For each of the 15 1 min PSVD measurements recorded using the Parsivel[2] disdrometer, Equation (6) was applied to compute the corresponding rainfall intensity value, and the mean I_P value was then obtained by averaging these 15 values.

To assess the agreement between the two measurement methods, the (I_T, I_P) pairs corresponding to each fixed sampling position were plotted in Figure 7, in which the solid line indicates the perfect agreement (1:1), while the dashed lines indicate the ±20% error boundaries.

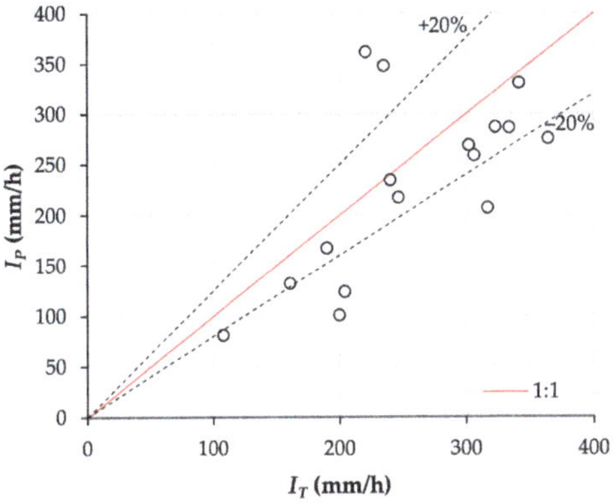

Figure 7. Comparison between simulated rainfall intensity measured with rain collectors (I_T) and that measured using a Parsivel[2] disdrometer (I_P).

Overall, the Parsivel[2] disdrometer appears to underestimate rainfall intensity compared with the rain collectors. Indeed, 87.5% of the measures lie below the 1:1 line (Figure 7), and 43.75% of the points exhibit errors greater than or equal to 20%. This result is consistent with those found in literature [81], where a general underestimation of rainfall intensity using the Parsivel[2] is detected and variance between identical Parsivel[2] devices increases significantly with precipitation intensity. Furthermore, many authors [79,84,105–107] have reported that underestimation in rainfall intensity can be attributed to instrumental errors, including margin fallers—raindrops pass only partially through the sensor's laser beam and are thus misclassified as smaller drops—and double detection—a single large drop is erroneously registered as two or more separate smaller drops.

However, it is important to note that the measurement surfaces are not identical: the disdrometer measures intensity over a rectangular area of 54 cm^2, whereas the rain collectors cover circular areas of 92.76 cm^2 (p_v) and 96.07 cm^2 (p_c).

Therefore, the observed discrepancies between I_T and I_P can be explained by the expected underestimation under extreme rainfall intensity conditions, the differences in the real measurement surfaces, and the fair spatial uniformity exhibited by the simulator.

To further deepen the analysis on the differences between the two measurement methods, spatial distribution of rainfall intensity obtained using both rain collectors (I_T) and a Parsivel[2] (I_P) is shown in Figure 8.

The coherent spatial overlay of the position of the rain collectors and the disdrometer, combined with consistent color coding, enables immediate visual comparison between the two measurement methods. This facilitates the identification of areas with similar rainfall intensities, the detection of extreme events, and a first evaluation of the spatial distribution captured by the disdrometer. Specifically, similar rainfall intensity values were recorded in only a few positions and, compared with the rain collector data, the intensities measured with the disdrometer appear less uniformly distributed.

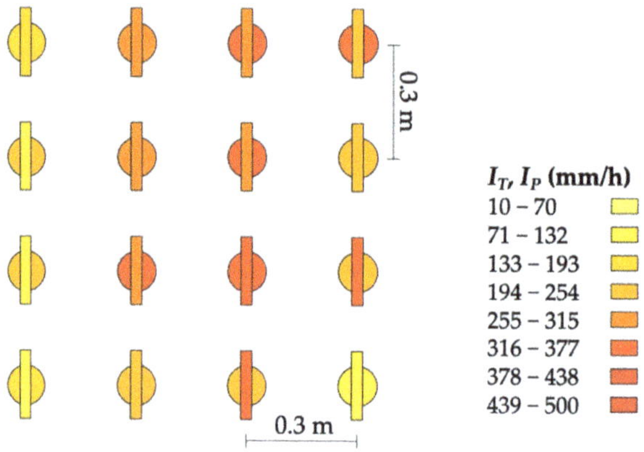

Figure 8. Spatial distribution of simulated rainfall intensity in a 1.2 m × 1.2 m sampling surface, obtained using rain collectors (I_T) and a Parsivel[2] disdrometer (I_P).

To quantitatively assess the uniformity of the simulated rainfall based on the full dataset from both rain collectors and disdrometric measurements, the *CU* coefficient was calculated using Equation (1). Notwithstanding similar mean rainfall intensities, equal to 255.8 mm/h (rain collectors) and 229.6 mm/h (disdrometer), the volumetric method yielded a *CU* of 75.7%, while the disdrometric method resulted in a lower *CU* of 61.1%.

According to the ASABE classification [56], the RS in the 1.2 m × 1.2 m surface would be rated as fair (70% < *CU* < 80%) when evaluated using rain collectors and poor (60% < *CU* < 70%) when evaluated using the disdrometer.

These discrepancies are consistent with findings reported in literature, which highlight that the number, spatial arrangement, and physical characteristics of rainfall collectors, such as their size and shape, directly influence the representation of rainfall distribution over a plot surface [55].

3.2. Rainfall Kinetic Power and Momentum Spatial Distribution

To evaluate the uniformity of the rainfall energy-related parameters, i.e., the rainfall kinetic power (P_n) and momentum (M), derived by applying Equations (8) and (9) to the disdrometer raw PSVD, their spatial distribution was analyzed. Figure 9 shows, for each sampled position, the mean P_n and M values of fifteen 1 min PSVD.

Overall, the distribution of energetic parameters (Figure 9) mirrors the pattern previously observed for rainfall intensity (Figure 8), displaying a non-uniform spatial distribution within the wetted area of 1.2 m × 1.2 m. The rainfall momentum exhibits a higher degree of spatial uniformity compared to kinetic power, likely due to the quadratic dependence of the latter on fall velocity. Indeed, the uneven distribution of rainfall intensity reflects the non-uniform spatial pattern of the drop size distribution and the associated fall velocities. Thus, this variability in raindrop fall velocity is further amplified in the calculation of kinetic power.

Specifically, the RS discharges a mean P_n of 1.54 J/m^2/s and a mean M of 0.39 N/m^2 in the 1.2 m x 1.2 m sampled area, registering a *CU* coefficient, respectively, equal to 61.5% and 65.0%. These values reflect a level of spatial variability comparable to that observed for rainfall intensity (I_P) (CU = 61.1%). Therefore, when assessed through disdrometric methods, the simulator exhibits *poor* uniformity [56] across the three metrics (I_P, P_n, M).

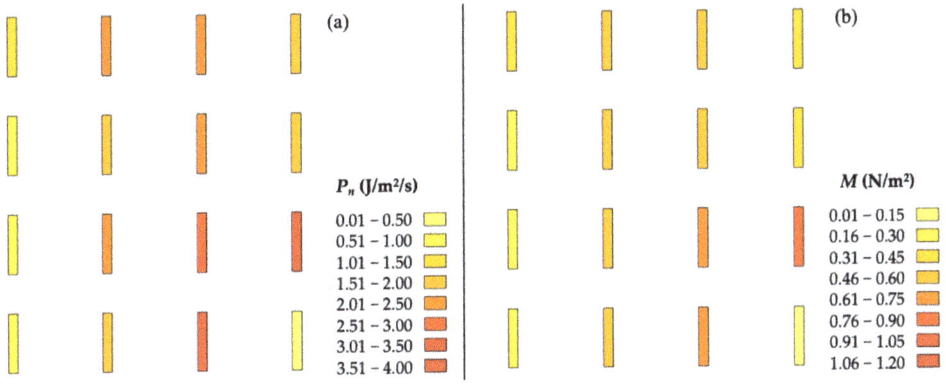

Figure 9. The spatial distribution of the rainfall kinetic power (P_n) (**a**) and Momentum (M) (**b**) obtained using a Parsivel[2] disdrometer.

3.3. Raindrop Fall Velocity

Given the 10 m drop falling height of the employed RS, raindrops are expected to reach terminal velocities comparable to those observed under natural precipitation conditions. To verify this, the weighted mean fall velocity for each diameter class, v_P, measured with the Parsivel[2] disdrometer was compared with the velocities $V_{(D)}$ calculated using empirical relationships from literature. Specifically, Equations (2) [89,90] and Equation (3) [97] represent typical fall velocities under natural rainfall conditions, while Equation (4) [49] was developed to estimate drop velocities in laboratory settings.

Figure 10 shows the comparison between the measured v_P, derived from raw PSVDs using Equation (10), and the calculated $V_{(D)}$ relationships. In particular, Figure 10a displays v_P versus $V_{(D)}$, with a 1:1 reference line and $\pm 20\%$ deviation bands, allowing visual assessment of the agreement between measured and predicted velocities. Figure 10b illustrates the variation in both v_P and $V_{(D)}$ as a function of drop diameter.

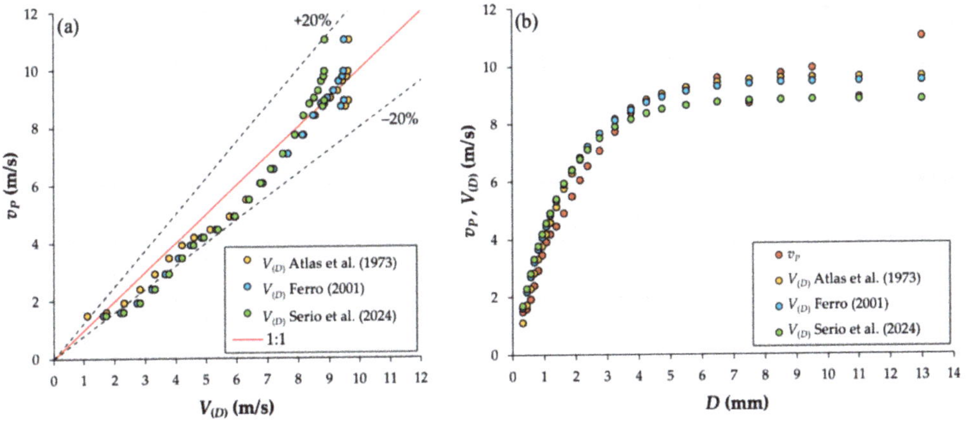

Figure 10. Comparison between weighted mean velocity measured with the Parsivel[2] disdrometer, v_P, and calculated, $V_{(D)}$, raindrop fall velocity (**a**); variation in both v_P and $V_{(D)}$ as a function of drop diameter, D (**b**).

Generally, the agreement between experimental measurements and empirical relationships demonstrates the

Effectiveness of the RS system in generating drops that fall at velocities consistent with terminal conditions, as evidenced by the close alignment between measured and predicted values (Figure 10a,b). However, the Parsivel^2 tends to underestimate fall velocities compared with all the tested empirical relationships, consistently with what has been observed in natural precipitation events [79].

Moreover, Equation (4) results applicable for predicting the fall velocity of simulated raindrops under the present experimental conditions, despite having been previously validated only for drip-type simulators with limited drop heights [49] (Figure 10a,b). The points of Atlas et al. [89] fall entirely within the ±20% error band, while those of Ferro [99] and Serio et al. [49] show greater errors, between 2 m/s and 4 m/s (Figure 10a).

Furthermore, for smaller diameters ($D < 3$ mm), the Parsivel^2 shows a general tendency to slightly underestimate the fall velocity compared with the values calculated using literature-based relationships. Conversely, for larger diameters ($D > 3$ mm), all the empirical models accurately describe the velocity of the simulated raindrops (Figure 10b), with greater uncertainty for diameter values above 6 mm. These results are consistent with the findings by Luo et al. [108], who reported that the Parsivel^2 exhibited a slight negative bias in mean fall velocity for the 1.375–2.75 mm diameter range, while showing a positive bias for drops exceeding 5.5 mm in diameter.

It is worth noticing also that the empirical relationships tend toward an asymptotic value of approximately 9–10 m/s (Figure 10a,b), which is consistent with the maximum velocities observed in nature for drops of around 6–8 mm. In contrast, the Parsivel^2 disdrometer recorded both unrealistically large drop diameters (up to 13 mm) and a maximum weighted mean fall velocity of 11.02 m/s (Figure 10a,b). These outliers may be attributed to coalescence phenomena induced by the simulation system, instrumental errors such as overlapping drops within the laser beam being interpreted as a single drop, or the limited fall height, which may prevent the onset of breakup processes typically observed in larger natural raindrops [109].

It is also important to consider that the velocity comparison in Figure 10 is presented per fixed diameter class, without accounting for the relative abundance of drops within each class, as the goal was to assess whether the agreement between measured and calculated velocities holds consistently across the full range of drop sizes. However, such an evaluation must be weighted by the actual frequency of occurrence of each diameter class. Specifically, in the present dataset, approximately 99.9% of all simulated raindrops fall within the diameter range of 0.31 mm to 6.50 mm. Consequently, discrepancies observed at larger diameters may have limited practical relevance, given their negligible contribution to the overall drop population.

3.4. Broader Implications and Future Applications

The RS characterized here is being currently applied to investigate the erosive effects of precipitation on stone-based heritage surfaces [20,102]. Nevertheless, the comprehensive characterization presented in this study using volumetric methods and disdrometric technics extends its potential far beyond this specific research field. Indeed, the detailed assessment of rainfall intensity, kinetic energy, and momentum, as well as their spatial distribution and the information regarding fall velocity provide a complete description of the simulated rainfall regime, therefore allowing the use of this simulator for a wide range of hydrological and erosive research topics.

Specifically, we recommend the 1.2 m × 1.2 m as the maximum sample size for experiments requiring the highest level of accuracy and fully controlled rainfall parameters. When testing samples that exceed this size, the characteristics for the area beyond this 1.2 m × 1.2 m plot must be evaluated relying only on the calibration data derived from the rain collectors, which cover the larger area, or spatially interpolating the disdrometer data to the unmeasured zones, while acknowledging the associated uncertainties that this estimation process introduces.

Moreover, since the selected sampling procedure can substantially influence the results of the simulator characterization, standardization of sampling procedures is of paramount importance and must combine different characterization methodologies for a more robust and comprehensive analysis.

However, in light of the results presented in this work, future research should aim to enhance the spatial uniformity of simulated rainfall and to standardize the protocol for a more robust and reliable comparison between events, thereby enhancing the overall reproducibility and applicability of the simulator for comparative studies.

4. Conclusions

This study provides a comprehensive characterization of a laboratory RS originally designed to reproduce the erosive effects of high-intensity rainfall on heritage materials. The integrated use of volumetric and disdrometric techniques allowed for a detailed evaluation of the simulator's hydrological and energetic performance, confirming its ability to realistically simulate the erosive effects of precipitation by achieving natural-like drop velocities and, therefore, kinetic energy and momentum.

The volumetric calibration, performed using different collector types and sampling durations, demonstrated that the simulator delivers consistent average rainfall intensities, but exhibits marked spatial variability over the entire wetted area. However, the *CU* coefficient improved up to nearly 80% within the central area.

Conversely, the analysis of the spatial distribution of rainfall intensity, kinetic power, and momentum obtained with a Parsivel^2 disdrometer within the same central area revealed poor uniformity.

Disdrometric measurements also confirmed the simulator's ability to generate raindrops able to achieve velocities comparable to those predicted by empirical relationships for natural rainfall, confirming its reliability in reproducing realistic fall dynamics even under limited falling height conditions.

Future research should aim to enhance the spatial uniformity of simulated rainfall and provide deeper insight into the physical realism of simulated precipitation, verifying whether the measured drop size distribution fits commonly used probabilistic functions such as gamma, Weibull, or lognormal.

Author Contributions: Conceptualization, F.G.C. and M.F.-R.; Methodology, R.C., M.A.S., F.G.C. and M.F.-R.; Software, G.B.-S.; Validation, R.C., M.A.S., F.G.C. and M.F.-R.; Formal Analysis, R.C., G.B.-S. and A.O.-M.; Investigation, R.C.; Resources, G.B.-S., A.O.-M. and M.F.-R.; Data Curation, R.C. and M.F.-R.; Writing—Original Draft Preparation, R.C.; Writing—Review and Editing, R.C., M.A.S., G.B.-S., V.F. and M.F.-R.; Visualization, R.C., M.A.S., F.G.C. and V.F.; Supervision, G.B.-S. and M.F.-R.; Project Administration, R.C., G.B.-S. and M.F.-R.; Funding Acquisition, G.B.-S., F.G.C. and M.F.-R. All authors have read and agreed to the published version of the manuscript.

Funding: Financial support is acknowledged from Spanish MCIN/AEI/10.13039/501100011033 and "ERDF A way of making Europe" by the European Union under project grant PID2023–147116OB-I00 (Nano shield 2.0) and by the "SiciliAn MicronanOTecH Research And Innovation Center" "SAMOTHRACE" (MUR, PNRR-M4C2, ECS_00000022, CUP B73C220081001), spoke 3—Università degli Studi di Palermo "S2-COMMs—Micro and Nanotechnologies for Smart & Sustainable Communities".

Data Availability Statement: The original contributions presented in this study are included in the article. Further inquiries can be directed to the corresponding authors.

Acknowledgments: The author R. C. gratefully acknowledges the hospitality and support provided by the University of León, specifically the Department of Chemistry and Applied Physics at the Industrial Engineering School, during his research stay. Moreover, the author R.C. also acknowledges the participation in the Erasmus+ Program (zero grant modality) of the University of Palermo (D.R. n. 848/2025). The authors want to thank Ángel Raga–Martín for his help in the experiments.

Conflicts of Interest: The authors declare no conflicts of interest.

List of Symbols and Units

The following Symbols and Units are used in this manuscript:

Symbol	Description	Unit
A	Rain collector sampling area	m^2
a_n	Coefficient of Equation (3)	cm
CU	Christiansen uniformity coefficient	%
D	Raindrop diameter	mm
D_i	Mean diameter of the *i*-th diameter class	mm
h	Drop falling height	m
I_i	Rainfall intensity of a single rain collector	mm/h
I_P	Rainfall intensity measured by the Parsivel^2 disdrometer	mm/h
I_S	Mean rainfall intensity	mm/h
I_T	Mean rainfall intensity across the treatments T1, T2, and T3	mm/h
L	Length of the Parsivel^2 laser beam	mm
M	Rainfall momentum	N/m^2
N	Total number of drops in 1-min sampling time by the Parsivel^2 disdrometer	(-)
n_c	Number of collectors	(-)
n_D	Number of diameter classes for which the disdrometer recorded velocities	(-)
n_i	Number of drops in the *i*-th diameter class	(-)
n_{ij}	Number of drops in the *i*-th dimeter class and *j*-th velocity class	(-)
Pn	Rainfall kinetic power	$J/m^2/s$
t	Sampling time for the disdrometric characterization	s
t_S	Time of the simulation for the volumetric characterization	h
$V_{(D)}$	Terminal fall velocity of raindrops according to literature relationships	m/s
v_j	Mean fall velocity if the *j*-th velocity class	m/s
V_n	Coefficient of Equation (3)	m/s
Vol	Water volume sampled by rain collectors	L
v_P	Mean raindrop fall velocity recorded by Parsivel^2 for each *i*-th diameter class	m/s
W	Width of the Parsivel^2 laser beam	mm
σ_{eff}	Effective sampling area of the Parsivel^2 disdrometer	mm^2
ρ	Water density	Kg/m^3

Abbreviations

The following abbreviations are used in this manuscript:

ANOVA	Analysis of variance
ASABE	American Society of Agricultural and Biological Engineers
ASTM	Advancing Standards Transforming Markets
df	Degree of freedom
GIS	Geographical iformation system
LPM	Laser Precipitation Monitor
MQ	Mean square
Parsivel	Particle Size Velocity

p_c	Plastic cups
PSVD	Particle Size and Velocity Distribution
p_v	Plastic vessels
RS	Rainfall simulator
SQ	Sum of square
T	Treatment for the volumetric characterization of the rainfall simulator

References

1. Cerdá, A. Simuladores de Lluvia y Su Aplicación a La Geomorfología: Estado de La Cuestión. *Cuad. Investig. Geográfica* **1999**, *25*, 45–84. [CrossRef]
2. Dunkerley, D. Rainfall Drop Arrival Rate at the Ground: A Potentially Informative Parameter in the Experimental Study of Infiltration, Soil Erosion, and Related Land Surface Processes. *Catena* **2021**, *206*, 105552. [CrossRef]
3. Meyer, L.D. Rainfall Simulators for Soil Conservation Research. In *Soil Erosion Research Methods*; Lal, R., Ed.; Soil & Water Conservation Societ: Ankeny, IA, USA, 1988; pp. 75–95.
4. Moore, I.D.; Hirschi, M.C.; Barfield, B.J. Kentucky Rainfall Simulator. *Trans. ASAE* **1983**, *26*, 1085–1089. [CrossRef]
5. Humphry, J.B.; Daniel, T.C.; Edwards, D.R.; Sharpley, A.N. A Portable Rainfall Simulator for Plot-Scale Runoff Studies. *Appl. Eng. Agric.* **2002**, *18*, 199–204. [CrossRef]
6. Lora, M.; Camporese, M.; Salandin, P. Design and Performance of a Nozzle-Type Rainfall Simulator for Landslide Triggering Experiments. *Catena* **2016**, *140*, 77–89. [CrossRef]
7. Meyer, L.D. Rainfall Simulators for Soil Erosion Research. In *Soil Erosion Research Methods*; Lal, R., Ed.; Routledge: Abingdon, UK, 1994; pp. 83–104.
8. Iserloh, T.; Ries, J.B.; Arnáez, J.; Boix-Fayos, C.; Butzen, V.; Cerdà, A.; Echeverría, M.T.; Fernández-Gálvez, J.; Fister, W.; Geißler, C.; et al. European Small Portable Rainfall Simulators: A Comparison of Rainfall Characteristics. *Catena* **2013**, *110*, 100–112. [CrossRef]
9. Hamed, Y.; Albergel, J.; Pépin, Y.; Asseline, J.; Nasri, S.; Zante, P.; Berndtsson, R.; El-Niazy, M.; Balah, M. Comparison between Rainfall Simulator Erosion and Observed Reservoir Sedimentation in an Erosion-Sensitive Semiarid Catchment. *Catena* **2002**, *50*, 1–16. [CrossRef]
10. Iserloh, T.; Fister, W.; Seeger, M.; Willger, H.; Ries, J.B. A Small Portable Rainfall Simulator for Reproducible Experiments on Soil Erosion. *Soil Tillage Res.* **2012**, *124*, 131–137. [CrossRef]
11. Mhaske, S.N.; Pathak, K.; Basak, A. A Comprehensive Design of Rainfall Simulator for the Assessment of Soil Erosion in the Laboratory. *Catena* **2019**, *172*, 408–420. [CrossRef]
12. Serio, M.A.; Caruso, R.; Carollo, F.G.; Bagarello, V.; Ferro, V.; Nicosia, A. The Hydraulic Assessment of a New Portable Rainfall Simulator Using Different Nozzle Models. *Water* **2025**, *17*, 1765. [CrossRef]
13. Martínez-Murillo, J.F.; Nadal-Romero, E.; Regüés, D.; Cerdà, A.; Poesen, J. Soil Erosion and Hydrology of the Western Mediterranean Badlands throughout Rainfall Simulation Experiments: A Review. *Catena* **2013**, *106*, 101–112. [CrossRef]
14. Chouksey, A.; Lambey, V.; Nikam, B.; Aggarwal, S.; Dutta, S. Hydrological Modelling Using a Rainfall Simulator over an Experimental Hillslope Plot. *Hydrology* **2017**, *4*, 17. [CrossRef]
15. Roth, C.H.; Meyer, B.; Frede, H.G. A Portable Rainfall Simulator for Studying Factors Affecting Runoff, Infiltration and Soil Loss. *Catena* **1985**, *12*, 79–85. [CrossRef]
16. Ogden, C.B.; van Es, H.M.; Schindelbeck, R.R. Miniature Rain Simulator for Field Measurement of Soil Infiltration. *Soil Sci. Soc. Am. J.* **1997**, *61*, 1041–1043. [CrossRef]
17. Boulange, J.; Malhat, F.; Jaikaew, P.; Nanko, K.; Watanabe, H. Portable Rainfall Simulator for Plot-Scale Investigation of Rainfall-Runoff, and Transport of Sediment and Pollutants. *Int. J. Sediment Res.* **2019**, *34*, 38–47. [CrossRef]
18. Abudi, I.; Carmi, G.; Berliner, P. Rainfall Simulator for Field Runoff Studies. *J. Hydrol.* **2012**, *454–455*, 76–81. [CrossRef]
19. Simelane, M.P.Z.; Soundy, P.; Maboko, M.M. Effects of Rainfall Intensity and Slope on Infiltration Rate, Soil Losses, Runoff and Nitrogen Leaching from Different Nitrogen Sources with a Rainfall Simulator. *Sustainability* **2024**, *16*, 4477. [CrossRef]
20. Fernández-Raga, M.; Rodríguez, I.; Caldevilla, P.; Búrdalo, G.; Ortiz, A.; Martínez-García, R. Optimization of a Laboratory Rainfall Simulator to Be Representative of Natural Rainfall. *Water* **2022**, *14*, 3831. [CrossRef]
21. Richards, J.; Zhao, G.; Zhang, H.; Viles, H. A Controlled Field Experiment to Investigate the Deterioration of Earthen Heritage by Wind and Rain. *Herit. Sci.* **2019**, *7*, 51. [CrossRef]
22. Arunrat, N.; Sereenonchai, S.; Kongsurakan, P.; Hatano, R. Assessing Soil Organic Carbon, Soil Nutrients and Soil Erodibility under Terraced Paddy Fields and Upland Rice in Northern Thailand. *Agronomy* **2022**, *12*, 537. [CrossRef]

23. Bharali, B. Rate of Infiltration for Different Soil Textures Using Rainfall Simulator and Green–Ampt Model. *ISH J. Hydraul. Eng.* **2021**, *27*, 70–76. [CrossRef]
24. Ma, G.; Li, G.; Mu, X.; Hou, W.; Ren, Y.; Yang, M. Effect of Raindrop Splashes on Topsoil Structure and Infiltration Characteristics. *Catena* **2022**, *212*, 106040. [CrossRef]
25. Fernández-Raga, M.; Palencia, C.; Keesstra, S.; Jordán, A.; Fraile, R.; Angulo-Martínez, M.; Cerdà, A. Splash Erosion: A Review with Unanswered Questions. *Earth Sci. Rev.* **2017**, *171*, 463–477. [CrossRef]
26. Rodrigo-Comino, J. *Precipitation: Earth Surface Responses and Processes*; Elsevier: Amsterdam, Netherlands, 2021; ISBN 9780128226995.
27. Aksoy, H.; Unal, N.E.; Cokgor, S.; Gedikli, A.; Yoon, J.; Koca, K.; Inci, S.B.; Eris, E. A Rainfall Simulator for Laboratory-Scale Assessment of Rainfall-Runoff-Sediment Transport Processes over a Two-Dimensional Flume. *Catena* **2012**, *98*, 63–72. [CrossRef]
28. Saber, A.N.; Somjunyakul, P.; Ok, J.; Watanabe, H. Rainfall-Runoff Simulation of Radioactive Cesium Transport by Using a Small-Scale Portable Rainfall Simulator. *Water Air Soil Pollut.* **2019**, *230*, 226. [CrossRef]
29. Eckley, C.S.; Branfireun, B. Simulated Rain Events on an Urban Roadway to Understand the Dynamics of Mercury Mobilization in Stormwater Runoff. *Water Res.* **2009**, *43*, 3635–3646. [CrossRef]
30. Biswas, S.; Kranz, W.L.; Shapiro, C.A.; Snow, D.D.; Bartelt-Hunt, S.L.; Mamo, M.; Tarkalson, D.D.; Zhang, T.C.; Shelton, D.P.; van Donk, S.J.; et al. Effect of Rainfall Timing and Tillage on the Transport of Steroid Hormones in Runoff from Manure Amended Row Crop Fields. *J. Hazard. Mater.* **2017**, *324*, 436–447. [CrossRef]
31. Rončević, V.; Živanović, N.; Radulović, L.; Ristić, R.; Sadeghi, S.H.; Fernández-Raga, M.; Prats, S.A. Design, Calibration, and Performance Evaluation of a High-Fidelity Spraying Rainfall Simulator for Soil Erosion Research. *Water* **2025**, *17*, 1863. [CrossRef]
32. Clarke, M.A.; Walsh, R.P.D. A Portable Rainfall Simulator for Field Assessment of Splash and Slopewash in Remote Locations. *Earth Surf. Process Landf.* **2007**, *32*, 2052–2069. [CrossRef]
33. Lascelles, B.; Favis-Mortlock, D.T.; Parsons, A.J.; Guerra, A.J.T. Spatial and Temporal Variation in Two Rainfall Simulators: Implications for Spatially Explicit Rainfall Simulation Experiments. In *Earth Surface Processes and Landforms*; Wiley: Hoboken, NJ, USA, 2000; Volume 25, pp. 709–721.
34. Boulal, H.; Gómez-Macpherson, H.; Gómez, J.A.; Mateos, L. Effect of Soil Management and Traffic on Soil Erosion in Irrigated Annual Crops. *Soil Tillage Res.* **2011**, *115–116*, 62–70. [CrossRef]
35. Ries, J.B.; Iserloh, T.; Seeger, M.; Gabriels, D. Rainfall Simulations—Constraints, Needs and Challenges for a Future Use in Soil Erosion Research. *Z. Fur Geomorphol.* **2013**, *57*, 1–10. [CrossRef]
36. da Silva Luz, C.C.; de Almeida, W.S.; de Souza, A.P.; Schultz, N.; Anache, J.A.A.; de Carvalho, D.F. Simulated Rainfall in Brazil: An Alternative for Assesment of Soil Surface Processes and an Opportunity for Technological Development. *Int. Soil Water Conserv. Res.* **2024**, *12*, 29–42. [CrossRef]
37. Grismer, M. Standards Vary in Studies Using Rainfall Simulators to Evaluate Erosion. *Calif. Agric.* **2012**, *66*, 102–107. [CrossRef]
38. Carollo, F.G.; Serio, M.A.; Caruso, R. Rainfall Energy Characteristics of a Modified Kamphorst Simulator. In *Biosystems Engineering Promoting Resilience to Climate Change—AIIA 2024—Mid-Term Conference*; Springer: Berlin/Heidelberg, Germany, 2025; pp. 138–145. [CrossRef]
39. Guida, G.; Lucchese, A.; Nicosia, A.; Palmeri, V.; Pampalone, V.; Ferro, V. Effects of Bacteria Inoculation on Soil Hydrology and Erosion Processes of a Clay Loam Soil. In Proceedings of the 2024 IEEE International Workshop on Metrology for Agriculture and Forestry (MetroAgriFor), Padua, Italy, 29 October 2024; pp. 392–396. [CrossRef]
40. Ma, X.; Li, Z.; Ren, Z.; Xu, G.; Gao, H.; Xie, M.; Wang, P. Evaluating the Role of Hydrological and Sediment Connectivity in Runoff and Sediment Transfer on the Loess Plateau: An in-Situ Field Rainfall Experiment. *J. Hydrol.* **2025**, *659*, 133226. [CrossRef]
41. *ASTM D6459-19*; Advancing Standards Transforming Markets (ASTM) Test Method for Determination of Rolled Erosion Control Product (RECP) Performance in Protecting Hillslopes from Rainfall-Induced Erosion. 2019. [CrossRef]
42. Ricks, M.D.; Horne, M.A.; Faulkner, B.; Zech, W.C.; Fang, X.; Donald, W.N.; Perez, M.A. Design of a Pressurized Rainfall Simulator for Evaluating Performance of Erosion Control Practices. *Water* **2019**, *11*, 2386. [CrossRef]
43. Manning, C.; Faulkner, B.; Donald, W.N.; Perez, M.A. Comparison of Erosion Control Products Using an ASTM D6459 Rainfall Simulator: Insights and Suggestions. *J. Irrig. Drain. Eng.* **2023**, *149*, 04023017. [CrossRef]
44. Manning, C.J.W. Rainfall Simulator Construction and Evaluation of Erosion Control Practices. Master's Thesis, Auburn University: Auburn, Alabama, 2021.
45. Midha, V.-K.; S, S.-K.; Sharma, A. Biodegradable Geomeshes for Rainsplash Erosion Control. *J. Fiber Bioeng. Inform.* **2017**, *10*, 155–161. [CrossRef]
46. Johannsen, L.L.; Zambon, N.; Strauss, P.; Dostal, T.; Neumann, M.; Zumr, D.; Cochrane, T.A.; Klik, A. Impact of Disdrometer Types on Rainfall Erosivity Estimation. *Water* **2020**, *12*, 963. [CrossRef]

47. Meshesha, D.T.; Tsunekawa, A.; Haregeweyen, N. Application of an Optical Disdrometer to Characterize Simulated Rainfall and Measure Drop-Size Distribution. *Hydrol. Sci. J.* **2018**, *63*, 1574–1587. [CrossRef]
48. Koch, T.; Chifflard, P.; Aartsma, P.; Panten, K. A Review of the Characteristics of Rainfall Simulators in Soil Erosion Research Studies. *MethodsX* **2024**, *12*, 102506. [CrossRef] [PubMed]
49. Serio, M.A.; Carollo, F.G.; Caruso, R.; Ferro, V. Guidelines for the Energetic Characterization of a Portable Drip-Type Rainfall Simulator for Soil Erosion Research. *Water* **2024**, *16*, 2100. [CrossRef]
50. Carollo, F.G.; Caruso, R.; Ferro, V.; Serio, M.A. Characterizing the Kamphorst Rainfall Simulator for Soil Erosion Investigations. *J. Hydrol.* **2024**, *643*, 132025. [CrossRef]
51. Lazarus, R.R.; Wan Jaafar, W.Z.; Alengaram, U.J.; Hin, L.S. Overview of the Research Gaps in the Rainfall Simulator Study. *Soil Sci. Soc. Am. J.* **2023**, *87*, 1231–1248. [CrossRef]
52. Isidoro, J.M.G.P.; Green, D.; Iserloh, T.; de Lima, J.L.M.P.; Pattison, I.; Marzen, M.; Silveira, A.; Stirling, R. Towards Harmonization in the Use of Rainfall Simulators—On the Pursuit of Better and More Comparable Experimental Results. In Proceedings of the 18th Biennial Conference of the Euromediterranean Network of Experimental and Representative Basins, Portoferraio, Italy, 7–10 June 2022.
53. Kibet, L.C.; Saporito, L.S.; Allen, A.L.; May, E.B.; Kleinman, P.J.A.; Hashem, F.M.; Bryant, R.B. A Protocol for Conducting Rainfall Simulation to Study Soil Runoff. *J. Vis. Exp.* **2014**, *86*, e51664. [CrossRef]
54. Iserloh, T.; Isidoro, J.M.G.P.; de Lima, J.L.M.P.; Marzen, M.; de Lima, M.I.P.; Green, D.; Seeger, M.; Ries, J.B. Moving towards Harmonisation in Rainfall Simulation. In Proceedings of the EGU General Assembly 2021, online, 19–30 April 2021. EGU21-5667. [CrossRef]
55. Green, D.; Pattison, I. Christiansen Uniformity Revisited: Re-Thinking Uniformity Assessment in Rainfall Simulator Studies. *Catena* **2022**, *217*, 106424. [CrossRef]
56. American Society of Agricultural and Biological Engineers. *Design and Installation of Microirrigation Systems*; American Society of Agricultural and Biological Engineers: St. Joseph, MI, USA, 2014.
57. Esteves, M.; Planchon, O.; Lapetite, J.M.; Silvera, N.; Cadet, P. The 'EMIRE' Large Rainfall Simulator: Design and Field Testing. *Earth Surf. Process Landf.* **2000**, *25*, 681–690. [CrossRef]
58. Salem, H.M.; Meselhy, A.A. A Portable Rainfall Simulator to Evaluate the Factors Affecting Soil Erosion in the Northwestern Coastal Zone of Egypt. *Nat. Hazards* **2021**, *105*, 2937–2955. [CrossRef]
59. Luk, S.; Abrahams, A.D.; Parsons, A.J. Sediment Sources and Sediment Transport by Rill Flow and Interrill Flow on a Semi-Arid Piedmont Slope, Southern Arizona. *Catena* **1993**, *20*, 93–111. [CrossRef]
60. Kavian, A.; Mohammadi, M.; Cerda, A.; Fallah, M.; Abdollahi, Z. Simulated Raindrop's Characteristic Measurements. A New Approach of Image Processing Tested under Laboratory Rainfall Simulation. *Catena* **2018**, *167*, 190–197. [CrossRef]
61. Fernández-Raga, M.; Cabeza-Ortega, M.; González-Castro, V.; Peters, P.; Commelin, M.; Campo, J. The Use of High-Speed Cameras as a Tool for the Characterization of Raindrops in Splash Laboratory Studies. *Water* **2021**, *13*, 2851. [CrossRef]
62. Catari, G.; Latron, J.; Gallart, F. Assessing the Sources of Uncertainty Associated with the Calculation of Rainfall Kinetic Energy and Erosivity—Application to the Upper Llobregat Basin, NE Spain. *Hydrol. Earth Syst. Sci.* **2011**, *15*, 679–688. [CrossRef]
63. Angulo-Martínez, M.; Barros, A.P. Measurement Uncertainty in Rainfall Kinetic Energy and Intensity Relationships for Soil Erosion Studies: An Evaluation Using PARSIVEL Disdrometers in the Southern Appalachian Mountains. *Geomorphology* **2015**, *228*, 28–40. [CrossRef]
64. Assouline, S. Drop Size Distributions and Kinetic Energy Rates in Variable Intensity Rainfall. *Water Resour. Res.* **2009**, *45*, 1–7. [CrossRef]
65. Carollo, F.G.; Serio, M.A.; Pampalone, V.; Ferro, V. Advances in a Measurement Method of Rainfall Kinetic Power and Momentum Affecting Soil Erosion Processes. *Hydrol. Process* **2024**, *38*, e15172. [CrossRef]
66. Carollo, F.G.; Caruso, R.; Di Stefano, C.; Ferro, V.; Pampalone, V.; Serio, M.A. Measurement Method of Rainfall Energetic Characteristics Using the Weibull Drop Size Distribution. *J. Hydrol.* **2026**, *664*, 134368. [CrossRef]
67. Lanzinger, E.; Theel, M.; Windolph, H. Rainfall Amount and Intensity Measured by the Thies Laser Precipitation Monitor. In Proceedings of the TECO-2006, Geneva, Switzerland, 4–6 December 2006.
68. Garrett, T.J.; Fallgatter, C.; Shkurko, K.; Howlett, D. Fall Speed Measurement and High-Resolution Multi-Angle Photography of Hydrometeors in Free Fall. *Atmos. Meas. Tech.* **2012**, *5*, 2625–2633. [CrossRef]
69. Löffler-Mang, M.; Joss, J. An Optical Disdrometer for Measuring Size and Velocity of Hydrometeors. *J. Atmos. Ocean. Technol.* **2000**, *17*, 130–139. [CrossRef]
70. OTT Hydromet GmbH. *Operating Instructions—Present Weather Sensor OTT Parsivel*^{2}; OTT Hydromet GmbH: Kempten, Germany, 2017.

71. Schönhuber, M.; Lammer, G.; Randeu, W.L. One Decade of Imaging Precipitation Measurement by 2D-Video-Distrometer. *Adv. Geosci.* **2007**, *10*, 85–90. [CrossRef]
72. Kruger, A.; Krajewski, W.F. Two-Dimensional Video Disdrometer: A Description. *J. Atmos. Ocean. Technol.* **2002**, *19*, 602–617. [CrossRef]
73. Petrů, J.; Kalibová, J. Measurement and Computation of Kinetic Energy of Simulated Rainfall in Comparison with Natural Rainfall. *Soil Water Res.* **2018**, *13*, 226–233. [CrossRef]
74. Neumann, M.; Zumr, D.; Kavka, P.; Laburda, T.; Lolk Johannsen, L.; Zambon, N.; Dostál, T.; Strauss, P.; Klik, A. Characterization of an Artificially Generated Rainfall Used for a Soil Erosion Research. *Vodohospodářské Tech.-Ekon. Inf.* **2019**, *61*, 10. [CrossRef]
75. Gires, A.; Bruley, P.; Ruas, A.; Schertzer, D.; Tchiguirinskaia, I. Disdrometer Measurements under Sense-City Rainfall Simulator. *Earth Syst. Sci. Data* **2020**, *12*, 835–845. [CrossRef]
76. Cottenot, L.; Courtemanche, P.; Nouhou-Bako, A.; Darboux, F. A Rainfall Simulator Using Porous Pipes as Drop Former. *Catena* **2021**, *200*, 105101. [CrossRef]
77. Jadhao, V.G.; Pandey, A.; Mishra, S.K. Sediment Modeling Using Laboratory-Scale Rainfall Simulator and Laser Precipitation Monitor. *Environ. Res.* **2023**, *237*, 116859. [CrossRef]
78. Bosio, R.; Cagninei, A.; Poggi, D. Large Laboratory Simulator of Natural Rainfall: From Drizzle to Storms. *Water* **2023**, *15*, 2205. [CrossRef]
79. Angulo-Martínez, M.; Beguería, S.; Latorre, B.; Fernández-Raga, M. Comparison of Precipitation Measurements by OTT Parsivel ^{2} and Thies LPM Optical Disdrometers. *Hydrol. Earth Syst. Sci.* **2018**, *22*, 2811–2837. [CrossRef]
80. Frasson, R.P.d.M.; da Cunha, L.K.; Krajewski, W.F. Assessment of the Thies Optical Disdrometer Performance. *Atmos. Res.* **2011**, *101*, 237–255. [CrossRef]
81. Fernández-Raga, M.; Fraile, R.; Keizer, J.J.; Varela Teijeiro, M.E.; Castro, A.; Palencia, C.; Calvo, A.I.; Koenders, J.; Da Costa Marques, R.L. The Kinetic Energy of Rain Measured with an Optical Disdrometer: An Application to Splash Erosion. *Atmos. Res.* **2010**, *96*, 225–240. [CrossRef]
82. Caracciolo, C.; Prodi, F.; Battaglia, A.; Porcu', F. Analysis of the Moments and Parameters of a Gamma DSD to Infer Precipitation Properties: A Convective Stratiform Discrimination Algorithm. *Atmos. Res.* **2006**, *80*, 165–186. [CrossRef]
83. Marzuki; Randeu, W.L.; Kozu, T.; Shimomai, T.; Hashiguchi, H.; Schönhuber, M. Raindrop Axis Ratios, Fall Velocities and Size Distribution over Sumatra from 2D-Video Disdrometer Measurement. *Atmos. Res.* **2013**, *119*, 23–37. [CrossRef]
84. Saha, R.; Testik, F.Y. Assessment of OTT Parsivel^2 Raindrop Fall Speed Measurements. *J. Atmos. Ocean. Technol.* **2023**, *40*, 557–573. [CrossRef]
85. Adirosi, E.; Gorgucci, E.; Baldini, L.; Tokay, A. Evaluation of Gamma Raindrop Size Distribution Assumption through Comparison of Rain Rates of Measured and Radar-Equivalent Gamma DSD. *J. Appl. Meteorol. Climatol.* **2014**, *53*, 1618–1635. [CrossRef]
86. Raupach, T.H.; Berne, A. Correction of Raindrop Size Distributions Measured by Parsivel Disdrometers, Using a Two-Dimensional Video Disdrometer as a Reference. *Atmos. Meas. Tech.* **2015**, *8*, 343–365. [CrossRef]
87. Adirosi, E.; Baldini, L.; Lombardo, F.; Russo, F.; Napolitano, F.; Volpi, E.; Tokay, A. Comparison of Different Fittings of Drop Spectra for Rainfall Retrievals. *Adv. Water Resour.* **2015**, *83*, 55–67. [CrossRef]
88. Gunn, R.; Kinzer, G.D. The Terminal Velocity of Fall For Water Droplets in Stagnant Air. *J. Meteorol.* **1949**, *6*, 243–248. [CrossRef]
89. Atlas, D.; Srivastava, R.C.; Sekhon, R.S. Doppler Radar Characteristics of Precipitation at Vertical Incidence. *Rev. Geophys.* **1973**, *11*, 1–35. [CrossRef]
90. Atlas, D.; Ulbrich, C.W. Path- and Area-Integrated Rainfall Measurement by Microwave Attenuation in the 1–3 Cm Band. *J. Appl. Meteorol.* **1977**, *16*, 1322–1331. [CrossRef]
91. Beard, K.V. Terminal Velocity and Shape of Cloud and Precipitation Drops Aloft. *J. Atmos. Sci.* **1976**, *33*, 851–864. [CrossRef]
92. Serio, M.A.; Carollo, F.G.; Ferro, V. Raindrop Size Distribution and Terminal Velocity for Rainfall Erosivity Studies. A Review. *J. Hydrol.* **2019**, *576*, 210–228. [CrossRef]
93. Yoo, C.; Cho, E.; Na, W.; Kang, M.; Lee, M. Change of Rainfall–Runoff Processes in Urban Areas Due to High-Rise Buildings. *J. Hydrol.* **2021**, *597*, 126155. [CrossRef]
94. Hadidi, A. Proposal and Design a Comprehensive Framework to Provide Water and Energy from Rain and Precipitation. *Water Resour. Ind.* **2024**, *32*, 100263. [CrossRef]
95. Carollo, F.G.; Ferro, V. Modeling Rainfall Erosivity by Measured Drop-Size Distributions. *J. Hydrol. Eng.* **2015**, *20*, C4014006. [CrossRef]
96. Carollo, F.G.; Ferro, V.; Serio, M.A. Estimating Rainfall Erosivity by Aggregated Drop Size Distributions. *Hydrol. Process* **2016**, *30*, 2119–2128. [CrossRef]

97. Ferro, V. Tecniche Di Misura e Monitoraggio Dei Processi Erosivi. In *Quaderni di Idronomia Montana—Trasporto di Azcqua e Sedimenti a Scala di Versante*; Bagarello, V., Ferro, V., Giordano, G., Eds.; Editoriale Bios: Cosenza, Italy, 2001; Volume 21/2, pp. 63–128.
98. Laws, J.O. Measurements of the Fall-velocity of Water-drops and Raindrops. *Eos Trans. Am. Geophys. Union* **1941**, *22*, 709–721. [CrossRef]
99. Blanchard, D.C. The Behavior of Water Drops at Terminal Velocity in Air. *Eos Trans. Am. Geophys. Union* **1950**, *31*, 836–842. [CrossRef]
100. Epema, G.F.; Riezebos, H.T. Fall Velocity of Waterdrops at Different Heights as a Factor Influencing Erosivity of Simulated Rain. *Catena Suppl.* **1983**, *4*, 1–17.
101. Jayawardena, A.W.; Rezaur, R.B. Drop Size Distribution and Kinetic Energy Load of Rainstorms in Hong Kong. *Hydrol. Process* **2000**, *14*, 1069–1082. [CrossRef]
102. Rodríguez, I.; Ortiz, A.; Caldevilla, P.; Giganto, S.; Búrdalo, G.; Fernández-Raga, M. Comparison between the Effects of Normal Rain and Acid Rain on Calcareous Stones under Laboratory Simulation. *Hydrology* **2023**, *10*, 79. [CrossRef]
103. Si, L.; Xi, Y.; Wei, J.; Wang, H.; Zhang, H.; Xu, G.; Liu, Y. The Influence of Inorganic Salt on Coal-Water Wetting Angle and Its Mechanism on Eliminating Water Blocking Effect. *J. Nat. Gas. Sci. Eng.* **2022**, *103*, 104618. [CrossRef]
104. Martínez-García, R.; González-Campelo, D.; Fraile-Fernández, F.J.; Castañón, A.M.; Caldevilla, P.; Giganto, S.; Ortiz-Marqués, A.; Zelli, F.; Calvo, V.; González-Domínguez, J.M.; et al. Performance Study of Graphene Oxide as an Antierosion Coating for Ornamental and Heritage Dolostone. *Adv. Mater. Technol.* **2023**, *8*, 2300486. [CrossRef]
105. Tokay, A.; Wolff, D.B.; Petersen, W.A. Evaluation of the New Version of the Laser-Optical Disdrometer, OTT Parsivel^2. *J. Atmos. Ocean. Technol.* **2014**, *31*, 1276–1288. [CrossRef]
106. Fraile, R.; Castro, A.; Fernández-Raga, M.; Palencia, C.; Calvo, A.I. Error in the Sampling Area of an Optical Disdrometer: Consequences in Computing Rain Variables. *Sci. World J.* **2013**, *2013*, 369450. [CrossRef] [PubMed]
107. Park, S.-G.; Kim, H.-L.; Ham, Y.-W.; Jung, S.-H. Comparative Evaluation of the OTT PARSIVEL2 Using a Collocated Two-Dimensional Video Disdrometer. *J. Atmos. Ocean. Technol.* **2017**, *34*, 2059–2082. [CrossRef]
108. Luo, L.; Wang, L.; Huo, T.; Chen, M.; Ma, J.; Li, S.; Wu, J. Raindrop Size Distribution and Rain Characteristics of the 2017 Great Hunan Flood Observed with a Parsivel2 Disdrometer. *Atmosphere* **2021**, *12*, 1556. [CrossRef]
109. Pruppacher, H.R.; Klett, J.D. Cloud Particle Interactions-Collision, Coalescence, and Breakup. In *Microphysics of Clouds and Precipitation*; Springer: Dordrecht, The Netherlands, 1978; pp. 464–503.

water

MDPI

Article

Correlation Analysis of Geological Disaster Density and Soil and Water Conservation Prevention and Control Capacity: A Case Study of Guangdong Province

Yaping Lu [1,2,3], Jingcheng Fu [4] and Li Tang [5,*]

1 Guangzhou Urban Planning & Design Survey Research Institute Co., Ltd., Guangzhou 510060, China; luyaping@gzsghy.wecom.work
2 Collaborative Innovation Center for Natural Resources Planning and Marine Technology of Guangzhou, Guangzhou 510060, China
3 Guangdong Enterprise Key Laboratory for Urban Sensing, Monitoring and Early Warning, Guangzhou 510060, China
4 Nanjing Research Institute of Hydrology and Water Conservation Automation, Ministry of Water Resources, Nanjing 210012, China; fujingcheng@nsy.com.cn
5 College of Geological and Surveying Engineering, Taiyuan University of Technology, Taiyuan 030024, China
* Correspondence: li370336892@163.com; Tel.: +86-186-3661-9229

Abstract: This study investigates the spatial coupling between geohazard susceptibility and soil conservation capacity in Guangdong Province, China, using integrated spatial analysis and machine learning approaches. Through kernel density estimation, hotspot analysis, principal component analysis (PCA), and t-SNE clustering applied to 11,252 geohazard records and nine soil conservation factors, we identify three critical mechanisms: (1) Topographic steepness (LS factor) constitutes the primary control on geohazard distribution (r = 0.162, $p < 0.001$), with high-risk clusters concentrated in northeastern mountainous regions (Meizhou-Huizhou-Heyuan); (2) Vegetation coverage (C_mean) mediates rainfall impacts, exhibiting significant risk reduction (r = −0.099, $p < 0.001$) despite counterintuitive negative correlations with mean rainfall erosivity; (3) Soil conservation effectiveness depends on topographic context, reducing geohazard density in moderate slopes (Cluster 0: 527.04) but proving insufficient in extreme terrain (Cluster 2: LS = 20.587). The emerging role of rainfall variability (R_slope, r = 0.183) highlights climate change impacts.

Keywords: geological disaster; soil and water conservation; correlation; principal component; clustering

1. Introduction

Geological hazards are destructive surface changes induced by geological processes or human activities, constituting natural disasters that pose severe threats to both society and ecosystems [1,2]. Global disaster inventories confirm that high-relief terrains and coastal zones are particularly prone to landslides, debris flows, and subsidence, events that frequently result in substantial socio-economic losses across these vulnerable regions. Critically, the escalating pace of climate change—manifested through increased frequency and intensity of extreme weather events—combined with the intensification of land-use practices, is dramatically amplifying the exposure and susceptibility of communities to these pervasive hazards [3]. This heightened risk profile underscores an urgent need for innovative mitigation strategies to enhance resilience [4,5].

Soil–water conservation (SWC) serves as a foundational disaster mitigation strategy by enhancing slope stability through erosion control [6]. Previous studies have demonstrated that vegetation restoration, structural stabilizers (e.g., lime, cement, and fly ash), and terrain engineering (e.g., soil compaction) can significantly mitigate rainfall-induced slope failures [7–9]. However, regionally divergent disaster susceptibilities persist, reflecting spatial heterogeneities in SWC implementation efficacy [10,11]. Thus, quantifying spatial linkages between geohazard density and SWC effectiveness emerges as a critical research frontier for sustainable risk governance [12].

Guangdong Province, situated in southern China, features diverse topography characterized by complex landforms and a humid climate with abundant rainfall. The region experiences various geological disasters, including landslides, collapses, and debris flows. Due to significant terrain undulations, intricate geological conditions, and high-intensity precipitation, Guangdong faces perennial threats from geological hazards. In recent years, accelerated urbanization and infrastructure development have intensified human impacts on the geological environment, leading to increased frequency and risk of geological disasters. To mitigate these risks, soil and water conservation measures have been widely implemented across Guangdong. These include vegetation restoration, soil improvement, and conservation engineering projects, which collectively enhance ecological conditions and soil structure [13]. For instance, the following may be considered: Vegetative cover and slope surface treatment reduce rainfall-induced soil erosion [14]; Barrier facilities decelerate debris flow velocity and dissipate destructive energy [15]. Such interventions not only lower disaster occurrence probabilities but also improve regional ecological resilience. However, the effectiveness of conservation measures varies significantly across regions, influenced by local geological settings, climatic patterns, and anthropogenic activities.

In recent years, scholars worldwide have made significant achievements in the research on geological hazards and soil–water conservation. Studies primarily focus on the following aspects: (1). Spatial distribution and influencing factors of geological hazards: Technologies such as remote sensing and Geographic Information Systems (GIS) are used to analyze spatial distribution characteristics and key influencing factors [16–18] (e.g., topography, rainfall, soil types) of geological hazards. These studies provide crucial data support for revealing the mechanisms of geological hazard occurrence. (2). Effectiveness evaluation of soil–water conservation measures [19,20]: Through field observations and experimental studies, the effectiveness of various soil–water conservation measures (e.g., vegetation restoration, engineering barriers) in reducing soil erosion and controlling geological hazards is evaluated. These studies provide a scientific basis for promoting and optimizing soil–water conservation practices. (3). Relationship between geological hazards and soil–water conservation [21,22]: Some scholars employ statistical analysis to explore the relationship between geological hazard frequency and soil–water conservation measures, finding that conservation factors (e.g., vegetation coverage [23], soil retention capacity [24]) significantly influence hazard occurrence. Despite these substantial accomplishments, most studies concentrate on single factors or localized regions, lacking systematic, global-scale spatial analysis of the relationship between geological hazard density and soil–water conservation capacity. Systematic research remains relatively scarce for regions like Guangdong Province, characterized by complex geological conditions and diverse hazard types. In particular, studies on the spatial distribution of geological hazard density and its correlation with soil–water conservation capacity are insufficient. The complexity of geological hazards and the diversity of conservation measures in Guangdong necessitate further research to uncover intrinsic relationships, providing scientific foundations for disaster prevention and mitigation.

This study aims to elucidate the spatial distribution characteristics of geological hazard density in Guangdong Province through spatial analysis methods and explore its relationship with soil and water conservation capacity. The specific objectives include: (1). Reveal spatial distribution patterns of geological hazard density: Utilize spatial hotspot analysis to uncover the spatial distribution laws of geological hazard density within the study area and identify high-risk zones. (2). Investigate spatial correlations between geological hazard density and soil and water conservation factor: Quantify the spatial correlation between geological hazard density and soil and water conservation factor, and identify key influencing factors. The findings will provide critical insights for geological hazard prevention and mitigation in Guangdong Province and other similar regions.

2. Materials and Methods

2.1. Study Area

Guangdong Province is located in the Pearl River Basin in southern China. The province features complex topography, primarily consisting of coastal plains, hills, and mountains. It experiences a humid subtropical monsoon climate, characterized by abundant annual rainfall. The average annual precipitation exceeds 1500 mm, with uneven distribution concentrated predominantly in the summer rainy season. Major rivers, including the Pearl River (Zhu Jiang), East River (Dong Jiang), and West River (Xi Jiang), form an extensive river network. This hydrological system exerts a crucial impact on regional soil–water conservation and the occurrence of geological hazards [25].

Guangdong Province exhibits a complex and diverse geological environment, primarily composed of granite, metamorphic rocks, and sedimentary rocks, representing distinct geological formations. The terrain features significant topographic relief, with mountainous areas accounting for over 76% of the total land area. Owing to complex topography and highly variable climatic conditions, the region is prone to frequent geological hazards, with landslides, rockfalls, and debris flows being the most prevalent types. During torrential rain seasons (April–September), intense rainfall and elevated soil saturation levels significantly increase the susceptibility to catastrophic geological events, particularly in mountainous and rural slope-cutting zones [26].

2.2. Data Source and Type

The spatial distribution data of geological hazard sites in Guangdong Province was obtained from the Resource and Environment Science and Data Center of the Chinese Academy of Sciences (https://www.resdc.cn/DataSearch.aspx) (accessed on 31 March 2024), encompassing 11,252 hazard sites across seven categories: collapses, sinkholes, debris flows, land subsidence, ground fissures, landslides, and unstable slopes [27].

The dataset of soil conservation service preventing soil water erosion in China (1992–2019) is available at Science Data Bank (https://cstr.cn/31253.11.sciencedb.07135) (accessed on 31 March 2024) [28]. This dataset includes nine zip files (".rar"). All the data in the zip files are raster data (".tif"). The details about every zip file we used were as follows (Figure 1).

"C1992–2019.rar": It contains the mean value and changing rate of the C-factor from 1992 to 2019, which were named "C_mean.tif" and "C_slope.tif", respectively.

"K_300.rar": It contains the K-factor data, named "K_300.tif" (300 m × 300 m resolution).

"LS_300.rar": It contains the LS-factor data, named "LS_300.tif" (300 m × 300 m resolution).

"P_300.rar": It contains the P-factor data, named "P_300.tif" (300 m × 300 m resolution).

"R1992–2019.rar": The mean value and changing rate of the R-factor in China from 1992 to 2019, named "R_mean.tif" and "R_slope.tif", respectively. The resolution of the data is 1 km × 1 km.

"SC 1992–2019.rar": The mean value and changing rate of soil conservation service (300 m × 300 m resolution).

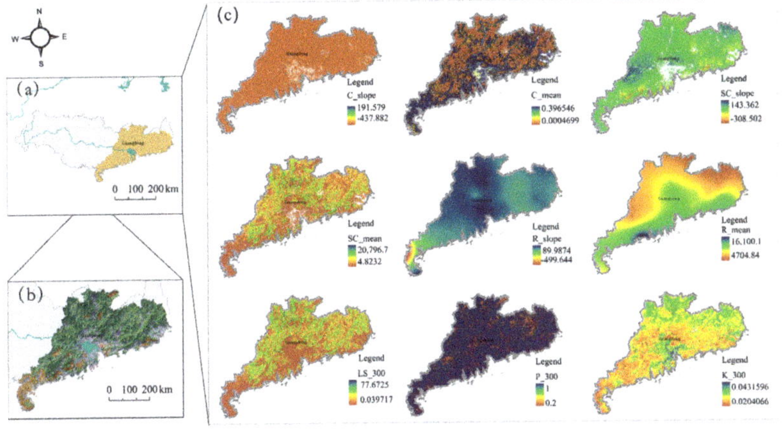

Figure 1. Study Area and Dataset. (**a**) geographical location of Guangdong Province; (**b**) geomorphic features of Guangdong Province; (**c**) spatial distribution of the dataset.

Which, C is the vegetation cover and management factor (dimensionless); K is the soil erodibility factor (t h MJ^{-1} mm^{-1}); LS is the topographic factor (dimensionless) with L being the slope length factor and S being the slope factor; P is the support practice factor (dimensionless); R is the rainfall erosivity factor (MJ mm ha^{-1} h^{-1} a^{-1}); SC is the soil conservation service (t ha^{-1} a^{-1}) is defined as soil retention, which is soil water erosion prevented by vegetation and practice measures.

2.3. Method

(1) Using the Kernel Density Estimation (KDE) method [29,30], we generated a density distribution map of geohazard sites via ArcGIS Pro 3.0.0 to determine their spatial density distribution.

$$\text{Gridcode} = \frac{1}{nh}\sum_{i=1}^{n} K\left(\frac{x - x_i}{h}\right) \tag{1}$$

where Gridcode represents the estimated density value at location x; n denotes the total number of geohazard sites; h is the bandwidth (smoothing parameter); K signifies the kernel function, with commonly used types including the Gaussian kernel and Uniform kernel; x_i indicates the location of the i-th geohazard site.

The bandwidth (search_radius) selection mechanism strictly follows Silverman's rule of thumb for spatial correction algorithms. The tool automatically calculates the optimal radius based on the spatial distribution of the input points (11,252 disaster points) using the following formula:

$$h = 0.9 \times min\left(\hat{\sigma}, \frac{IQR}{1.34}\right) \times n^{-\frac{1}{5}} \tag{2}$$

where $\hat{\sigma}$ is the standard deviation of the nearest neighbor distances, IQR is the interquartile range, and n = 11,252.

(2) Using ArcGIS Pro 3.0.0 and geohazard distribution data of Guangdong Province, we applied spatial statistical analysis to calculate the z-score and p-value for each geohazard site [31]. The z-score measures the local clustering intensity of geohazard sites, while the p-value (derived from the z-score based on the standard normal distribution) evaluates

the statistical significance. Confidence intervals further determine the reliability range of the results. This approach ultimately identified significant hot spot and cold spot clusters within Guangdong Province, revealing the spatial distribution characteristics and their statistical significance of geohazards.

The steps to identify significant geological hazard clusters using the Z-score are as follows:

1. Calculate the local density: Compute the kernel density value x_i for each location in Guangdong Province;

2. Calculate the global statistics:

$$\overline{x} = \frac{1}{n}\sum_{i=1}^{n} x_i \tag{3}$$

$$s = \sqrt{\frac{1}{n-1}\sum_{i=1}^{n}(x_i - \overline{x})^2} \tag{4}$$

3. Calculate Z-score:

$$z_i = \frac{x_i - \overline{x}}{S} \tag{5}$$

where x is the observed value; $\overline{x}$ is the average value.

4. Significance test for the Z-score of individual locations (Hotspot Analysis):

If $z_i \geq 2.58$, location i is a significant hotspot at the 99% confidence level (i.e., the hazard density at this location is far above the average level);

If $1.96 \leq z_i < 2.58$, location i is a hotspot at the 95% confidence level;

If $-1.96 \leq z_i < 1.96$, location i is not significant (i.e., the hazard density is close to the average level);

If $-2.58 \leq z_i < -1.96$, location i is a significant cold spot at the 95% confidence level;

If $z_i \leq -2.58$, location i is a significant cold spot at the 99% confidence level (i.e., the hazard density at this location is far below the average level).

(3) By overlaying soil and water conservation indices with geohazard density distribution data, we calculated the correlations and significance levels between geohazard density and each factor. Principal Component Analysis (PCA) [32,33] was employed to extract dominant influencing factors and quantify their impact on geohazard density.

The input data includes 11,252 observation samples and nine soil and water conservation factor variables (C_mean, C_slope, K_300, LS_300, P_300, R_mean, R_slope, SC_mean, SC_slope). The analysis process is as follows:

1. Data Standardization: Use Z-standardization to eliminate the effects of dimensions:

$$Z = \sigma(x - \mu) \tag{6}$$

where μ is the mean of the variable, and σ is the standard deviation.

2. Correlation Matrix Calculation: Construct a 9×9 Pearson correlation matrix R

$$\boldsymbol{R} = \begin{bmatrix} r_{11} & \cdots & r_{19} \\ \vdots & \ddots & \vdots \\ r_{91} & \cdots & r_{99} \end{bmatrix} \tag{7}$$

$$r_{ij} = \mathrm{corr}(x_i, x_j) = \frac{\sum_{k=1}^{n}(x_{ki-}\overline{x_i})}{\sqrt{\sum_{k=1}^{n}(x_{ki-}\overline{x_i})^2 \sum_{k=1}^{n}\left(x_{kj-}\overline{x_j}\right)^2}} \tag{8}$$

3. Eigen Decomposition: Solve for the eigenvalues and eigenvectors of the correlation matrix. The eigenvalues λ satisfy the following:

$$|R - \lambda I| = 0 \tag{9}$$

where R is the correlation matrix and I is the identity matrix.

4. Component Selection: Retain the principal components based on the Kaiser criterion (eigenvalues > 1).

5. Factor Rotation: Apply Varimax rotation to optimize the structure of the loading matrix.

(4) KMO Test: Conduct the Kaiser–Meyer–Olkin (KMO) test to quantitatively evaluate the sampling adequacy and variable intercorrelation strength, ensuring data suitability for factor analysis. The KMO calculation formula is as follows:

1. Calculate the partial correlation matrix:

$$P = -\frac{\mathbf{R}^{-1}}{\sqrt{diag(\mathbf{R}^{-1}) \times diag(\mathbf{R}^{T})}} \tag{10}$$

where, R^{-1} is the inverse of matrix R mathbit R^{T} is the transpose of matrix $\times$ R, and the diagonal elements are set to 0.

2. Calculate the KMO value:

$$KMO = \frac{\sum_{i \neq j} r_{ij}^2}{\sum_{i \neq j} r_{ij}^2 + \sum_{i \neq j} p_{ij}^2} \tag{11}$$

(5) Bartlett's Test of Sphericity: Perform Bartlett's test to statistically verify the presence of significant systematic associations among variables by rejecting the null hypothesis of an identity correlation matrix ($p < 0.01$).

$$\chi^2 = \left[(n-1) - \frac{2p+5}{6}\right] \times (-ln|\mathbf{R}|) \tag{12}$$

where p represents the number of variables, which is 9. $|\mathbf{R}|$ values are the determinants of the correlation matrix.

(6) t-Distributed Stochastic Neighbor Embedding (t-SNE) is a nonlinear dimensionality reduction method particularly suitable for high-dimensional data visualization. This method retains the local structure features of the data by optimizing the Kullback–Leibler divergence between the probability distributions in the low-dimensional and high-dimensional spaces. The principle is as follows:

1. High-Dimensional Space Similarity: For high-dimensional data points xi and xj, the conditional probability is defined as:

$$p_{j|i} = \frac{exp\left(-\|x_i - x_j\|^2/2\sigma_i^2\right)}{\sum_{k \neq i} exp\left(-\|x_i - x_k\|^2/2\sigma_i^2\right)} \tag{13}$$

where σ_i is the variance of the Gaussian distribution centered at x_i, controlled by the perplexity parameter.

$$Prep(P_i) = 2^{H(P_i)},\ H(P_i) = -\sum_j p_{j|i} log_2 p_{j|i} \tag{14}$$

2. Low-Dimensional Space Similarity: In the low-dimensional embedding space, similarity qij is measured using a Student's t-distribution:

$$q_{ij} = \frac{\left(1 + \|y_i - y_j\|^2\right)^{-1}}{\sum_{k \neq l} \left(1 + \|y_k - y_l\|^2\right)^{-1}} \quad (15)$$

3. Optimization Objective Function: The objective is to minimize the Kullback–Leibler divergence between pairwise distributions:

$$KL(P\|Q) = \sum_i \sum_j p_{ij} log \frac{p_{ij}}{q_{ij}} \quad (16)$$

Using gradient descent optimization, the partial derivative with respect to y_i is as follows:

$$\frac{\delta KL}{\delta y_i} = 4 \sum (p_{ij} - q_{ij})(y_i - y_j)\left(1 + \|y_i - y_j\|^2\right)^{-1} \quad (17)$$

In this study, t-SNE is used to reduce the dimensionality of the standardized soil conservation factors (including eight indicators such as terrain ruggedness, vegetation cover index, and rainfall erosivity) from a high-dimensional space to a two-dimensional visualization space. The t-distribution characteristic of t-SNE effectively handles the long-tail distribution features present in this study (such as the slope distribution), avoiding the "crowding problem" and enhancing the visualization of geomorphological features. The parameters are set as follows:

Perplexity = 30, balancing local and global structures;

Learning rate = 200, ensuring the convergence of the optimization process;

Early exaggeration = 12, enhancing cluster separation;

Number of iterations (n_iter) = 1000, ensuring sufficient convergence of the algorithm.

(7) Subsequently, K-means clustering was employed to delineate characteristic zones using the following:

1. Core formula of K-means clustering:

Objective function:

$$J(k) = \sum_{i=1}^{k} \sum_{x \in C_i} \|x - \mu_i\|^2 \quad (18)$$

where k is the number of clusters, C_i is the set of samples in the i-th cluster, μ_i is the centroid of the i-cluster, and $\|x - \mu_i\|$ is the Euclidean distance from the sample point x to the centroid μ_i. The centroid μ_i is calculated as follows:

$$\mu_i = \frac{1}{|C_i|} \sum_{x \in Ci} x \quad (19)$$

The Euclidean distance is calculated as follows:

$$d(x, y) = \sqrt{\sum_{j=1}^{p} (x_j - y_j)^2} \quad (20)$$

where x, y are two p-dimensional sample points (where $p = 9$ soil and water conservation factors), and x_j, y_j are the values of the j-th variable.

Cluster number selection criteria:

① Elbow method: The rate of decrease in $J(k)$ as k increases:

$$\Delta(k) = \left| \frac{J(k) - J(k+1)}{J(k+1)} \right| \quad (21)$$

when $\Delta(k)$ is significantly smaller than $\Delta(k+1)$, choose k as the optimal value.

2. Kruskal–Wallis test (overall difference test):

$$H = \left(\frac{12}{N(N+1)} \sum_{i=1}^{k} \frac{R_i^2}{n_i} \right) - 3(N+1) \quad (22)$$

where N is the total sample size, k is the number of clusters, R_i is the rank sum of the i-th cluster, and n_i is the sample size of the i-th cluster.

3. Mann–Whitney U test (pairwise comparison):

$$U = R_1 - \frac{n_1(n_1+1)}{2} \quad (23)$$

where R_1 is the rank sum of the first group.

3. Results

3.1. Distribution Characteristics of Geohazard Density

Figure 2a displays the spatial distribution of geohazard types across Guangdong Province, indicating province-wide coverage with a particularly dense concentration of both variety and quantity in cities like Guangzhou and Shenzhen. Figure 2b illustrates regional geohazard intensity, where Areas (0.001–282.841) represent widespread low-intensity zones. Areas (1979.884–2825.644) denote concentrated high-intensity zones, primarily in Meizhou, Huizhou, Heyuan, and the surrounding regions. Figure 2c identifies geohazard hot spots and cold spots using hot spot analysis: Deep red areas (99% confidence level) cluster in Meizhou, Huizhou, and northern Heyuan. Orange areas (95% confidence level) show broader distribution, including peri-urban zones of Qingyuan City.

3.2. Correlation Between Geological Hazard Density and Soil Conservation Factors

Figure 3 illustrates the correspondence between kernel density values and nine erosion factors (C_mean, C_slope, K_300, LS_300, P_300, R_mean, R_slope, SC_mean, SC_slope). Correlation analysis demonstrates significant relationships between geological hazard density and soil–water conservation factors ($p < 0.05$) (Figure 4, Tables 1 and 2). Key correlations include: strong negative correlations with mean rainfall erosivity (r = −0.247) and mean vegetation coverage (C_mean = −0.099); strong positive correlations with topographic factor (LS_300 = 0.162), mean soil conservation capacity (SC_mean = 0.116), and rate of change in rainfall erosivity (R_slope = 0.183); while soil erodibility (K_300 = 0.019) and rate of change in soil conservation capacity (SC_slope = 0.067) exhibited slight yet statistically significant positive correlations.

Table 1. Correlation Matrix.

	grid_code	C_mean	C_slope	K_300	LS_300	P_300	R_mean	R_slope	SC_mean	SC_slope
grid_code	1	−0.099	−0.018	0.019	0.162	0.007	−0.247	0.183	0.116	0.067
C_mean	−0.099	1	0.269	0.065	−0.549	−0.348	0.256	−0.237	−0.566	0.109
C_slope	−0.018	0.269	1	0.073	−0.137	−0.119	−0.059	0.002	−0.167	−0.126
K_300	0.019	0.065	0.073	1	−0.012	−0.154	−0.346	0.281	−0.047	0.023
LS_300	0.162	−0.549	−0.137	−0.012	1	0.258	−0.252	0.22	0.954	−0.262
P_300	0.007	−0.348	−0.119	−0.154	0.258	1	0.011	0.018	0.248	−0.098
R_mean	−0.247	0.256	−0.059	−0.346	−0.252	0.011	1	−0.244	−0.05	−0.345
R_slope	0.183	−0.237	0.002	0.281	0.22	0.018	−0.244	1	0.194	0.085
SC_mean	0.116	−0.566	−0.167	−0.047	0.954	0.248	−0.05	0.194	1	−0.422
SC_slope	0.067	0.109	−0.126	0.023	−0.262	−0.098	−0.345	0.085	−0.422	1

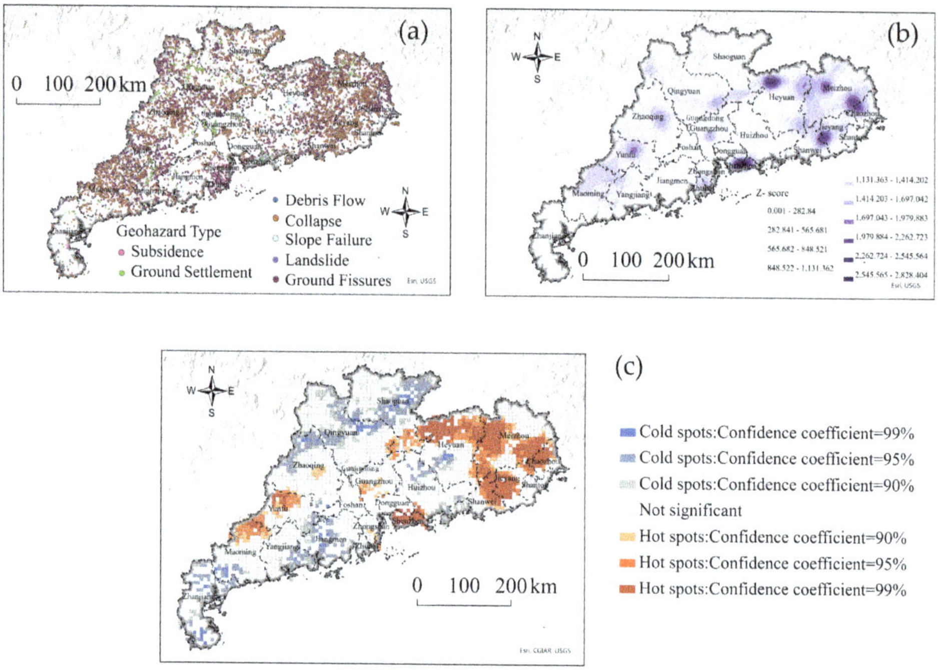

Figure 2. Spatial Distribution of Geohazard Types in Guangdong Province. (**a**) geohazard type of Guangdong Province; (**b**) Z-score distribution; (**c**) cold and hot spots distribution.

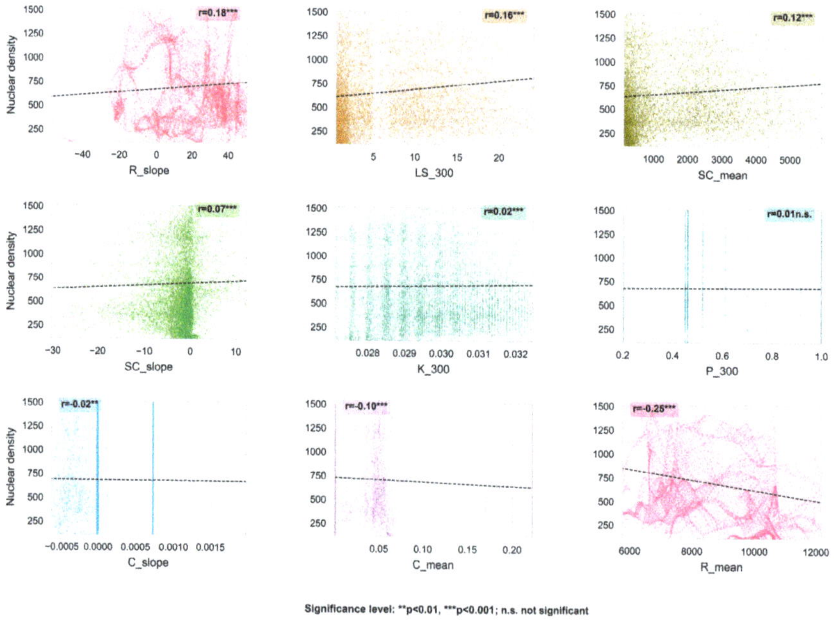

Figure 3. The correspondence between kernel density values and nine erosion factors.

Figure 4. Correlation Diagram between Soil Conservation Factors and Geological Hazard Kernel Density.

Table 2. Correlation and Significance Analysis Results.

Factor	Correlation	*p*-Value	Significance	Interpretation
R_mean	−0.247	0	***	$p < 0.001$
R_slope	0.183	0	***	$p < 0.001$
LS_300	0.162	0	***	$p < 0.001$
SC_mean	0.116	0	***	$p < 0.001$
C_mean	−0.099	0	***	$p < 0.001$
SC_slope	0.067	0	***	$p < 0.001$
K_300	0.019	0.0006	***	$p < 0.001$
C_slope	−0.018	0.0019	**	$p < 0.01$
P_300	0.007	0.2469	n.s.	$p \geq 0.10$

Notes: Significance (*p*-Value): *p*-Value < 0.001, "***"; *p*-Value < 0.01, "**"; not significant, "n.s.".

3.3. *Principal Components of Soil and Water Conservation Affecting Nuclear Density*

The results of the factor analysis suitability test indicate that the Bartlett's test of sphericity yielded a statistic of 153,644.1 ($p < 0.001$), strongly rejecting the null hypothesis of variable independence and supporting the suitability of the data for dimensionality reduction analysis (Table 3 and Figure 5). However, the KMO test value was 0.47, below the recommended threshold of 0.6, indicating weak commonality information among the soil and water conservation factors. Specifically, there is significant informational independence between variables, such as a strong direct association between LS_300 (topographical factor) and SC_mean (soil conservation capacity) ($\rho = 0.97$, $\rho^2 = 0.94$), whereas the association between vegetation cover (C_mean) and rainfall erosivity (R_mean) is mainly mediated by topography ($r = -0.55$ vs. $\rho = -0.13$). This complex interaction pattern reflects the spatial heterogeneity characteristics of the soil and water conservation system in Guangdong Province.

Table 3. The results of the factor analysis suitability test.

Variable Pair	r_{ij}	r_{ij}^2	p_{ij}	p_{ij}^2
C_mean vs. C_slope	0.27	0.07	0.19	0.04
C_mean vs. K_300	0.06	0.00	0.19	0.04
C_mean vs. LS_300	−0.55	0.30	0.30	0.09
C_mean vs. P_300	−0.35	0.12	−0.29	0.09
C_mean vs. R_mean	0.26	0.07	0.41	0.17
C_mean vs. R_slope	−0.24	0.06	−0.13	0.02
C_mean vs. SC_mean	−0.57	0.32	−0.39	0.15
C_mean vs. SC_slope	0.11	0.01	−0.08	0.01
C_slope vs. K_300	0.07	0.01	−0.04	0.00
C_slope vs. LS_300	−0.14	0.02	0.11	0.01
C_slope vs. P_300	−0.12	0.01	−0.04	0.00
C_slope vs. R_mean	−0.06	0.00	−0.10	0.01
C_slope vs. R_slope	0.00	0.00	0.09	0.01
C_slope vs. SC_mean	−0.17	0.03	−0.13	0.02
C_slope vs. SC_slope	−0.13	0.02	−0.29	0.09
K_300 vs. LS_300	−0.01	0.00	−0.15	0.02
K_300 vs. P_300	−0.15	0.02	−0.07	0.00
K_300 vs. R_mean	−0.35	0.12	−0.38	0.14
K_300 vs. R_slope	0.28	0.08	0.26	0.07
K_300 vs. SC_mean	−0.05	0.00	0.12	0.02
K_300 vs. SC_slope	0.02	0.00	−0.12	0.01
LS_300 vs. P_300	0.26	0.07	0.20	0.04
LS_300 vs. R_mean	−0.25	0.06	−0.64	0.41
LS_300 vs. R_slope	0.22	0.05	−0.05	0.00
LS_300 vs. SC_mean	0.95	0.91	0.97	0.94
LS_300 vs. SC_slope	−0.26	0.07	0.39	0.15
P_300 vs. R_mean	0.01	0.00	0.16	0.02
P_300 vs. R_slope	0.02	0.00	−0.02	0.00
P_300 vs. SC_mean	0.25	0.06	−0.18	0.03
P_300 vs. SC_slope	−0.10	0.01	−0.09	0.01
R_mean vs. R_slope	−0.24	0.06	−0.04	0.00
R_mean vs. SC_mean	−0.05	0.00	0.61	0.37
R_mean vs. SC_slope	−0.35	0.12	−0.05	0.00
R_slope vs. SC_mean	0.19	0.04	0.08	0.01
R_slope vs. SC_slope	0.08	0.01	0.16	0.03
SC_mean vs. SC_slope	−0.42	0.18	−0.48	0.23
Total		2.90		3.25
KMO = 2.90/(2.90 + 3.25) = 0.47				

3.3.1. Results of Principal Component Analysis (PCA)

Principal Component Analysis (PCA) reveals that soil and water conservation factors explain 83.13% of the total variance across five principal components (Figure 6, Tables 4 and 5). PC1 (31.42% variance) has high positive loadings from topography (LS_300, 0.54) and soil conservation services (SC_mean, 0.55), and a high negative loading from vegetation cover (C_mean, −0.46). PC2 (19.28% variance) is driven positively by rainfall erosivity (R_mean, 0.60) and negatively by soil erodibility (K_300, −0.50) and soil conservation change rate (SC_slope, −0.41). Subsequent PCs capture complex interactions: PC3 (13.83%) shows inverse vegetation cover (C_slope, 0.59) and soil conservation change (SC_slope, −0.56); PC4 (9.80%) reflects synergy between engineering measures (P_300, 0.59) and vegetation change (C_slope, 0.68); PC5 (8.80%) indicates opposite responses in rainfall erosion (R_slope, −0.57) and engineering measures (P_300, −0.55).

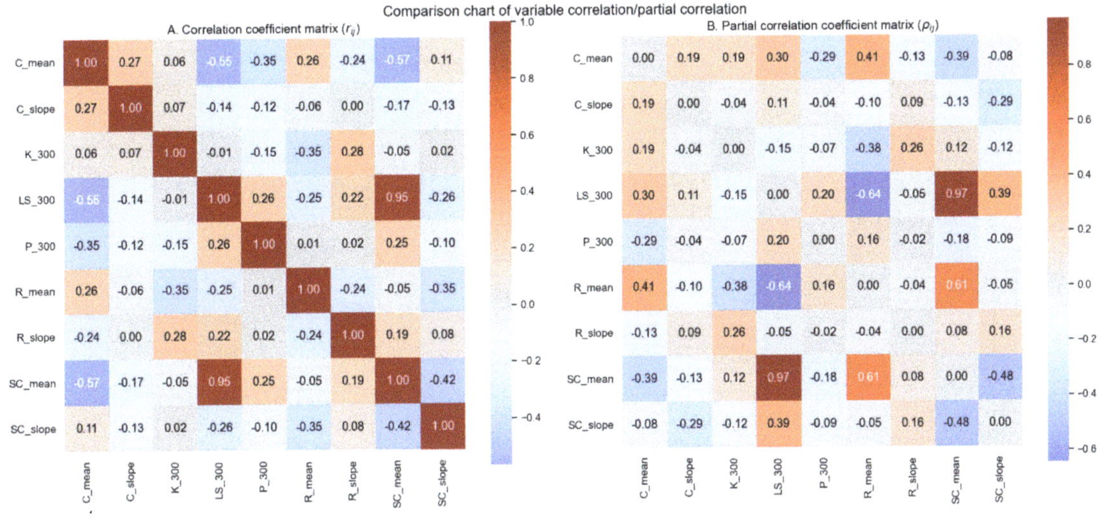

Figure 5. Comparison chart of variable correlation and partial correlation.

Table 4. Principal Component Loadings Table.

	PC1	PC2	PC3	PC4	PC5	PC6	PC7	PC8	PC9
C_mean	−0.462	0.067	0.233	−0.072	0.008	0.029	0.754	0.386	−0.067
C_slope	−0.158	−0.054	0.591	0.683	0.119	−0.234	−0.054	−0.291	−0.017
K_300	−0.01	−0.498	0.355	−0.237	−0.276	0.605	0.067	−0.35	0.022
LS_300	0.544	−0.014	0.116	−0.015	0.289	−0.041	0.383	−0.052	0.675
P_300	0.262	0.167	−0.304	0.588	−0.546	0.314	0.253	0.068	−0.028
R_mean	−0.134	0.596	0.09	−0.291	−0.378	−0.196	0.123	−0.566	0.135
R_slope	0.193	−0.428	0.127	−0.128	−0.565	−0.624	0.036	0.195	0.009
SC_mean	0.548	0.112	0.176	−0.122	0.183	−0.051	0.268	−0.14	−0.717
SC_slope	−0.201	−0.407	−0.557	0.105	0.18	−0.22	0.355	−0.509	−0.079

Table 5. Principal Component Variance Table.

Principal Component	Variance Explained	Cumulative Variance
PC1	0.314249	0.314249
PC2	0.192757	0.507006
PC3	0.13828	0.645286
PC4	0.098034	0.74332
PC5	0.087963	0.831283
PC6	0.079334	0.910616
PC7	0.052394	0.96301
PC8	0.035187	0.998197
PC9	0.001803	1

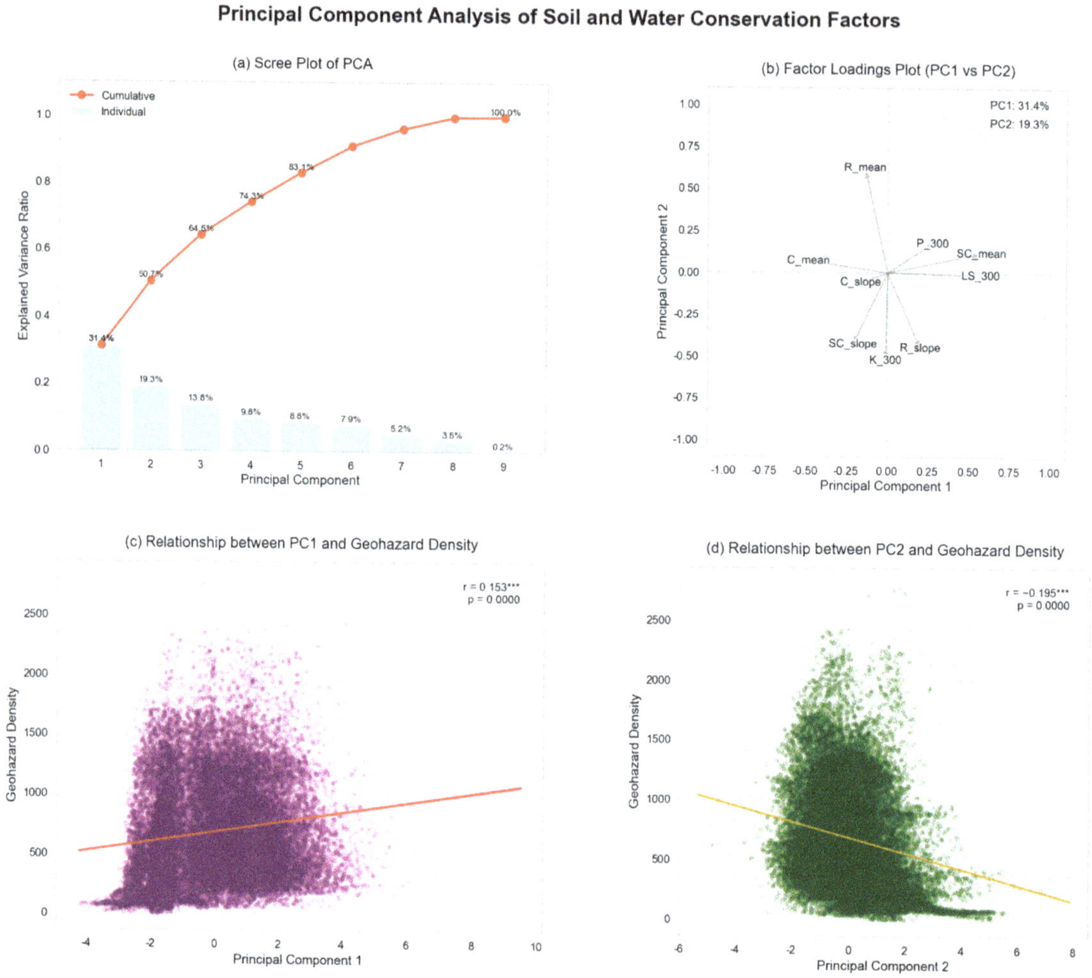

Figure 6. Principal Component Analysis of Soil and Water Conservation Factors.

3.3.2. t-SNE Dimensionality Reduction and Clustering Analysis

Cluster analysis identified five feature regions (contour coefficient = 0.246, Calinski–Harabasz index = 7046.5) (Figure 7, Table 6): Cluster 1 exhibits the highest disaster density (672.59), where low vegetation coverage (0.026) and steep terrain (topographic factor = 11.305) dominate risk factors, with moderate soil conservation capacity (2462.48) insufficient to offset topographic disadvantages. Cluster 2 shows extreme terrain (topographic factor = 20.587), driving high disaster density (578.11), where even the strongest soil conservation capacity (5454.60) remains inadequate. Clusters 3 and 0 exhibit similar vegetation coverage and terrain factors, but they differ in hazard levels (600.42 vs. 527.04). Cluster 3 is characterized by high soil conservation capacity (781.42), while Cluster 0 is influenced by high rainfall erosivity (9863.08). Cluster 4 represents the sole low-risk zone (disaster density = 101.84), where flat terrain (topographic factor = 1.306) and high vegetation coverage (0.206) synergistically mitigate the impact of maximum rainfall erosivity (10,482.20).

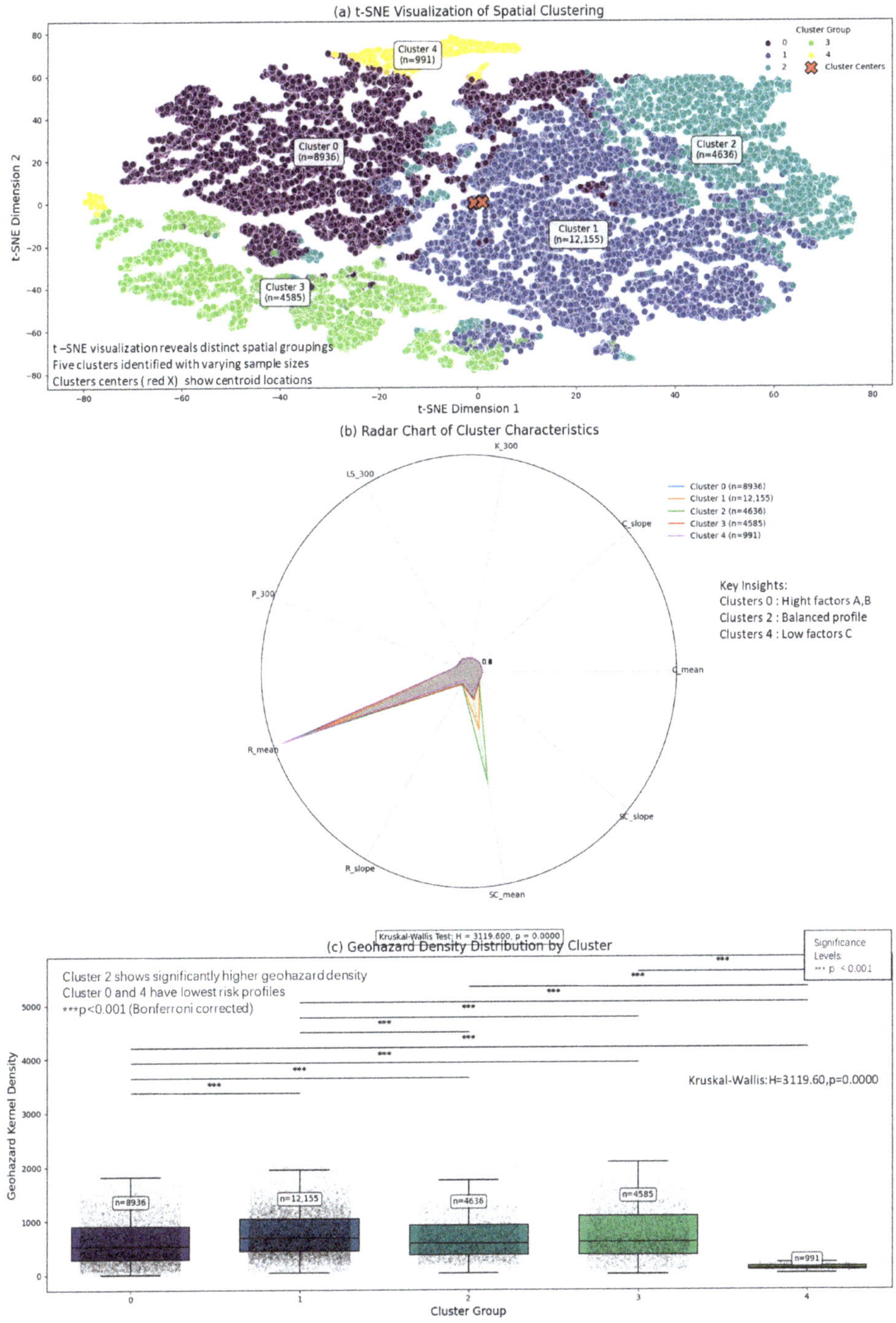

Figure 7. Cluster analysis identified five feature regions.

Table 6. Clustering Feature Table.

Cluster	Disaster Density (Median)	Vegetation Coverage (Mean)	Terrain Factor (Mean)	Rainfall Erosivity (Average)	Soil Conservation (Mean)
0	527.04	0.197	3.339	9863.08	725.937
1	672.59	0.026	11.305	7730.72	2462.48
2	578.112	0.032	20.587	9617.99	5454.6
3	600.42	0.197	3.531	8542.3	781.424
4	101.835	0.206	1.306	10482.2	307.082

The spatial distribution of clusters (Figure 8) and corresponding geomorphic characteristics (Figure 9) reveal critical patterns: Clusters 1 and 3 share identical geomorphic composition (marine deposits: 76.3% + fluvial-lacustrine deposits: 11.8% + alluvial deposits: 15.8%) yet exhibit different disaster densities (672.59 vs. 600.42). Cluster 2 displays extreme geomorphic homogeneity (marine deposits: 93.5%), with the maximum topographic factor (20.587) confirming uncontrollable risks in open marine terrain. Cluster 0 contains the highest proportion of alluvial deposits (32.7%), driving its heightened sensitivity to rainfall erosion (9863.08) and resulting in moderate disaster density (527.04). Cluster 4 is enriched with gentle continental slopes (42.3% occurrence) and lacustrine plains, where flat terrain (topographic factor = 1.306) and dense vegetation (coverage = 0.206) completely suppress the strongest rainfall erosivity (10,482.20), achieving minimal disaster density (101.84).

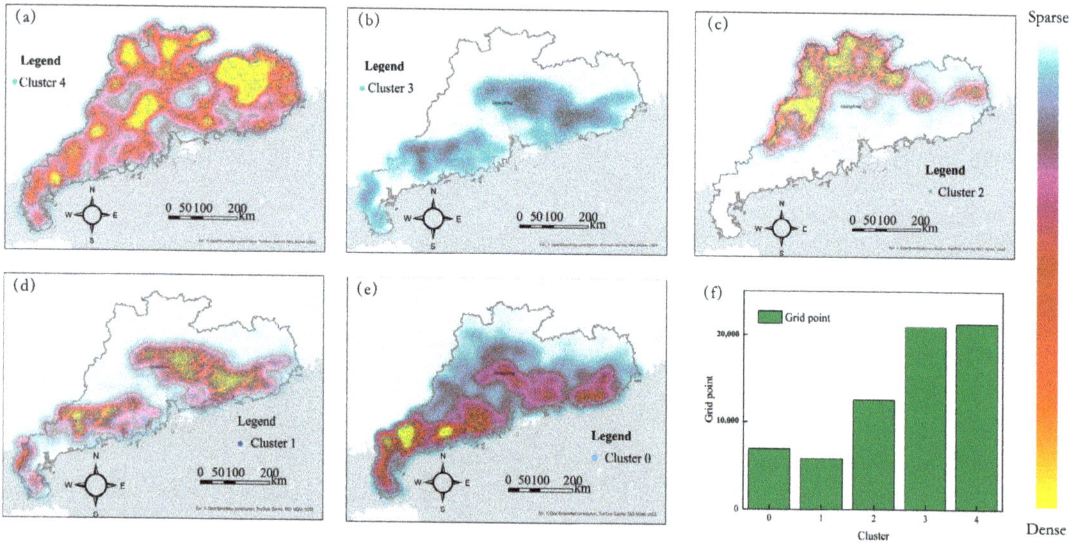

Figure 8. Spatial Distribution of Clusters. (**a**) the spatial distribution of cluster 4; (**b**) the spatial distribution of cluster 3; (**c**) the spatial distribution of cluster 2; (**d**) the spatial distribution of cluster 1; (**e**) the spatial distribution of cluster 0; (**f**) the grid point of clusters.

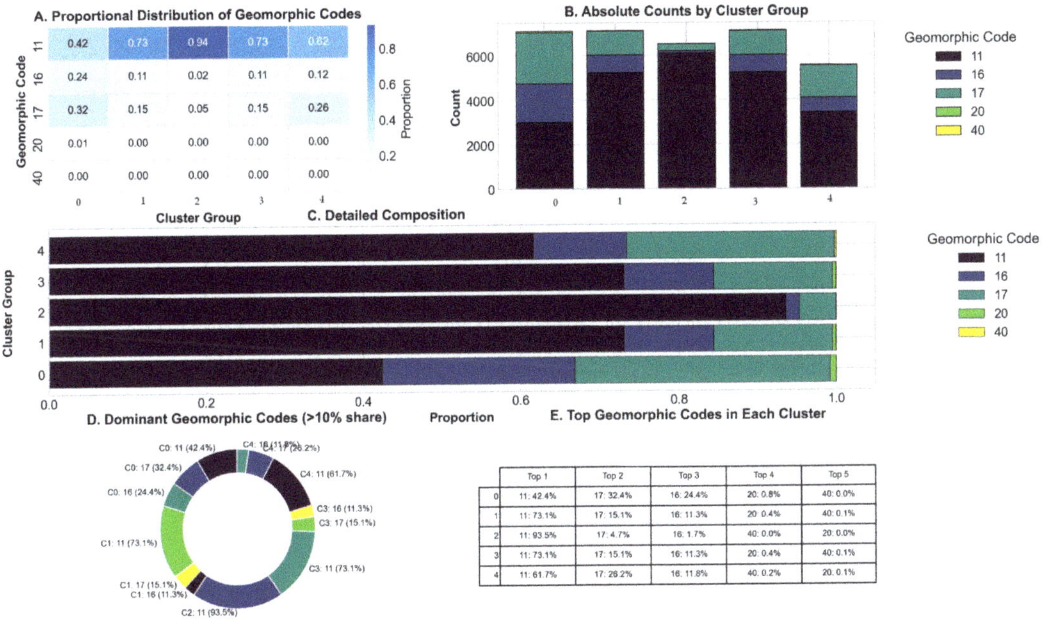

Figure 9. Spatial Distribution Characteristics of Clustering across Different Geomorphological Types (Code 11 (Marine deposits): Dominated by wave and current action in coastal environments; Code 16 (Pluvial-lacustrine deposits): Fluvial-lacustrine transitional zones with periodic flooding; Code 17 (Alluvial deposits): River channel and floodplain sedimentation; Code 20 (Arid pluvial deposits): Ephemeral flood deposits in dry regions; Code 40 (Gentle continental slopes): Transitional slope environments between continental shelf and deep basin.).

4. Discussion

This study presents a comprehensive spatial analysis of geohazard density and soil conservation capacity in Guangdong Province. Our findings reveal fundamental relationships as follows:

Comprehensive analysis indicates that Guangdong Province exhibits diverse and widespread geohazard types, with higher intensity observed in eastern and northern regions (e.g., Meizhou and Qingyuan). Geohazard hotspots are primarily concentrated in areas surrounding Meizhou and Huizhou, while cold spots dominate cities such as Maoming and Yangjiang.

The correlation between the kernel density of geological disasters and soil and water conservation factors reveals two core mechanisms. (1) Natural protection effect: A high average rainfall (R_mean) represents a normal level, where a higher average rainfall promotes vegetation growth, and dense vegetation (C_mean) significantly inhibits geological disasters, confirming its ecological barrier function. This is consistent with existing research, which indicates that vegetation itself is beneficial for slope stability [34]. Additionally, for slopes affected by rainfall, the application of vegetation significantly increases the safety factor of the slope [35]. (2) Risk-driven pattern: Steep terrain (LS_300) directly triggers disasters, while the positive correlation of soil conservation (SC_mean) is an interesting phenomenon. This reflects that areas with stronger soil conservation capacity often have higher geological disaster risk levels (positive correlation). This positive correlation is the result rather than the cause [36]. It reveals that in high-risk areas (due to steep terrain), more management efforts (soil and water conservation projects) have been made, thereby increasing soil conservation. Therefore, high soil conservation is an indicator or marker of the strong management needs and actions in high-risk areas. In practice, we cannot use the level of soil conservation alone as an evaluation standard for geological disaster susceptibility. A comprehensive evaluation, including actual management measures, is needed.

Although the KMO value (0.47) is lower than the conventional threshold, the spatial heterogeneity of soil and water conservation factors it reflects is precisely the essential characteristic of the disaster system in Guangdong Province. A key warning is that the increasing trend of rainfall erosivity (R_slope) becomes an emerging source of risk, highlighting the potential threat of climate change. Extreme rainfall induces geological disasters, becoming a challenge for geological disaster prevention and control under climate change conditions [37,38]. The weak impact of soil characteristics (K_300) indicates that regional disaster differentiation is mainly dominated by terrain and climate. The five PCA dimensions form an integrated framework for soil–water conservation dynamics. PC1 and PC2 dominate (50.7% combined variance), revealing core drivers: PC1 highlights the conflict between topography-driven conservation efforts and inadequate vegetation management; PC2 underscores high rainfall erosivity coupled with degrading soil conservation capacity. This structure provides a quantitative basis for spatial differentiation of geological hazard risks, prioritizing regions where natural forces and conservation lapses intersect.

The partitioning results reveal three core patterns: (1) Absolute dominance of topography: Clusters 1–2 demonstrate that steep terrain is a sufficient condition for disasters, even with strong management measures (Cluster 2), the risk cannot be completely mitigated [39]; (2) Vegetation-topography synergistic barrier: Cluster 4 validates that flat terrain combined with dense vegetation can completely inhibit disasters, highlighting the combined necessity of landform modification and ecological restoration; (3) Dual-path mechanism in medium-risk areas: Cluster 0 and Cluster 3 demonstrate considerable similarity across multiple environmental indicators (e.g., vegetation coverage, terrain factor, soil conservation), but exhibit significant divergence in geomorphic composition. As the fundamental basis of surface materials, geomorphic composition governs the efficiency of water–soil interactions. The heightened erosion sensitivity in Cluster 0 arises from the triple hybrid of marine, fluvial-lacustrine, and alluvial deposits (98%), compounded by the amplification effect of arid deposits, leading to persistent erosion. Conversely, Cluster 3's elevated disaster density is primarily attributed to the abrupt destabilization tendency inherent to marine-dominated sediments—despite lower erosion sensitivity, individual disaster events likely generate larger-scale impacts. The similarity in environmental parameters is superseded by geomorphic distinctions: under comparable vegetation and terrain conditions, disaster typology (gradual erosion versus catastrophic failure) more effectively explains disaster density variation than total erosion volume. This analysis substantiates the principle that geological foundations dictate disaster mechanisms. Future investigations should examine event-based rainfall records and historical disaster patterns (e.g., prevalence of large-scale landslides or liquefaction events in Cluster 3). Regarding mitigation strategies: Cluster 0 necessitates distributed interventions (e.g., vegetated swales); Cluster 3 requires prioritized monitoring of critical rainfall thresholds (e.g., installation of piezometers).

The geomorphology-disaster feedback reveals three universal mechanisms: (1) Absolute dominance of topography: In marine sedimentation zones (Clusters 1–2), disaster risk is inherently determined by coastal geomorphic configuration (e.g., cliff inclination, sediment consolidation), where engineering interventions (e.g., Cluster 2's conservation capacity of 5454.60) offer only marginal risk reduction due to fundamental landscape constraints.

(2) Hydrological sensitivity effect: The alluvial system (Cluster 0) exhibits amplified feedback to rainfall erosion dynamics, with runoff coefficient increases during storm events, mandating adaptive rain–flood management frameworks integrating real-time hydrological monitoring. (3) Geomorphology–ecology synergistic barrier: Slope transition zones (Cluster 4) develop efficient disaster buffering through geometric–ecological coupling—where gentle gradients and multilayer vegetation coverage (canopy + understory) jointly dissipate kinetic energy, establishing a replicable regional management paradigm for hazard mitigation efficacy.

5. Conclusions

Based on 11,252 geohazard records (Resource and Environment Science and Data Center) and multi-decadal soil conservation datasets (Science Data Bank, 1992–2019), this study employed Spatial Clustering (Kernel Density Estimation + Z-score hotspot analysis), Multivariate Statistics (PCA with KMO and Bartlett's $p < 0.01$ validation), and Machine-Learning Partitioning (t-SNE and DBSCAN clustering) to analyze geohazard-environment linkages across Guangdong. Four principal conclusions emerge from this research:

1. Spatial risk stratification: Guangdong exhibits distinct geohazard zones: high-risk mountainous northeast (Clusters 1–2), medium-risk transitional areas (Clusters 0, 3), and low-risk coastal plains (Cluster 4). This stratification provides a targeted management basis.
2. Key controlling factors: Geohazard susceptibility is primarily governed by topographic steepness (LS factor), modulated by vegetation coverage (C_mean) and rainfall variability (R_slope). Soil conservation measures show location-dependent effectiveness, achieving optimal results when combined with vegetation restoration on moderate slopes.
3. Climate change impacts: Increasing rainfall variability (R_slope) emerges as a significant emerging risk factor, particularly given climate projections for intensified precipitation extremes in South China.
4. Management implications: High-risk zones: Prioritize engineering-slope stabilization combined with deep-rooted vegetation; Medium-risk zones: Implement vegetation-based solutions with continuous cover monitoring; All zones: Establish early-warning systems triggered by R_slope thresholds.

Author Contributions: Conceptualization, L.T. and Y.L.; methodology, J.F.; software, L.T. and Y.L.; validation, L.T., Y.L. and J.F.; formal analysis, L.T., Y.L. and J.F.; investigation, L.T., Y.L. and J.F.; writing—original draft preparation, L.T., Y.L. and J.F.; writing—review and editing, L.T., Y.L. and J.F.; visualization, L.T., Y.L. and J.F.; supervision, L.T., Y.L. and J.F.; project administration, L.T., Y.L. and J.F. All authors have read and agreed to the published version of the manuscript.

Funding: This research received no external funding.

Data Availability Statement: The original contributions presented in this study are included in the article material. Further inquiries can be directed to the corresponding author.

Acknowledgments: The authors have reviewed and edited the output and take full responsibility for the content of this publication.

Conflicts of Interest: Author Y.L. was employed by the company Guangzhou Urban Planning & Design Survey Research Institute Co., Ltd. The remaining authors declare that the research was conducted in the absence of any commercial or financial relationships that could be construed as a potential conflict of interest.

References

1. Froude, M.J.; Petley, D.N. Global fatal landslide occurrence from 2004 to 2016. *Nat. Hazards Earth Syst. Sci.* **2018**, *18*, 2161–2181. [CrossRef]
2. Herrera, G.; Mateos, R.M.; García-Davalillo, J.C.; Grandjean, G.; Poyiadji, E.; Maftei, R.; Filipciuc, T.-C.; Auflič, M.J.; Jež, J.; Podolszki, L.; et al. Landslide databases in the Geological Surveys of Europe. *Landslides* **2018**, *15*, 359–379. [CrossRef]
3. United Nations Office for Disaster Risk Reduction (UNDRR). *Sendai Framework for Disaster Risk Reduction 2015–2030*; United Nations: Geneva, Switzerland, 2015; p. 9.
4. Gariano, S.L.; Guzzetti, F. Landslides in a changing climate. *Earth-Sci. Rev.* **2016**, *162*, 227–252. [CrossRef]
5. Race, D.; Gentle, P.; Mathew, S. Living on the margins: Climate change impacts and adaptation by remote communities living in the Pacific Islands, the Himalaya and desert Australia. *Clim. Risk Manag.* **2023**, *40*, 100503. [CrossRef]
6. Borrelli, P.; Robinson, D.A.; Fleischer, L.R.; Lugato, E.; Ballabio, C.; Alewell, C.; Meusburger, K.; Modugno, S.; Schütt, B.; Ferro, V.; et al. An assessment of the global impact of 21st century land use change on soil erosion. *Nat. Commun.* **2017**, *8*, 2013. [CrossRef]
7. Wang, L.; Liu, Q.; Li, X.A.; Qin, B.; Hong, B.; Shi, J. Effect of nanosilica on the hydrological properties of loess and the microscopic mechanism. *Sci. Rep.* **2024**, *14*, 13692. [CrossRef]
8. Das, P.; Patwa, D.; G, V.; Bharat, T.V. Influencing factors on the simulation of rainfall-induced landslide prediction based on case study. *Bull. Eng. Geol. Environ.* **2022**, *81*, 194. [CrossRef]
9. Zhang, X.; Huang, T.; Ge, Z.; Man, T.; Huppert, H.E. Infiltration characteristics of slurries in porous media based on the coupled Lattice-Boltzmann discrete element method. *Comput. Geotech.* **2025**, *177*, 106865. [CrossRef]
10. Sun, X.; Liu, G.; Zhao, T.; Tang, L.; Han, X.; Peng, W. Application of a geomorphic restoration method for landslide susceptibility mapping along the rapidly uplifting section of the upper Jinsha river, South-Western China. *Bull. Eng. Geol. Environ.* **2025**, *84*, 42. [CrossRef]
11. Xue, Y.; Miao, F.S.; Wu, Y.P.; Dias, D.; Li, L. Combing soil spatial variation and weakening of the groundwater fluctuation zone for the probabilistic stability analysis of a riverside landslide in the Three Gorges Reservoir area. *Landslides* **2023**, *20*, 1013–1029. [CrossRef]
12. Peng, J.; Sun, P.; Igwe, O.; Li, X.A. Loess caves, a special kind of geo-hazard on loess plateau, northwestern China. *Eng. Geol.* **2018**, *236*, 79–88. [CrossRef]
13. Li, X.; Yu, K.; Xu, G.; Li, P.; Li, Z.; Shi, P.; Jia, L. Exploring ecosystem service dynamics and drivers in the upper and middle Yellow River Basin under large-scale ecological restoration. *Ecol. Eng.* **2025**, *217*, 107643. [CrossRef]
14. Wang, Y.N.; Li, S.K.; Li, Z.Y.; Garg, A. Exploring the application of the MICP technique for the suppression of erosion in granite residual soil in Shantou using a rainfall erosion simulator. *Acta Geotech.* **2023**, *18*, 3273–3285. [CrossRef]
15. Ren, S.; Zhang, P.; Man, T.; Galindo-Torres, S.A. Numerical assessments of the influences of soil-boulder mixed flow impact on downstream facilities. *Comput. Geotech.* **2023**, *153*, 105055. [CrossRef]
16. Thiery, Y.; Malet, J.P.; Sterlacchini, S.; Puissant, A.; Maquaire, O. Landslide susceptibility assessment by bivariate methods at large scales: Application to a complex mountainous environment. *Geomorphology* **2007**, *92*, 38–59. [CrossRef]
17. Wang, S.; Lin, X.; Qi, X.; Li, H.; Yang, J. Landslide susceptibility analysis based on a PSO-DBN prediction model in an earthquake-stricken area. *Front. Environ. Sci.* **2022**, *10*, 912523. [CrossRef]
18. Sun, X.; Chen, J.; Han, X.; Bao, Y.; Zhan, J.; Peng, W. Application of a GIS-based slope unit method for landslide susceptibility mapping along the rapidly uplifting section of the upper Jinsha River, South-Western China. *Bull. Eng. Geol. Environ.* **2020**, *79*, 533–549. [CrossRef]
19. Hou, Y.; Hu, N.; Teng, C.; Zheng, L.; Zhang, J.; Gong, Y. Monitoring and Evaluation of Ecological Restoration Effectiveness: A Case Study of the Liaohe River Estuary Wetland. *Sustainability* **2025**, *17*, 2973. [CrossRef]
20. Taylor, A.C.; Fletcher, T.D. Nonstructural urban stormwater quality measures: Building a knowledge base to improve their use. *Environ. Manag.* **2007**, *39*, 663–677. [CrossRef]
21. Wang, L.; Li, X.A.; Zheng, Z.Y.; Zheng, H.; Ren, Y.B.; Chen, W.J.; Lei, H.N. Analysis of the slope failure mechanism a under tunnel erosion environment in the south-eastern Loess Plateau in China. *Catena* **2022**, *212*, 106039. [CrossRef]
22. Yeh, H.F.; Lee, C.H. Soil water balance model for precipitation-induced shallow landslides. *Environ. Earth Sci.* **2013**, *70*, 2691–2701. [CrossRef]
23. Sun, H.; Li, W.; Scaioni, M.; Fu, J.; Guo, X.; Gao, J. Influence of spatial heterogeneity on landslide susceptibility in the transboundary area of the Himalayas. *Geomorphology* **2023**, *433*, 108723. [CrossRef]
24. Liu, J.; Fu, Z.; Zhou, L.; Feng, G.; Wang, Y.; Luo, W. Non-Tectonic Geohazards of Guangdong Province, China, Monitored Using Sentinel-1A/B from 2015 to 2022. *Sensors* **2024**, *24*, 5449. [CrossRef] [PubMed]
25. Cui, Q.L.; Shen, S.L.; Xu, Y.S.; Wu, H.N.; Yin, Z.Y. Mitigation of geohazards during deep excavations in karst regions with caverns: A case study. *Eng. Geol.* **2015**, *195*, 16–27. [CrossRef]

26. Liu, X.; Chen, H. Regional assessment on ecological risk of ecosystems under natural hazards: An application in Guangdong Province (SE China). *Nat. Hazards* **2020**, *100*, 205–229. [CrossRef]
27. Data Center for Resources and Environmental Sciences, Chinese Academy of Sciences (RESDC). *Spatial Distribution Dataset of Geological Hazard Points in China (Collated Through 2019)*; RESDC: Beijing, China, 2021. Available online: https://www.resdc.cn/DataSearch.aspx (accessed on 31 March 2024).
28. Li, J.; He, H.; Zeng, Q.; Chen, L.; Sun, R. Dataset of Soil Conservation Capacity Preventing Water Erosion in China (1992–2019) [Dataset]. ScienceDB. 2023. Available online: https://www.scidb.cn/en/detail?dataSetId=3b2e9ea9d65c4c92b6357519a7ddc6dd (accessed on 31 March 2024).
29. Olsen, K.F.; Lindrup, R.M.H.; Morup, M. Think Global, Adapt Local: Learning Locally Adaptive K-Nearest Neighbor Kernel Density Estimators. In Proceedings of the 27th International Conference on Artificial Intelligence and Statistics (AISTATS), Valencia, Spain, 2–4 May 2024; p. 29.
30. Xu, M.; Ishibuchi, H.; Gu, X.; Wang, S. Dm-KDE: Dynamical kernel density estimation by sequences of KDE estimators with fixed number of components over data streams. *Front. Comput. Sci.* **2014**, *8*, 563–580. [CrossRef]
31. Zhu, A.X.; Wang, R.; Qiao, J.; Qin, C.-Z.; Chen, Y.; Liu, J.; Du, F.; Lin, Y.; Zhu, T. An expert knowledge-based approach to landslide susceptibility mapping using GIS and fuzzy logic. *Geomorphology* **2014**, *214*, 128–138. [CrossRef]
32. Sun, X.; Yu, C.; Li, Y.; Rene, N.N. Susceptibility Mapping of Typical Geological Hazards in Helong City Affected by Volcanic Activity of Changbai Mountain, Northeastern China. *ISPRS Int. J. Geo-Inf.* **2022**, *11*, 344. [CrossRef]
33. Li, L.; Wang, L. Modified Method in Group Comparison Matrixes Consistency Check and Its Application in Building Material Evaluation. In Proceedings of the International Symposium on Ecological Environment and Technology of Concrete, Beijing, China, 1–4 August 2011; pp. 425–431.
34. Lann, T.; Bao, H.; Lan, H.X.; Zheng, H.; Yan, C.G.; Peng, J.B. Hydro-mechanical effects of vegetation on slope stability: A review. *Sci. Total Environ.* **2024**, *926*, 171691. [CrossRef]
35. Razali, I.H.; Taib, A.M.; Abd Rahman, N.; Hasbollah, D.Z.A.; Dan, M.F.M.; Ramli, A.B.; Ibrahim, A. Slope Stability Analysis of Riverbank in Malaysia with the Effects of Vegetation. *Phys. Chem. Earth* **2023**, *129*, 103334. [CrossRef]
36. Huang, Z.; Wang, X.; Yuan, M.H.; Duan, W.H.; Xia, J.Y.; Li, J.; Zhao, Y.F. Differential Response of Ecosystem Service to Restoration Methods and Restoration Time in Shallow Landslide-Prone Areas. *Land Degrad. Dev.* **2025**, *36*, 3051–3062. [CrossRef]
37. Shi, S.Y.; Zhou, J.; Chen, X.J.; Xu, Y.J.; Tao, J. Risk Assessment of Geological Disasters Triggered by Heavy Rainfall in Mountainous Areas: A Case Study in Hebei Province, China. *Appl. Ecol. Environ. Res.* **2025**, *23*, 5443–5460. [CrossRef]
38. da Silva, R.C.; Marengo, J.A.; Lemes, M.R. Analysis of Extreme Rainfall and Landslides in the Metropolitan Region of the Paraiba do Sul River Valley and North Coast of Sao Paulo, Brazil. *Theor. Appl. Climatol.* **2024**, *155*, 3927–3949. [CrossRef]
39. Guo, H.Y.; Martínez-Graña, A.M. Correlation between soil moisture change and geological disasters in E'bian area (Sichuan, China). *Appl. Sci.* **2024**, *14*, 6685. [CrossRef]

water

MDPI

Article

Spatial Distribution, Temporal Behaviour, and Trends of Rainfall Erosivity in Central Italy Using Coarse Data

Francesca Todisco, Alessio Massimi Alunno * and Lorenzo Vergni

Department of Agricultural, Food and Environmental Science, University of Perugia, 06124 Perugia, Italy; francesca.todisco@unipg.it (F.T.); lorenzo.vergni@unipg.it (L.V.)
* Correspondence: alessio.massimialunno@unipg.it

Abstract: This study examines the spatio-temporal dynamics of rainfall erosivity, R, in the Umbria region (central Italy), based on a 20-year dataset of 30 min precipitation records from 54 stations. Using the RUSLE2 framework, models of varying complexity were evaluated to estimate the R-factor: the original model (Model A), and models based solely on event rainfall depth h_e or daily rainfall depth h_d. All the models show consistency in the spatial and temporal patterns of the R-factor: higher erosivity is observed in the southern and northwestern areas, while summer contributes the most to annual erosivity due to the high average intensity of rainfall events. Trend analyses indicate stationarity across most stations. Compared to Model A (mean R-factor: 1840 MJ mm ha^{-1} h^{-1} y^{-1}), the models based on h_e underestimate the R-factor by about 15%, whereas the R-factor derived from the h_d-dependent model is almost equivalent. The estimate from Model A is also approximately 20% higher than that of a previous study conducted on a more limited dataset. The most likely reason for this difference appears to be the formula used for estimating the R-factor. The study highlights the practicality of simplified models, which offer a viable alternative in contexts where high-resolution precipitation data are unavailable. It also demonstrates the benefits of denser station networks and longer observation periods, particularly in regions characterised by complex terrains.

Keywords: RUSLE2; rainfall runoff erosivity factor; runoff models; erosion models; hydrological processes seasonality; zoning of hydrological processes

1. Introduction

Soil is a vital, non-renewable resource essential for agricultural productivity, ecosystem services, and global sustainability. Its role in carbon sequestration, water regulation, and biodiversity maintenance underpins numerous ecological and economic processes [1]. However, unsustainable land-use practices, such as deforestation, overgrazing, and inappropriate agricultural methods, have led to widespread soil degradation, compromising both food security and ecosystem resilience. Among the various degradation processes, water erosion is particularly significant, as it not only diminishes soil fertility but also accelerates desertification, alters hydrological cycles, and contributes to downstream sedimentation [2].

Water erosion involves the detachment and transport of soil particles through chemical and mechanical mechanisms. This process is influenced by a combination of interrelated factors, including climatic conditions, soil characteristics, topography, vegetation cover, and land management practices [3]. Climatic changes, particularly in regions prone to intense rainfall events, amplify erosion risks by increasing runoff potential and the erosive force of precipitation [4]. In the Mediterranean region, a hotspot for erosion studies, the interplay of steep topographies, erodible soils, and intense rainfall creates a scenario of heightened vulnerability [5–7].

Rainfall aggressiveness, a key driver of erosion, is determined by the intensity and duration of precipitation events. This parameter is quantified through the concept of rainfall erosivity, which measures the kinetic energy of rainfall and its ability to generate runoff [8,9]. The interaction of raindrop impact with soil particles initiates the detachment process, while the subsequent overland flow facilitates particle transport [10,11]. The R-factor, central to models like the Universal Soil Loss Equation (USLE) and its successors (RUSLE and RUSLE-2), captures this erosive potential, combining the effects of kinetic energy and maximum 30 min rainfall intensity [12]. These models have become fundamental tools for assessing soil erosion risk, offering insights that extend beyond soil conservation to support flood and landslide risk assessments [13–16].

The USLE framework estimates annual soil loss (A) using the equation:

$$A = R \times K \times L \times S \times C \times P,$$

where

- A: average annual soil loss ($t\ ha^{-1}$);
- R: rainfall erosivity factor ($MJ\ mm\ ha^{-1}\ h^{-1}$);
- K: soil erodibility factor ($t\ h\ ha^{-1}\ MJ^{-1}\ mm^{-1}$);
- L: slope length factor;
- S: slope steepness factor;
- C: cropping and management factor;
- P: erosion control practices factor.

Accurate estimation of the R-factor requires high-temporal-resolution rainfall data, usually recorded at intervals of 1 to 60 min. These data allow for precise results based on the accumulated erosivity of individual rainstorm events [17]. However, in many regions, such data are scarce, requiring the formulation of empirical models that relate erosivity to readily available rainfall metrics, such as daily, monthly, or annual totals [18–20].

Rainfall erosivity studies across Europe demonstrate significant spatial and temporal variability. In Slovenia, Mikos et al. [21] used high-resolution precipitation data to assess erosion risks, while in Spain's Ebro catchment, Angulo-Martínez et al. [22] highlighted the influence of extreme events on sediment transport. Mici'c Ponjiger et al. [23] produced rainfall erosivity maps for the Western Balkans region based on RUSLE and RUSLE2 modelling and ERA5 data, highlighting the presence of well-defined spatial patterns. Petroselli et al. [20], for a small area in central Italy, conducted a comparative evaluation of several simplified formulas that can be used to estimate annual rainfall erosivity in the absence of high-resolution rainfall datasets. In Switzerland, Meusburger et al. [24] produced detailed erosivity maps that informed national conservation strategies. At a continental scale, advancements in geospatial analysis have enabled estimates of rainfall erosivity across Africa and Europe using satellite-derived precipitation data and geostatistical models [25,26]. In the Algarve region of Portugal, these approaches have been further refined by integrating secondary variables like elevation, thus improving spatial predictions of erosivity [27].

In Italy, a comprehensive assessment based on 10 years of pluviographic data from 386 meteorological stations produced a high-resolution (500 m) national map of rainfall erosivity [28]. The study revealed distinct spatio-temporal patterns, underscoring the importance of localised assessments for effective land management. The broader applications of these findings extend to natural hazard mitigation, with rainfall erosivity data aiding in the prediction of floods and landslides caused by high-intensity precipitation events. In Europe, a detailed study [29] leveraging a decade of pluviographic data from over 1500 meteorological stations led to the creation of a high-resolution (500 m) map of rainfall erosivity across the continent. This work highlighted significant spatial variability in erosivity values, shaped by climatic and geographical factors. This framework supports policymakers and stakeholders in implementing sustainable land-use practices. However, previous studies have certain shortcomings, such as the limited length of the historical series used and the density of weather stations, even in areas characterised by high spatial variability in precipitation.

The present study examines rainfall erosivity estimation using four models of varying complexity, focusing on the region of Umbria in central Italy. The reference Model A utilises high-resolution semi-hourly data to compute the R-factor with exceptional accuracy, capturing the detailed dynamics of individual rainfall events. While ideal for focused research, the computational demands and data requirements of Model A limit its applicability in regions lacking detailed records.

Model B introduces a methodological simplification by employing monthly parameters (α_m, β_m) specific to each meteorological station, striking a balance between accuracy and practicality. Model C further reduces computational complexity by segmenting the year into three seasonal periods (winter, humid, and summer) and applying constant parameters for each station. This approach lowers data requirements while preserving the robustness of erosivity estimates. Finally, Model D is designed for contexts where only daily rainfall data are available, establishing a relationship between daily cumulative rainfall (h_d) and daily erosivity (R_d).

The first objective is to analyse the spatiotemporal variability of the R-factor at the regional scale. The resulting isoerosivity maps are essential for land-use planning and soil conservation, as they identify areas of heightened vulnerability, support the prioritisation of mitigation strategies, and contribute to sustainable land management [30]. The second objective is to define and test simplified models to facilitate R-factor estimation through less detailed input data. The resulting models are useful for size remediation and protection solutions for soil conservation at the local scale, and also for when long-time series of detailed meteorological data are unavailable. Finally, by integrating high-resolution precipitation data, simplified modelling approaches, and advanced geospatial techniques, enhance the planning, programming, and management of the soil erosion protection policy.

Spatial resolution is a critical factor in this context. The results obtained are only truly valuable when they can be effectively integrated into areas of a relatively limited size, such as the Umbria region. In larger areas, the use of low-resolution spatial scales can provide general guidance for policy development; however, these scales are insufficient for delivering the detailed and precise information required for more targeted and actionable decision-making. For this reason, a further objective of the study was to compare the R-factor map obtained in this work, based on a dense network of stations, with the one derived from larger-scale studies [29].

Finally, the statistical trend analysis of annual erosivity series allowed for completing the study with valuable insights into the presence of potential long-term changes in erosion risk.

2. Materials and Methods

2.1. Materials

The study refers to the Umbria region, which extends across central peninsular Italy with a surface area of 8465 km^2. The region is characterised by a predominantly temperate climate, with cool winters and warm summers, although local variability is influenced by the region's morphology, which includes both hilly and mountainous areas. The average annual precipitation is approximately 857 mm with a relevant interannual variability. Precipitation is primarily concentrated in spring and autumn (40.5%), followed by winter (34.4%). Although summer is the driest season, it still contributes 25.1%. The distribution and intensity of rainfall play a crucial role in erosion processes and the management of regional water resources.

The dataset used for the analysis was defined based on semi-hourly precipitation time series provided by the Regional Hydrological Service, encompassing a total of 94 meteorological stations. Overall, the initial database contained 29,779,498 semi-hourly records, corresponding to approximately 1700 years of observations. However, not all stations in the initial dataset were included in the final analysis. Stations with excessively short historical series were excluded, as well as years with incomplete months characterised by more than 5% of missing data. This filtering process was carried out to avoid underestimation of the annual R parameter.

The final dataset, obtained after these operations, includes 54 meteorological stations (as shown in Figure 1 and Table 1), each with a sufficiently long historical precipitation series. To enhance the representativeness of the areas near the regional boundaries, stations from Verghereto (ID54) and Castel Cellesi (ID12), located in neighbouring areas outside of Umbria, were also included in the analysis. The density of the monitoring network, with one station every 156.8 km^2 (equivalent to a grid of approximately 12.5 km × 12.5 km), ensures high spatial coverage, which is crucial for an accurate representation of erosive dynamics at the regional scale.

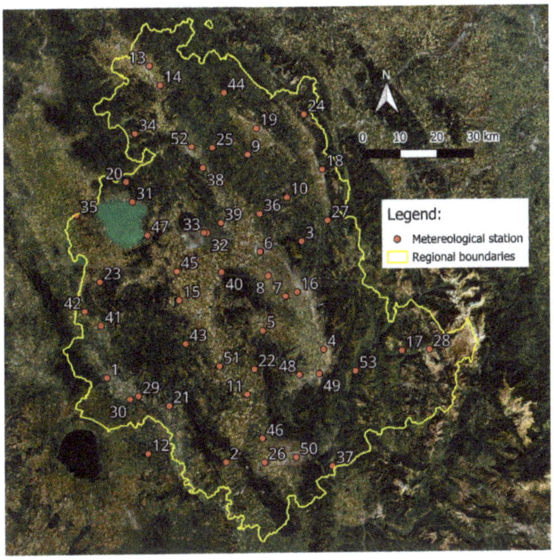

Figure 1. Localization of the stations in the territory. Numbers represent station's IDs.

The altitude of the stations ranges from 109 to 1087 m above sea level (mean: 382 m a.s.l.). Of the total number of monitored stations, 46% are in hilly areas (300–700 m a.s.l.), 42% in flat areas (0–300 m a.s.l.), and 13% are situated in mountainous regions (>700 m a.s.l.). Two stations are located at altitudes exceeding 1000 m a.s.l.

Table 1. Identification codes (ID), name, and altitude above sea level (m a.s.l.).

ID	Metereological Station	Elevation (m a.s.l.)	ID	Metereological Station	Elevation (m a.s.l.)
1	Allerona	153.2	28	Norcia	691.3
2	Amelia	319.5	29	Orvieto	311.4
3	Armenzano	707.7	30	Orvieto Scalo	109.3
4	Azzano	235.2	31	Passignano sul Trasimeno	327.8
5	Bastardo	331.5	32	Perugia Fontivegge	308.5
6	Bastia Umbra	203.3	33	Perugia Santa Giuliana	433.8
7	Bevagna	211.5	34	Petrelle	341.9
8	Cannara	191.4	35	Petrignano del Lago	303.9
9	Carestello	517.8	36	Pianello	231.5
10	Casacastalda	730.1	37	Piediluco	370.0
11	Casigliano	272.5	38	Pierantonio	239.0
12	Castel Cellesi	362.8	39	Ponte Felcino	191.6
13	Cerbara	310.0	40	Ponte Nuovo Torgiano	170.0
14	Città di Castello	304.0	41	Ponte Santa Maria	240.1
15	Compignano	240.2	42	Ponticelli	244.9
16	Foligno	220.0	43	Ripalvella	453.1
17	Forsivo	963.2	44	San Benedetto Vecchio	729.4
18	Gualdo Tadino	596.5	45	San Biagio della Valle	257.0
19	Gubbio	471.5	46	San Gemini	297.1
20	La Cima	790.8	47	San Savino	260.0
21	Lago Corbara	128.0	48	San Silvestro	379.9
22	Massa Martana	327.9	49	Spoleto	353.2
23	Moiano	270.3	50	Terni	123.0
24	Monte Cucco	1087.2	51	Todi	326.1
25	Montelovesco	634.1	52	Umbertide	305.7
26	Narni Scalo	108.9	53	Vallo di Nera	306.7
27	Nocera Umbra	535.7	54	Verghereto	1069.2

2.2. Methods

2.2.1. Isoerosivity Maps

Isoerosivity maps are generated through a zonation process based on cluster analysis of the R parameter, which is estimated using the following models. The visualisation is achieved through Inverse Distance Weighting (IDW) interpolation of the mean annual R values recorded at individual stations.

2.2.2. Model A

According to the definition by Wischmeier and Smith [8], the calculation of the rainfall erosivity factor (expressed in MJ mm h^{-1} ha^{-1} y^{-1}), requires precipitation data with a temporal resolution of at least semi-hourly intervals. Given the limited availability of such high-resolution data, simplified calculation models or those requiring precipitation data with lower temporal resolution are often used.

In this study, several models were parameterized using as reference data the erosivity calculated with the RUSLE 2 model [31], referred to in the text as Model A, as follows:

$$R_e = E_e \times I_{30}, \quad (1)$$

where R_e (MJ mm h^{-1} ha^{-1}) denotes the erosivity of single erosive event; E_e (MJ ha^{-1}) is the total energy of the event; and I_{30} (mm h^{-1}) is the maximum semi-hourly intensity during the event e.

In the rainfall dataset, following the methodology of Wischmeier and Smith [8], individual events are identified as rainy periods preceded and followed by at least six non-rainy hours. Each event is then subdivided into elementary events i, for which the specific energy e_i (MJ ha^{-1} mm^{-1}) is calculated as follows:

$$e_i = \left(1 - 0.72 \times e^{(-0.082 \times I_i)}\right), \quad (2)$$

where I_i (mm h^{-1}) denotes the erosivity of single erosive event.

The total energy of the event is given by the summation of the individual e_i values as follows:

$$E_e = \sum_{i=1}^{k} e_i, \quad (3)$$

where k is the total number of elementary events within the rainfall event.

R_e, calculated using Equation (1), serves as the reference value for subsequent evaluations and model parameterization. Specifically, the annual value (R_y) is defined as the summation of the erosivity of all individual events occurring between 1 January and 31 December:

$$R_y = \sum_{e=1}^{n} R_e, \quad (4)$$

where n is the total number of erosive events in the given year.

The mean annual erosivity R for each station is considered as the multi-year average:

$$R = \left(\sum_{y=1}^{N} R_y\right) \Big/ N, \quad (5)$$

where N is the total number of years analysed.

2.2.3. Simplified Models for Erosivity Estimation Based on the Event Rainfall Depth h_e

Accurate estimation of rainfall erosivity through derived USLE models necessitates semi-hourly precipitation records. However, reconstructing long-term historical series with this level of detail remains impractical. While gridded datasets (e.g., E-OBS or ERA5-Land) offer a potential alternative, their temporal resolution is typically inadequate for this purpose. Consequently, it becomes essential to develop models that simplify the estimation of the R factor using rainfall data with lower temporal resolution, which are more widely available.

To reduce the computational burden of calculating R_e in terms of the required temporal resolution of the data, a literature-based model was evaluated [32–35]. In this model, the rainfall erosivity index for a single event is assumed to be proportional to the power of the rainfall depth of the event h_e (mm):

$$\left[R_e = \alpha \times h_e^{\beta}\right]_{t,s}, \quad (6)$$

where α is the scale factor; β is the model parameter; t is the temporal reference; and s is the spatial reference.

The parameters α and β assume different values based on the selected temporal reference t and spatial reference s.

If the temporal reference t is not specified, the parameters are determined using the sample of all events available in the historical series of station s. Conversely, when t is specified (e.g., the month of January, the period from June to September, etc.), the following procedure is applied.

The sample of all events occurring during the specified *t* periods in the historical series is extracted (e.g., all events in January). For each event, h_e and R_e are calculated using the procedure described for Model A. The parameters α and β are then determined through regression analysis between the two variables. This procedure is repeated for each t and for each spatial reference.

Model B

With reference to Equation (6), the erosivity of a single event R_e is calculated by taking the temporal reference as the month t = m and the spatial scale as the individual station s. With S representing the number of stations and M the number of months m in a year, $S \times M$ parameter pairs $\{\alpha_{m,s}, \beta_{m,s}; m = 1, \ldots M; s = 1, \ldots S\}$ are determined and employed in Equation (6) to estimate the erosivity R_e of rainfall events with depth h_e in the month m and station s. As in the case of Model A, the annual erosivity R_y and the average annual erosivity R_m are subsequently calculated using Equations (4) and (5).

Although it represents a simplification compared to the estimation procedures of USLE and RUSLE-2, Model B remains highly complex due to the variability, and consequently, the large number of parameters required.

Model C

With reference to Equation (6), the erosivity of a single event R_e is calculated by taking the temporal reference as the period p, while the spatial scale remains focused on individual stations. With S representing the number of stations and P the number of periods p in a year, $S * P$ parameter pairs $\{\alpha_{p,s}, \beta_{p,s}; p = 1, \ldots P; s = 1, \ldots S\}$ are determined using the procedure described for Model A, and employed in Equation (6) to estimate the erosivity R_e of rainfall events with depth h_e for period p and station s. The total number of periods is three: 1, the winter period (January, February, March, December); 2, the humid period (April, May, October, November), and 3, the summer period (June, July, August, September). As in the case of Model A, the annual erosivity R_y and the average annual erosivity R_p are subsequently calculated using Equations (4) and (5).

2.2.4. Simplified Model for Erosivity Estimation Based on the Daily Rainfall Depth h_d (Model D)

In this model (Model D), the daily erosivity R_d is expressed as a function of the daily rainfall depth h_d (cumulative rainfall from 00:00 to 23:30 on day d). R_d is derived from the daily cumulative erosivity of individual rainfall events occurring on day d. Events are defined based on Wischmeier's six-hour criterion, with additional consideration given to events spanning two days. For example, an event starting at 22:00 on day d and ending at

02:00 on day d + 1 is divided into two separate events: the first covering the time interval 22:00–23:30 on day d, and the second covering 00:00–02:00 on day d + 1.

Model D functions were obtained by interpolating the scatter plots of the pairs (R_d,h_d) using third-degree polynomial functions. Different relationships were derived for each combination of temporal scale (period, p) and spatial division (regional subareas, z; z = 1, ... Z). The sub-zones were identified at the regional level based on the zoning represented in Figure 2a. In the northwestern and southern areas, "red-orange" zones (z = r), 27 stations with an annual average R lower than the regional average (R = 1863.6 MJ mm ha^{-1} h^{-1} y^{-1}) are located, while in the northern, eastern, and southwestern areas, "blue-green zone", (z = b), 27 stations with R values higher than the regional average are located. The temporal resolution remains that of the three periods p = 1, ... P as for Model C. This approach resulted in Z × P functions.

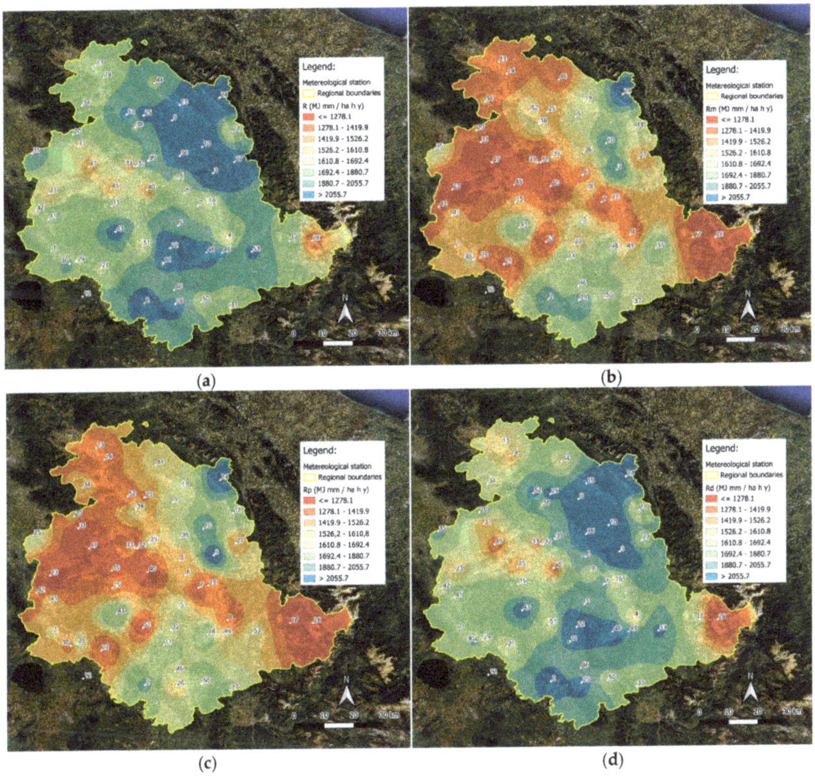

(c) (d)

Figure 2. Multi-year mean erosivity maps of the Umbria Region achieved using: (**a**), the RUSLE-2 model (Model A); (**b**), power model (6) with constant parameters α and β for each month and station (Model B); (**c**), power model (6) with constant parameters α and β for each period and station (Model C); (**d**), third-degree polynomial function between daily erosivity R_d and daily rainfall h_d (Model D).

2.2.5. Trend Analyses

The Mann–Kendall test [36,37] was employed to evaluate the presence of trends within the dataset. The null hypothesis posits the absence of a trend in the population from which the dataset is drawn, while the alternative hypothesis suggests the presence of a monotonic increasing or decreasing trend. Being a non-parametric test, it does not rely on the distribution of the variable under analysis and is less sensitive to outliers in the time series [19].

While the Mann–Kendall test is effective in detecting monotonic trends, it does not quantify their magnitude. To address this limitation, the non-parametric Theil–Sen estimator [38,39] was utilised to estimate the slope of regression lines interpolating the data.

The Zyp [40] and Kendall [41] packages in the R-statistical software (https://www.r-project.org/) were employed to calculate the Theil–Sen estimator and to determine the *p*-value associated with the Mann–Kendall test, respectively [19].

3. Results

The analysis of the 54 stations revealed that the duration of the historical series (column 2, Table 2) varies between a minimum of 15 years, recorded at the Armenzano station (ID 3), and a maximum of 24 years. The average length of the series per station is over 21.5 years. The extensive temporal range of the data provides a robust statistical foundation for the analysis, enabling the calculation of means and percentiles with high stability and reliability.

Table 2. For each station: ID (1); series length, years (2); average annual event count (3); mean h_e, mm (4); 95th percentile of h_e, mm (5); multi-year average of annual rainfall depth, mm (6); multi-year average percentage of rainfall during the winter period, % (7); multi-year average percentage of rainfall during the humid period, % (8); multi-year average percentage of rainfall during the summer period, % (9); mean R_e, MJ mm ha^{-1} h^{-1} (10); 95th percentile of R_e, MJ mm ha^{-1} h^{-1} (11); multi-year average annual erosivity, MJ mm ha^{-1} h^{-1} y^{-1} (12); multi-year average percentage of erosivity during the winter period, % (13); multi-year average percentage of erosivity during the humid period, % (14); multi-year average percentage of erosivity during the summer period, % (15).

1 ID	2 Series Length	3 Mean n° of Distinct Events	4 Mean h_e	5 95th Percentile h_e	6 P	7 P_w	8 P_h	9 P_s	10 Mean R_e	11 95th Percentile R_e	12 R	13 R_w	14 R_h	15 R_s
1	21	133.7	6.0	24.4	804.7	36.7	40.0	23.3	13.7	50.5	1823.7	14.8	39.2	46.0
2	19	146.8	6.4	27.2	940.4	37.9	40.2	21.9	16.1	74.6	2421.3	16.9	39.3	43.8
3	15	125.5	9.0	34.6	1130.4	36.3	40.7	22.9	21.9	103.9	2738.2	23.2	36.1	40.6
4	24	128.2	5.9	24.3	754.3	31.3	41.5	27.2	12.1	51.9	1567.4	14.7	34.8	50.6
5	23	127.1	6.7	27.2	845.0	34.7	40.1	25.3	16.9	71.1	2113.0	17.4	34.7	47.9
6	21	129.2	5.8	23.3	748.7	33.9	40.5	25.6	12.8	48.4	1655.3	15.7	32.8	51.5
7	24	144.0	5.1	21.9	734.4	32.1	41.1	26.8	10.0	41.6	1452.5	13.8	34.4	51.8
8	24	126.8	6.1	25.0	778.5	33.2	40.6	26.2	14.5	55.0	1904.5	16.2	34.7	49.1
9	24	162.9	6.2	27.0	1011.7	37.5	39.1	23.4	12.6	61.6	2087.2	17.7	32.4	49.9
10	23	134.1	7.5	29.3	1009.6	33.3	39.7	26.9	17.7	82.1	2361.5	16.5	32.0	51.5
11	24	135.4	6.2	26.0	839.8	34.1	41.6	24.3	16.0	67.6	2159.1	14.6	36.9	48.5
12	19	127.8	5.9	24.4	757.6	38.2	41.5	20.4	14.0	54.7	1885.0	15.4	44.5	40.1
13	21	140.6	5.7	23.6	806.5	34.4	40.8	24.8	11.1	45.8	1563.4	14.5	36.4	49.1
14	23	139.1	6.0	24.7	838.0	35.8	40.9	23.3	11.0	52.1	1525.8	17.6	34.8	47.6
15	24	139.1	5.4	22.5	755.1	33.3	39.3	27.4	12.8	51.0	1780.0	12.5	33.5	54.0
16	23	129.1	5.8	24.1	752.5	31.9	41.1	27.0	12.3	51.4	1624.2	14.1	36.6	49.3
17	24	143.2	6.1	24.4	869.5	32.1	40.7	27.2	11.7	44.2	1667.7	11.1	31.0	58.0
18	20	142.3	7.4	30.0	1057.2	34.5	40.2	25.3	14.4	66.3	2014.7	15.3	31.7	53.0
19	23	146.0	6.5	27.0	955.6	35.6	39.5	24.9	14.0	59.3	2051.1	15.1	30.7	54.2
20	23	152.9	5.6	24.2	845.6	32.3	41.3	26.4	12.4	54.6	1850.9	14.6	34.7	50.7
21	20	125.3	6.4	24.8	805.3	36.9	41.0	22.1	13.1	57.7	1635.6	18.1	36.4	45.4
22	19	143.8	6.1	26.5	881.6	34.3	40.2	25.5	14.2	62.3	2055.7	14.7	34.4	50.9
23	17	181.8	4.2	19.0	769.2	33.1	42.7	24.2	8.5	33.0	1620.7	12.1	40.5	47.5
24	23	171.7	7.7	32.9	1327.3	33.4	41.3	25.3	14.6	64.7	2522.2	13.9	35.3	50.8
25	23	133.2	6.2	24.6	830.8	31.1	41.4	27.4	14.3	56.7	1936.2	10.8	33.7	55.5
26	24	133.7	6.6	26.7	872.8	34.2	40.3	25.5	18.1	72.2	2399.1	13.5	36.4	50.1
27	22	139.8	6.8	27.4	946.1	34.8	38.5	26.7	13.3	58.0	1841.2	16.4	30.1	53.5
28	19	170.2	4.6	19.4	778.9	36.0	39.3	24.7	5.9	24.8	1004.4	16.3	40.9	42.8
29	22	125.6	6.8	27.0	855.4	37.3	39.5	23.2	15.3	62.2	1957.0	15.3	32.1	52.6
30	23	133.3	5.6	23.9	758.3	36.1	39.6	24.3	12.0	47.6	1634.8	14.5	32.4	53.0

Table 2. *Cont.*

1	2	3	4	5	6	7	8	9	10	11	12	13	14	15
ID	Series Length	Mean n° of Distinct Events	Mean h_e	95th Percentile h_e	P	P_w	P_h	P_s	Mean R_e	95th Percentile R_e	R	R_w	R_h	R_s
31	21	129.2	5.7	22.6	739.9	34.1	40.7	25.2	11.1	50.3	1432.4	13.7	34.5	51.9
32	24	156.6	5.7	25.4	893.2	35.8	39.8	24.4	12.0	50.4	1841.7	15.6	30.2	54.2
33	21	139.4	4.8	19.8	674.0	33.8	41.8	24.4	8.9	38.5	1247.0	13.5	39.0	47.5
34	23	119.0	6.5	25.4	781.6	33.4	39.0	27.6	13.6	56.8	1649.5	13.6	32.0	54.4
35	23	130.2	6.7	25.1	870.0	32.5	40.1	27.3	14.8	57.2	1931.5	12.4	32.8	54.8
36	20	141.5	6.3	25.2	897.9	33.6	38.4	28.1	13.2	58.4	1929.4	13.5	27.2	59.3
37	21	142.8	6.9	28.3	993.5	36.3	41.2	22.6	14.7	71.7	2104.2	16.7	39.5	43.8
38	20	135.0	6.5	27.8	879.7	36.5	38.7	24.9	13.4	58.3	1852.8	17.5	30.1	52.4
39	19	133.3	6.3	25.8	852.0	35.3	39.0	25.8	14.2	60.9	1940.6	14.7	31.0	54.2
40	22	134.6	5.4	22.4	728.8	33.9	39.8	26.4	10.4	46.0	1404.5	15.3	33.2	51.4
41	24	147.8	5.2	22.1	772.0	35.9	39.6	24.5	12.4	48.1	1881.6	15.2	34.1	50.7
42	17	150.0	5.1	22.4	769.5	34.7	41.9	23.4	10.9	43.7	1650.6	11.9	40.3	47.8
43	24	127.5	7.2	27.2	917.0	36.0	41.4	22.6	16.9	74.1	2186.3	17.2	35.1	47.8
44	24	134.3	6.2	25.0	828.6	31.6	40.9	27.5	12.4	49.0	1624.1	12.2	31.4	56.4
45	24	128.1	5.4	21.3	691.8	32.8	40.6	26.7	11.2	47.2	1449.6	12.7	34.9	52.4
46	19	122.5	6.9	26.1	839.2	33.9	40.6	25.5	18.4	78.8	2188.5	12.9	45.1	42.0
47	20	133.1	5.2	21.4	699.4	32.3	43.2	24.5	9.7	41.2	1316.4	14.1	36.0	49.9
48	22	133.7	6.8	27.6	900.8	33.0	40.2	26.8	17.9	75.6	2375.0	14.7	30.4	54.8
49	21	135.2	6.9	28.0	927.6	34.2	40.3	25.5	14.2	75.2	1912.7	16.7	37.2	46.2
50	19	125.6	6.6	24.8	828.6	33.7	40.9	25.4	17.0	57.9	2095.2	13.6	42.7	43.7
51	22	133.9	5.8	24.3	773.6	33.8	41.4	24.8	12.3	55.1	1660.8	14.8	37.3	47.8
52	21	172.9	5.2	22.7	900.2	36.3	38.7	25.1	11.4	42.6	2005.9	15.6	33.2	51.2
53	23	145.7	6.4	26.7	930.2	33.9	39.5	26.6	13.7	54.3	2001.9	14.1	31.6	54.3
54	20	149.9	7.8	31.0	1166.8	31.7	42.7	25.5	14.1	63.0	2097.0	12.3	35.6	52.1
mean	21.6	139.2	6.2	25.3	857.7	34.4	40.5	25.2	13.4	57.1	1863.6	14.9	35.0	50.2
min	15	119.0	4.2	19.0	674.0	31.1	38.4	20.4	5.9	24.8	1004.4	10.8	27.2	40.1
max	24	181.8	9.0	34.6	1327.3	38.2	43.2	28.1	21.9	103.9	2738.2	23.2	45.1	59.3
CV	0.10	0.10	0.14	0.12	0.14	0.05	0.03	0.07	0.21	0.24	0.18	0.14	0.11	0.08

3.1. Model A

The dataset of identified rainfall events includes a total of over 175,000 events. On average, approximately 139 events occur per year, with a maximum of 181 events at the Moiano station (ID 23) and a minimum at Petrelle (ID 34). The number of events is not correlated with either annual cumulative rainfall (r = 0.3) or event volume (r = 0.33).

The average annual number of events (column 3, Table 2) varies moderately (CV = 9.7%) between stations, suggesting differences in local rainfall regimes. Notable variations in the mean event rainfall depth, h_e (column 4, Table 2), are observed, with stations such as Armenzano (ID 3) (9.0 mm) characterised by high volumes and Moiano (ID 23) (4.2 mm) showing values reduced by up to 50%. The 95th percentile (column 5, Table 2), indicative of extreme events, ranges from 19.0 mm (Moiano, ID 23) to 34.6 mm (Armenzano, ID 3). The latter, located in the south and north-west, shows a higher predisposition to extreme events, likely due to specific topographical or microclimatic conditions.

The multi-year average annual rainfall, P (column 6, Table 2), varies significantly, with minimum values at Perugia Santa Giuliana (ID 33) (674.0 mm) and maximum values at Monte Cucco (ID 24) (1327.3 mm). Rainfall is unevenly distributed throughout the year, with the contribution during the humid period (P_h, column 8 in Table 2) representing the main share, ranging regionally from 38.4% to 43.2%, highlighting the greater importance of precipitation during transitional seasons compared to summer and winter periods.

The summer period (P_s, column 9 in Table 2) is the driest, with percentages ranging from 20.4% to 28.1%. Pianello (ID 36) shows the highest summer contribution, indicating less concentrated rainfall in transitional seasons. Winter precipitation (P_w, column 7 in Table 2) contributes one-third of the annual total (34.4%).

The average event erosivity (column 10, Table 2) and the 95th percentile (column 11, Table 2) reveal a differentiated potential risk of soil erosion across various areas. The mean R_e value ranges from a minimum of 5.9 MJ mm ha^{-1} h^{-1} (Norcia, ID 28) to a maximum of 21.9 MJ mm ha^{-1} h^{-1} (Armenzano, ID 3). The 95th percentile of R_e, reflecting the erosive capacity of the most intense events, again highlights Armenzano (ID 3) as the station with the highest value (103.9 MJ mm ha^{-1} h^{-1}).

The multi-year average erosivity (R, column 12, Table 2) was determined for the 54 stations using the Model A procedure and exhibits significant spatial variability. Stations such as Armenzano (ID 3) and Amelia (ID 2) stand out for high values, indicative of greater erosive potential, while others, such as Norcia (ID 28) and Perugia Santa Giuliana (ID 33), show lower values.

Although erosivity shows significant variations, its distribution across periods does not mirror that of rainfall regimes, which follow an opposite trend. The highest values are observed during the dry period (R_s, column 15 in Table 2), with an average multi-year erosivity of 50.2%. This is particularly noteworthy, considering that summer has the lowest rainfall among the analysed periods. This phenomenon is attributed to the frequency of intense storms, which significantly increase erosivity.

Winter erosivity (R_w, column 13 in Table 2) is the lowest among the analysed periods, likely due to the nature of typical winter meteorological events, often characterised by less intense precipitation. The multi-year average for the period is 14.9%, significantly lower than the average rainfall for the same interval, which stands at 34.4%.

The humid period, on the other hand, contributes 35% of the annual erosivity (R_h, column 13 in Table 2). This value reflects the transitional role of this period, positioned between the extremes of the dry and winter periods.

An interesting case is represented by the Pianello station (ID 36). During summer, the station records the highest regional erosivity value (59.3%), while during the humid period, it shows the lowest regional erosivity value (27.2%).

Figure 2a shows the isoerosivity map obtained. These values, calculated using Model A, are reported in column 12 of Table 2. IDW determines the cell values by applying a linearly weighted combination of the R values associated with a set of sample points, represented here by the 54 stations. The weight of each point is a function of the inverse distance, and the surface to be interpolated corresponds to the spatially dependent variable, in this case, R.

The southern and northwestern areas, are characterised by higher annual mean erosivity values (2116, 2738, and 1881 MJ mm ha^{-1} h^{-1} y^{-1}, respectively, for the mean, maximum, and minimum) compared to the northern, eastern, and southwestern areas, where rainfall erosivity is generally lower (1591, 1852, and 1004 MJ mm ha^{-1} h^{-1} y^{-1}, respectively, for the mean, maximum, and minimum). The regional annual mean value is 1863.6 MJ mm ha^{-1} h^{-1} y^{-1}, with a standard deviation of 336.4 MJ mm ha^{-1} h^{-1} y^{-1}.

3.2. Model B

The model was applied starting from the same dataset of individual rainfall events determined and used for the application of Model A (over 175,000 events).

With reference to Equation (6), the erosivity of a single event R_e was calculated considering the temporal reference as the month t = m = 1, . . ., 12, and the spatial scale as the specific station s.

The 54 × 12 pairs ($\alpha_{m,s}$, $\beta_{m,s}$) are displayed in Figure 3, where the trends of the two parameters throughout the 12 months of the year for the 54 stations can be observed. In both cases, a summer peak is evident during July and August, which tends to decrease towards the Gaussian extremes of the winter months; spring and autumn months exhibit intermediate values. For the same rainfall depth in an event h_e, summer events are thus characterised by a higher degree of erosivity compared to events occurring in the other periods of the year. The monthly values of α and β for each station are presented in Tables A1 and A2 in the Appendix A.

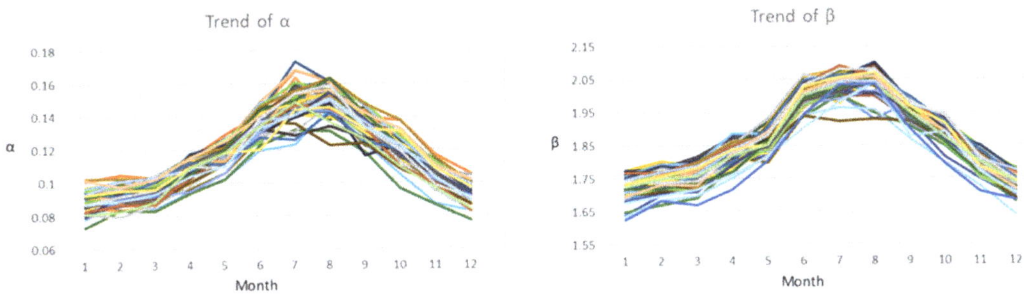

Figure 3. Monthly values of the parameters {αm, βm; m = 1, . . ., 12} of Model B, (6), for the 54 stations. Each station is represented by a different colour.

Figure 2b shows the isoerosivity map determined through IDW interpolation of the mean annual values, R_m, for each station, calculated using Equations (4)–(6) with the pairs ($\alpha_{m,s}$, $\beta_{m,s}$) given in Figure 3.

In this case, the northwestern and southern areas are characterised by higher values of mean annual erosivity (2299.0, 2298.9, 1446.7 MJ mm ha^{-1} h^{-1} y^{-1}, respectively, mean, maximum, and minimum) compared to the northern, eastern, and southwestern areas, where erosivity values are generally lower (1305.6, 1682.5, 756.0 MJ mm ha^{-1} h^{-1} y^{-1}, respectively, mean, maximum, and minimum). The regional mean annual value is 1531.2 MJ mm ha^{-1} h^{-1} y^{-1}, with a standard deviation of 318.3 MJ mm ha^{-1} h^{-1} y^{-1}.

Model B operates at the event scale temporal resolution, where the parameters α and β are kept constant across the months. This makes the procedure less computationally demanding compared to the semi-hourly scale required by Model A (USLE and RUSLE-2), although despite simplifying the process, Model B remains highly complex due to the variability and number of parameters involved.

3.3. Model C

The analysis of the temporal variability of the parameters {α_m, m = 1, . . . 12} e {β_m, m = 1, . . . 12}, shown in Figure 3, reveals a periodic pattern in the annual dynamics, with a maximum value in the summer months, a minimum in the winter months, and intermediate values corresponding to the spring and autumn months. Based on this, a data analysis was conducted for the selection and grouping of months with homogeneous erosivity values. A statistical analysis based on normalisation and clustering of the α and β data showed that these parameters can be considered statistically constant within three distinct periods, p (with p = 1, . . ., 3): summer period p_s from June to September; the humid period p_h, April, May, October and November; the winter period p_w, January, February, March and December. The period values of α and β for each station are presented in Table A3 in the Appendix A.

In addition, he percentage of the total annual rainfall height P_y relative to the three periods (Table 2) are substantially different with an average value at the regional scale of P_s = 211.7 mm, P_w = 286.8 mm and P_h = 339.1 mm, while the corresponding means R_s = 935.5 MJ mm ha^{-1} h^{-1} y^{-1}, R_w = 277. 7 MJ mm ha^{-1} h^{-1} y^{-1}, R_h = 652.3 MJ mm ha^{-1} h^{-1} y^{-1} show an opposite behaviour with the higher value in summer and the lower in winter period.

Figure 2c shows the isoerosivity map determined through IDW interpolation of the mean annual values, R_p, for each station, calculated using Equations (4)–(6) with the pairs (α_p, β_p).

The data shown in Figure 2c highlight a substantial agreement with the pattern obtained using RUSLE-2 (Model A, Figure 2a). Areas located in the south and north-west are characterised by significantly higher average annual erosivity values compared to other regions. Specifically, the average, maximum, and minimum erosivity are 1699.2, 2272.4, and 1334.7 MJ mm ha^{-1} h^{-1} y^{-1}, respectively. In contrast, the areas in the north, east, and south-west show lower average annual values, with an average of 1304.2 MJ mm ha^{-1} h^{-1} y^{-1}, a maximum of 1675.3 MJ mm ha^{-1} h^{-1} y^{-1}, and a minimum of 772.3 MJ mm ha^{-1} h^{-1} y^{-1}. The regional mean value of annual erosivity is 1499.1 MJ mm ha^{-1} h^{-1} y^{-1}, with a standard deviation of 294.0 MJ mm ha^{-1} h^{-1} y^{-1}, reflecting territorial variability.

In Model C, the complexity related to the calculations and the procedures required is further reduced compared to previous models, as the parameters α and β are considered constant across different periods. Like Model B, Model C requires a temporal resolution of the data at the scale of individual rainfall events.

3.4. Model D

The conducted analysis reveals a clear distinction in the levels of average annual erosivity across different geographical areas and periods. Thus, six functional relationships between the daily rainfall and erosivity were obtained, as shown in Table 3.

Table 3. Relationships between the daily rainfall (h_d) and erosivity (R_d) for each period and sub-zone processed using the D model.

Equation	P, Z	
$R_d = 0.00080848 \cdot h_d^3 - 0.03067570 \cdot h_d^2 + 1.35937424 \cdot h_d - 1.92473511$	p = 1; z = r	(7)
$R_d = -0.00022788 \cdot h_d^3 + 0.05795112 \cdot h_d^2 + 0.99900288 \cdot h_d - 2.06564377$	p = 2; z = r	(8)
$R_d = -0.00060982 \cdot h_d^3 + 0.18384798 \cdot h_d^2 + 0.56087557 \cdot h_d - 1.48459328$	p = 3; z = r	(9)
$R_d = -0.00032134 \cdot h_d^3 + 0.05897403 \cdot h_d^2 + 0.04415188 \cdot h_d + 0.09640815$	p = 1; z = b	(10)
$R_d = -0.00024928 \cdot h_d^3 + 0.07843005 \cdot h_d^2 + 0.47441833 \cdot h_d - 0.80569572$	p = 2; z = b	(11)
$R_d = -0.00036687 \cdot h_d^3 + 0.17375462 \cdot h_d^2 + 0.53993365 \cdot h_d - 1.04597592$	p = 3; z = b	(12)

The "blue-green zones" (Figure 2d) exhibit significantly higher erosivity values, with an annual average of 2040.3 MJ mm ha^{-1} h^{-1} y^{-1}, a maximum value of 2830.8 MJ mm ha^{-1} h^{-1} y^{-1}, and a minimum of 1698.5 MJ mm ha^{-1} h^{-1} y^{-1}.

Conversely, the "red-orange zones" (Figure 2d) show lower erosivity levels. In these areas, the average value stands at 2116.7 MJ mm ha^{-1} h^{-1} y^{-1}, with a maximum of 2093.8 MJ mm ha^{-1} h^{-1} y^{-1} and a minimum of 1698.8 MJ mm ha^{-1} h^{-1} y^{-1}.

Considering the entire study region, the average annual erosivity is 1915.4 MJ mm ha^{-1} h^{-1} y^{-1}, with a standard deviation of 328.9 MJ mm ha^{-1} h^{-1} y^{-1}.

3.5. Model Validation

To validate the obtained models, the meteorological station of Trestina (43.358 N, 12.237 E), not included in the database of Table 1, was considered. For this station, half-hourly precipitation data are available for the period 2007–2024.

For the application of Models B and C, Equation (6) was used with the parameters α and β derived from the nearest station (Città di Castello, ID14, approximately 9 km away). For Model D, Equations (7)–(9) were applied, as the Trestina station falls within the red zone. The estimated annual values (R_m, R_p, and R_d, corresponding to Models B, C, and D, respectively) were compared with the annual R values of Model A. The results are presented in Figure 4.

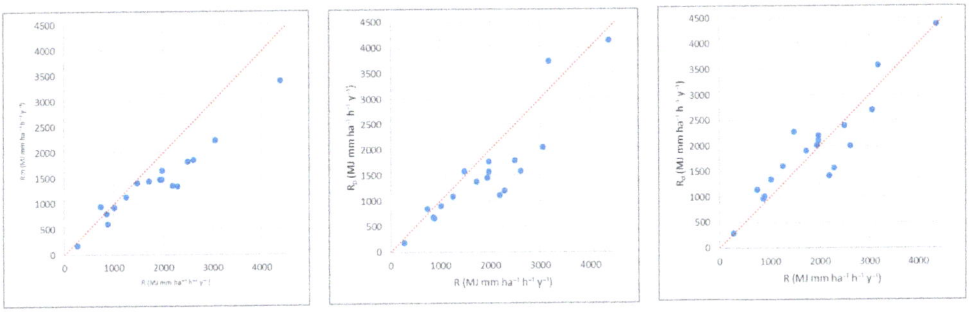

(a) (b) (c)

Figure 4. Scatter plot comparing annual cumulative erosivity values (MJ mm ha^{-1} h^{-1} y^{-1}) at the Trestina station obtained from different models: (**a**) Model B (R_m) vs. Model A (R); (**b**) Model C (R_p) vs. Model A (R); (**c**) Model D (R_d) vs. Model A (R).

As shown, while Models B and C exhibit a general agreement with the reference values, their annual estimates tend to show a slight underestimation. Model D demonstrates a stronger correlation, with values more evenly distributed around the 1:1 line, maintaining this pattern even in the presence of extreme values.

The comparison of annual values between Model A and Model B results in a MAPE of 24.4%, while the difference between Model A and Model C yields a MAPE of 23.3%. The comparison between Model A and Model D shows a slightly lower MAPE of 19.9%.

3.6. Trend Analyses

The trend analysis was conducted on the annual erosivity, R_y, as well as the annual rainfall, P, for each of the 54 stations. The historical series analysed are robust, spanning a total of 1167 years. Their duration ranges from a minimum of 15 years, observed in a single case at the Armenzano station (ID 3), to a maximum of 24 years. The average series length per station exceeds 21.5 years (Column 2, Table 2).

The results of the analysis indicate the absence of significant trends in annual erosivity in most meteorological stations, both for the multi-year average rainfall series and for those related to multi-year average erosivity. For most stations, neither rainfall events nor erosive events exhibit a predictable or consistent pattern over time.

A single exception was identified at the Nocera Umbra (ID 27), where a significant trend in rainfall was observed, with a Theil–Sen slope of 13.5 per year. Regarding erosivity, significant trends were observed in four stations: Armenzano (ID 3), Moiano (ID 23), Nocera Umbra (ID 27), and Ponte Felcino (ID 39). In these cases, the Theil–Sen slope values were 89.5, 109.0, 61.0, and 90.3, respectively.

4. Discussion

4.1. Comparison Between Simplified Models (B, C, D) and Reference Model (A)

Figure 2 presents isoerosivity maps generated through a zonation process based on clustering analysis. These maps depict the multi-year average annual erosivity values of the R parameter, calculated using different predictive models. Despite considering other explanatory factors, such as altitude (r = 0.24), latitude (r = 0.14), and longitude (r = 0.17), their correlations with erosivity were found to be insufficiently significant to provide a representative contribution.

To enhance the clarity and interpretability of the results, each map was constructed using uniform intervals, which were defined based on percentiles corresponding to specific erosivity ranges across the models. Adopting a uniform scale highlighted the peculiarities and differences among the various methods for calculating erosivity. The general pattern is consistently observed in all isoerosivity maps, although predictive models tend to underestimate the final values produced by RUSLE2 (Model A).

The statistics of the multi-year average annual erosivity values calculated using Models A, B, C, and D for individual stations are presented in Table 4. Regional values from the analysis by Panagos et al. [29] are also presented for comparison purposes.

Table 4. Statistics of mean annual erosivity (MJ mm ha^{-1} h^{-1} y^{-1}) at the Umbria region's 54 meteorological stations. The mean absolute percentage error, MAPE, and the Pearson correlation coefficient, r, refer to the comparison between the reference (model A) and all the other R estimates at the station level. The description of the models follows that of Figure 2. R_d is distinguished based on the regional mean value into R_{db} (>R_{mean}) and R_{dr} (<R_{mean}), while $R_{Panagos}$ is derived from [29].

	Model					
	A	**B**	**C**	**D**	**D**	
	R	**R_m**	**R_p**	**$R_{d\,r}$**	**$R_{d\,b}$**	**$R_{Panagos}$**
Minimum value	1004.4	756.0	772.3	992.5	879.2	1125.7
Maximum value	2738.2	2299.0	2272.4	2132.4	2694.4	2151.4
Mean value	1843.6	1531.2	1499.1	1626.3	2040.3	1518.7
Standard deviation	336.4	318.3	294.0	318.4	241.2	217.6
Coefficient of variation (%)	18.2	20.8	19.6	16.1	14.8	14.3
MAPE (%)		18.9	20.1	5.5	5.4	18.6
r		0.81	0.80	0.76	0.73	0.54

The minimum average value of R is observed at the Norcia station (ID 28), whereas Armenzano (ID 3) exhibits the maximum. These results are consistent for the erosivity estimates of Rm, Rp, and Rd. The highest variability in the estimates is found using Models B and C, whit CV = 20.8% and CV = 19.6%, (Table 4), respectively.

In general, although all models confirm spatial patterns similar to those provided by the reference model (Model A), the annual average erosivity shows a slight underestimation (≈15%) in the case of Models B and C, while it is more comparable in Model D (when averaging the values of $R_{d,r}$ and $R_{d,b}$).

With the increase in the degree of simplification applied to the model, the correlation between the model prediction and the expected value calculated by RUSLE2 (model A) decreases, from r = 0.81 and 0.80 for models B and C, to r = 0.76 and 0.73 for models D. Despite this variation, the explanatory power of the model relative to the reference data remains high considering the lower level of detail in the input data (from 30 min to daily time resolution). The stability and reliability of the Rm, Rp, and Rd estimates in representing the erosive phenomenon are further confirmed by the Mean Absolute Percentage Error (MAPE) between the model predictions and the expected values calculated using RUSLE2 (Model A) (Table 4). In this case, the best performance (i.e., the lowest MAPE) is achieved by Model D, with values around 5–6%, whereas Models B and C exhibit MAPE values of approximately 19–20%.

The simplified models prove to be sufficiently reliable not only at the annual and mean annual R-factor scales (Table 4) but also at the sub-annual level. This evaluation is demonstrated, by way of example, for Models B (R_m) and C (R_p).

The monthly distribution of the R and R_m are plotted in Figure 5a as a box plot to highlight potential seasonal trends and differences between the two estimates, with a particular focus on variability and extreme events characterising different months of the year. During the winter months (January, February, and December) the erosivity is generally low, with contained medians, relative compact distribution and similar values estimated by the two models. This pattern reflects the low intensity and variability of rainfall typical of the winter climate, which tends to have a lower potential erosive power [42].

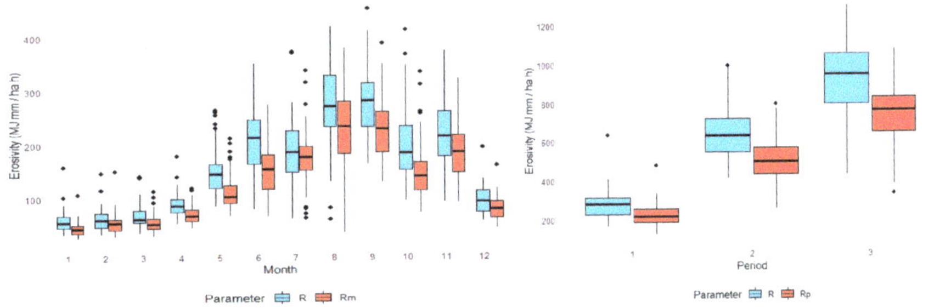

Figure 5. (**a**) Comparison of the monthly erosivity estimated by R (model A) and R_m (model B); (**b**) Comparison of the erosivity estimated by R and R_p (model C) across periods.

During the spring and summer, erosivity progressively increases, reaching the maximum in August, and then progressively decreases in autumn toward the winter values. This behaviour corresponds to a marked variability among stations starting from May. This variability remains almost unchanged for R until September; instead, Rm dispersion follows the erosivity trend with a maximum dispersion in August. The R presents a greater variability than Rm, with the difference diminishing in winter months when erosivity values are more similar. This seasonal trend confirms the system's greater stability in winter, where both models converge, and the heightened sensitivity to the variable summer climate. This behaviour could be attributed to the nature of the Rm parameter calculation, which incorporates reduced sensitivity to short-duration intense rainfall, characteristic of the summer period, better described by a parameter like I30 rather than event-scale cumulative rainfall (he).

Figure 5b compares the values of R and Rp across three main seasonal periods: 1, winter (January, February, March, and December); 2, humid (April, May, October, and November); and 3, summer (from June to September). The comparison confirms all the results discussed above relative to the R and Rm models with a progressive increase in intensity and variability of the rainfall erosivity from the winter to summer, with a particularly pronounced rise during spring and summer, coinciding with the intensification of rainfall phenomena.

These comparisons highlight the significance of factors such as the frequency and intensity of storm events in determining erosivity dynamics. The analysis of stations characterised by different seasonal distributions provides valuable insights into the influence of local factors on rainfall and erosivity properties. These results, enhanced by the integration of geomorphological data such as orography and microclimate, allow for a more accurate assessment of erosion risks and their potential environmental impacts.

Although models B and C offer a simplification in the calculation of erosivity at the event scale, they still require the knowledge of h_e, a variable that is typically not available in meteorological datasets. This significantly limits their practical utility. On the other hand, Todisco et al. [43], in a study based on 18 stations in the Umbria region from 2000 to 2010, demonstrated that the differences between h_e and h_d were not particularly pronounced, as they were found to be less than 5 mm in 95% of the cases analysed. Therefore, a more practical use of models B and C could involve substituting h_d for h_e. The impact of this solution, likely modest, will still require specific investigations.

The validation performed using the Trestina weather station (Figure 4) confirms the reliability of the results obtained through the application of the simplified models (Models B, C, and D). Among these, Model D represents the most effective trade-off between computational efficiency and accuracy. Specifically, the input data, characterised by a daily temporal resolution and a spatial scale at the sub-zone level (for Trestina, corresponding to the red zone), allow for a streamlined and efficient calculation process.

The literature offers several studies assessing the feasibility of obtaining reliable estimates of the R-factor through simplified empirical models. For certain regions of Italy, empirical approaches based on the Modified Fournier Index (MFI) have been shown to be both effective and easy to implement [19,20], as they require only monthly precipitation data as input. However, unlike these methodologies, the simplified models proposed in this study can be applied not only for the estimation of the annual R-factor but also for a (although approximate) assessment of erosivity at the event or daily scale.

4.2. Comparison with Panagos et al. (2015) [29]

The results obtained in this study were compared with the iso-erosivity maps developed in a previous work by Panagos et al. [29], which serve as a well-established reference in the field.

The erosivity map by Panagos et al. [29] (Figure 5b) was obtained through regional scale clipping operations applied to the map developed at the European scale. The erosivity value scale remains consistent with that presented in Figure 2. The comparison of individual values was conducted by sampling erosivity data from the Panagos map at the coordinates of the meteorological stations utilised in this study, using QGIS version 3.34.15.

The $R_{Panagos}$ are derived from a raster dataset developed at a European scale [29], which may imply a less detailed spatial resolution compared to Model A. The number of stations used at the regional scale is also smaller. Panagos et al. [29] analysed a total of 251 meteorological stations across Italy, with a spatial distribution that accounts for the country's main climatic conditions (Mediterranean, continental, and alpine) and different altitudinal zones. However, when considering the size of the Umbrian region in comparison to the national territory, only seven meteorological stations are representative of the region in Panagos' work, whereas this study analyses data from 54 stations. Additionally, the historical series used by Panagos has an average duration of 10 years, significantly shorter than the 21.5-year average in the present study, offering a broader temporal coverage. Furthermore, the use of different approaches for erosivity calculation (RUSLE for Panagos and RUSLE-2 for Model A) introduces another source of discrepancy between the two methods.

From the comparative analysis (Table 4), a relatively weak correlation ($r = 0.56$) emerged between the annual erosivity values estimated using the RUSLE-2 model, applied in the calculation of Model A, and those derived from the RUSLE model employed in Panagos's studies, $R_{Panagos}$.

The comparison between the R and $R_{Panagos}$ reveals a systematic tendency for the latter to provide lower estimates, indicating a potential intrinsic underestimation in [29]. This, in addition to the differences between the datasets used, may depend on the known tendency of the RUSLE method to underestimate precipitation energy compared to the RUSLE model2 [43].

Amelia (ID 2) recorded the highest erosivity value $R_{Panagos}$ = 2151.4 MJ mm ha^{-1} h^{-1} y^{-1} according to Panagos et al. [29], whereas Umbertide showed the lowest value $R_{Panagos}$ = 1125.7 MJ mm ha^{-1} h^{-1} y^{-1}). At the regional level, the average erosivity $R_{Panagos}$ = 1518.7 MJ mm ha^{-1} h^{-1} y^{-1}. The analysis of the distribution of values shows a general consistency across stations, although R highlights significant peaks for certain stations (ID 3, ID 24, and ID 43), while $R_{Panagos}$ is more uniform, exhibiting lower variability among stations. This behaviour suggests that the method used in Panagos et al. [29] reduces the impact of extreme values, likely due to differences in calculation methodology or the nature and density of the data used.

The differences between the results are particularly pronounced in some stations, such as ID3 and ID43, whereas in other cases, such as ID4, ID31, and ID44, there is an almost complete convergence between the two approaches. The general pattern confirms that Model A tends to produce higher erosivity values compared to those estimated by Panagos et al. [29].

Figure 6a compares the erosivity values (MJ mm ha^{-1} h^{-1}) across five models: R, $R_{Panagos}$, R_m, R_p, and R_d. Among them, R_d stands out with the highest median and the widest interquartile range (IQR), indicating greater variability and higher erosivity estimates. In contrast, $R_{Panagos}$ exhibits the lowest median and the narrowest IQR, reflecting more consistent and lower values. The R_m and R_p models show similar distributions, with R_p displaying slightly greater variability. The R model, though closer to R_d in median, reveals a broader spread compared to $R_{Panagos}$, indicating moderate variability. Outliers are present in all models, emphasising occasional extreme erosivity values.

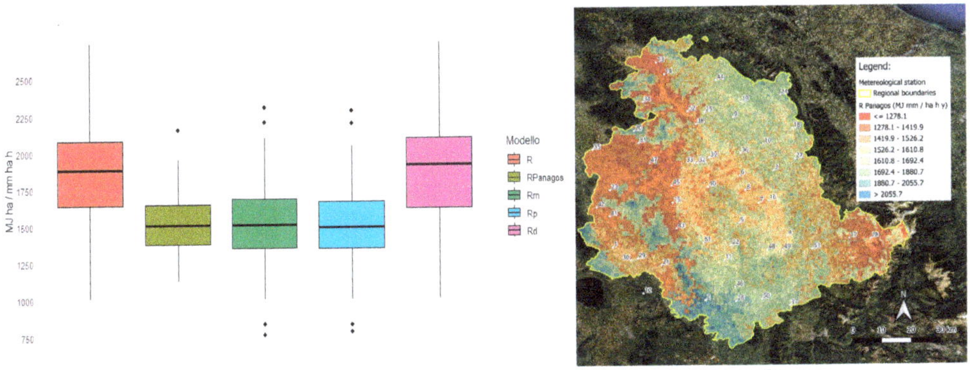

(a) (b)

Figure 6. (**a**) Comparison of multi-year average R, R_m, R_p, and R_d values with $R_{Panagos}$ [29]; (**b**) Multi-year average erosivity map of the Umbria Region, $R_{Panagos}$. Numbers represent station's IDs.

A direct comparison between R and $R_{Panagos}$ highlights their differences. The median and mean values of R are notably higher, with R also showing a larger IQR and more extended whiskers, capturing a wider range of data. Conversely, $R_{Panagos}$ demonstrates lower variability, with a compact box and shorter whiskers. While $R_{Panagos}$ has an outlier above the upper limit, R spans a broader range without evident outliers but includes a higher maximum value.

The greater variability in R likely stems from increased sensitivity in input data, influenced by the number of analysed stations or differing calculation criteria in RUSLE-2. By contrast, $R_{Panagos}$, derived using the RUSLE model, reflects less sensitivity to erosivity peaks, suggesting a more constrained methodology.

The isoerosivity map developed by Panagos [29] in Figure 6b is consistent with that obtained using Model A, despite systematic underestimations observed like in Models B and C. The dataset is projected onto a regular grid with a pixel resolution of $0.006° \times 0.006°$, which provides less detailed results compared to the models analysed in this study ($0.0009° \times 0.0009°$).

The areas with the lowest R-factor values are in the southeast and along the entire western strip, while erosivity values exceeding 2000 MJ mm ha^{-1} h^{-1} are found in the south and northeast. The central part of Umbria is characterised by average values.

5. Conclusions

This study aimed to assess rainfall erosivity in the Umbria region through the analysis of a twenty-year rainfall dataset. By exploring the spatial and temporal dynamics of erosive events, the research uncovered significant differences across areas and seasons, with implications for soil erosion risk. The R-factor was mapped at a 100 m resolution using IDW interpolation, leveraging extensive rainfall intensity data. The application of the RUSLE-2 model to a network of uniformly distributed meteorological stations demonstrated its ability to identify erosion-prone areas, supported by precise erosivity indices calculated from 30 min rainfall data.

Regional Zoning. Isoerosivity maps highlighted a clear spatial gradient, with higher rainfall erosivity values in the southern and north-western areas of Umbria. These patterns reflect the combined effects of topography, climate, and geomorphology, underscoring the importance of region-specific soil conservation plans.

Comparison of Medians, Model Variability, and Recommendation for Model D. Model A demonstrated high sensitivity to local erosivity variations, with values ranging from 1004.4 to 2738.2 MJ mm ha^{-1} h^{-1} y^{-1} and a standard deviation of 336.4 MJ mm ha^{-1} h^{-1} y^{-1}. Simplified models (B, C, and D) provided reliable yet slightly underestimated alternatives, making them suitable for contexts with limited data availability. Among these, Model D, which relies on daily rainfall data, emerged as the most practical solution for regions with restricted access to high-resolution pluviometric data. Striking a balance between computational simplicity and accuracy, it represents a robust choice for large-scale applications.

Importance of Station Density and Temporal Resolution. The reliability of Model A, confirmed by comparisons with prior studies (e.g., Panagos et al. [29]), highlights the significance of a dense meteorological network and long-term datasets. These factors enhance the detection of spatial and temporal variability, improving predictive accuracy and erosion management strategies.

Trend Analysis. The Mann–Kendall test applied to Model A revealed temporal stability in rainfall erosivity across most stations, with significant trends in only a few locations. This indicates a largely consistent pattern at the regional scale, though some areas may require intensified monitoring.

Seasonal Variability. The analysis identified pronounced seasonal differences in rainfall erosivity, emphasising the necessity of incorporating seasonal dynamics into erosion risk assessments to optimise soil protection measures during periods of higher erosivity.

By integrating high-resolution data with simplified models, this study provides a comprehensive framework for evaluating rainfall erosivity, contributing to sustainable land management and soil conservation in erosion-prone regions.

Author Contributions: Conceptualization, F.T. and A.M.A.; methodology, F.T.; software, L.V.; validation, L.V.; resources, F.T. and A.M.A.; formal analysis, data curation, A.M.A.; writing, A.M.A.; review and editing, F.T. and L.V.; visualisation, F.T., L.V. and A.M.A.; supervision, project administration, funding acquisition, F.T. All authors have read and agreed to the published version of the manuscript.

Funding: The study was financed by European Union-Next-GenerationEU-National Recovery and Resilience Plan (NRRP)–MISSION 4 COMPONENT 2, INVESTIMENT N. 1.1, CALL PRIN 2022 D.D. 104 02-02-2022–(Small reservoIrs restoration: Green blu-infrastructures to enHance rural area resilience To clImate chaNGe—SIGHTING) CUP N. J53D23002910006.

Data Availability Statement: The raw data used in the article can be requested at https://servizioidrografico.regione.umbria.it/.

Conflicts of Interest: The authors declare no conflicts of interest.

Abbreviations

The following abbreviations are used in this manuscript:

USLE	Universal Soil Loss Equation	R	Annual multi-year average erosivity using Model A
RUSLE	Revised Universal Soil Loss Equation	R_m	Annual multi-year average erosivity using Model B
MAPE	Mean Percent Average Error	R_p	Annual multi-year average erosivity using Model C
CV	Coefficient of Variation	R_d	Annual multi-year average erosivity using Model D
r	Pearson correlation coefficient	R_w	Winter multi-year average erosivity
QGIS	Quantum Geographic Information Science	R_h	Humid multi-year average erosivity
IDW	Inverse Distance Weight	R_s	Summer multi-year average erosivity
MFI	Modified Fournier Index	P	Annual multi-year average rainfall depth
h_e	Event rainfall depth	P_h	Humid multi-year average rainfall depth
R_e	Event erosivity	P_s	Summer multi-year average rainfall depth
		P_w	Winter multi-year average rainfall depth

Appendix A

Table A1. α values used in Equation (6) for each month and each station to calculate mean annual erosivity (R_m) in Model B.

ID Station	Month											
	1	2	3	4	5	6	7	8	9	10	11	12
1	0.094	0.097	0.098	0.115	0.126	0.136	0.159	0.160	0.147	0.118	0.110	0.099
2	0.101	0.105	0.103	0.115	0.119	0.145	0.168	0.162	0.139	0.138	0.116	0.105
3	0.098	0.100	0.100	0.112	0.121	0.122	0.152	0.151	0.142	0.129	0.106	0.103
4	0.087	0.089	0.093	0.110	0.112	0.142	0.153	0.144	0.142	0.117	0.102	0.095
5	0.090	0.097	0.099	0.112	0.116	0.146	0.142	0.161	0.141	0.123	0.109	0.095
6	0.089	0.094	0.097	0.105	0.115	0.131	0.150	0.153	0.142	0.119	0.103	0.095
7	0.097	0.098	0.103	0.112	0.121	0.139	0.157	0.155	0.140	0.124	0.111	0.101
8	0.092	0.095	0.096	0.112	0.119	0.140	0.146	0.160	0.132	0.112	0.101	0.092
9	0.091	0.093	0.097	0.105	0.112	0.133	0.146	0.150	0.137	0.122	0.105	0.095
10	0.082	0.089	0.094	0.096	0.106	0.127	0.144	0.145	0.121	0.115	0.098	0.087
11	0.093	0.095	0.100	0.114	0.122	0.143	0.173	0.161	0.137	0.118	0.110	0.095
12	0.094	0.099	0.105	0.117	0.124	0.145	0.146	0.138	0.144	0.126	0.112	0.098
13	0.082	0.085	0.092	0.105	0.120	0.138	0.156	0.154	0.137	0.122	0.100	0.090
14	0.085	0.084	0.086	0.103	0.114	0.135	0.164	0.143	0.132	0.116	0.101	0.090
15	0.087	0.089	0.093	0.104	0.115	0.127	0.147	0.141	0.126	0.113	0.096	0.090
16	0.088	0.089	0.095	0.100	0.112	0.132	0.141	0.139	0.133	0.113	0.103	0.092
17	0.078	0.090	0.084	0.094	0.102	0.119	0.123	0.145	0.123	0.104	0.087	0.083
18	0.087	0.091	0.092	0.110	0.113	0.126	0.153	0.143	0.128	0.118	0.100	0.095
19	0.086	0.091	0.094	0.111	0.111	0.132	0.145	0.149	0.129	0.121	0.103	0.092
20	0.092	0.098	0.099	0.116	0.129	0.141	0.155	0.153	0.145	0.126	0.109	0.095

Table A1. *Cont.*

ID Station	Month											
	1	2	3	4	5	6	7	8	9	10	11	12
21	0.094	0.097	0.101	0.115	0.120	0.138	0.142	0.161	0.132	0.126	0.111	0.098
22	0.092	0.094	0.100	0.114	0.117	0.143	0.147	0.155	0.133	0.129	0.113	0.101
23	0.091	0.097	0.095	0.110	0.119	0.134	0.155	0.142	0.140	0.128	0.109	0.096
24	0.073	0.084	0.083	0.093	0.102	0.123	0.132	0.131	0.118	0.096	0.086	0.077
25	0.084	0.086	0.090	0.106	0.112	0.127	0.147	0.145	0.130	0.116	0.100	0.089
26	0.091	0.092	0.097	0.105	0.124	0.149	0.168	0.161	0.142	0.124	0.111	0.100
27	0.081	0.079	0.087	0.098	0.113	0.129	0.131	0.148	0.134	0.112	0.095	0.083
28	0.084	0.094	0.097	0.104	0.111	0.139	0.146	0.145	0.129	0.117	0.100	0.091
29	0.097	0.100	0.099	0.118	0.120	0.137	0.159	0.148	0.137	0.133	0.108	0.099
30	0.094	0.096	0.102	0.115	0.126	0.142	0.141	0.150	0.139	0.123	0.114	0.099
31	0.096	0.096	0.095	0.116	0.123	0.129	0.144	0.148	0.116	0.123	0.113	0.100
32	0.086	0.089	0.091	0.103	0.119	0.142	0.156	0.156	0.137	0.119	0.101	0.090
33	0.095	0.094	0.096	0.107	0.119	0.135	0.127	0.134	0.127	0.121	0.104	0.093
34	0.082	0.087	0.088	0.100	0.111	0.137	0.135	0.123	0.124	0.108	0.098	0.086
35	0.091	0.097	0.097	0.106	0.122	0.143	0.143	0.143	0.137	0.125	0.110	0.095
36	0.096	0.100	0.101	0.110	0.121	0.131	0.158	0.145	0.138	0.127	0.113	0.097
37	0.082	0.089	0.092	0.095	0.106	0.129	0.144	0.144	0.133	0.116	0.101	0.091
38	0.092	0.096	0.096	0.114	0.113	0.133	0.155	0.150	0.135	0.128	0.110	0.101
39	0.095	0.097	0.104	0.114	0.119	0.144	0.148	0.154	0.140	0.131	0.112	0.096
40	0.097	0.101	0.099	0.112	0.119	0.142	0.149	0.154	0.141	0.123	0.112	0.097
41	0.097	0.102	0.096	0.111	0.120	0.148	0.155	0.163	0.146	0.127	0.108	0.101
42	0.096	0.098	0.099	0.109	0.120	0.147	0.159	0.157	0.148	0.128	0.112	0.097
43	0.080	0.093	0.092	0.108	0.120	0.138	0.141	0.155	0.138	0.117	0.106	0.093
44	0.083	0.087	0.087	0.104	0.111	0.133	0.147	0.145	0.133	0.109	0.099	0.083
45	0.086	0.092	0.095	0.111	0.121	0.135	0.158	0.156	0.135	0.115	0.106	0.090
46	0.103	0.103	0.102	0.116	0.122	0.147	0.155	0.163	0.147	0.138	0.118	0.100
47	0.094	0.093	0.101	0.113	0.111	0.127	0.126	0.142	0.124	0.120	0.109	0.097
48	0.092	0.095	0.097	0.105	0.120	0.144	0.150	0.163	0.142	0.123	0.108	0.097
49	0.093	0.101	0.103	0.108	0.118	0.137	0.141	0.152	0.136	0.121	0.109	0.099
50	0.102	0.098	0.102	0.115	0.125	0.140	0.154	0.156	0.142	0.133	0.110	0.102
51	0.087	0.092	0.096	0.110	0.109	0.139	0.154	0.159	0.131	0.119	0.098	0.090
52	0.093	0.095	0.096	0.109	0.118	0.120	0.144	0.146	0.132	0.131	0.108	0.097
53	0.087	0.089	0.093	0.104	0.110	0.133	0.141	0.151	0.133	0.123	0.103	0.090
54	0.081	0.083	0.088	0.109	0.118	0.134	0.149	0.136	0.126	0.109	0.097	0.084

Table A2. β values used in Equation (6) for each month and each station to calculate mean annual erosivity (R_m) in Model B.

ID Station	Month											
	1	2	3	4	5	6	7	8	9	10	11	12
1	1.694	1.719	1.727	1.805	1.887	1.975	2.048	2.057	1.981	1.858	1.792	1.713
2	1.717	1.736	1.727	1.843	1.859	2.014	2.072	2.051	1.968	1.936	1.820	1.730
3	1.779	1.801	1.786	1.798	1.815	2.028	1.978	2.066	1.922	1.874	1.804	1.781
4	1.733	1.762	1.793	1.842	1.898	1.993	2.017	2.058	1.942	1.928	1.804	1.777
5	1.764	1.790	1.804	1.857	1.859	2.007	2.037	2.058	1.980	1.924	1.840	1.777
6	1.757	1.757	1.794	1.865	1.872	1.994	2.027	2.040	1.981	1.909	1.815	1.756
7	1.696	1.728	1.740	1.819	1.838	2.003	1.999	1.999	1.922	1.865	1.766	1.708
8	1.762	1.780	1.817	1.875	1.881	1.988	2.019	2.042	1.997	1.915	1.829	1.759
9	1.689	1.724	1.724	1.814	1.797	1.943	2.047	2.025	1.933	1.863	1.767	1.699
10	1.761	1.770	1.797	1.829	1.858	1.991	2.038	2.055	1.975	1.905	1.828	1.763

Table A2. *Cont.*

ID Station	Month 1	2	3	4	5	6	7	8	9	10	11	12
11	1.758	1.766	1.800	1.851	1.920	2.032	2.075	2.062	1.957	1.910	1.833	1.757
12	1.679	1.711	1.739	1.812	1.860	2.002	2.004	1.929	1.965	1.900	1.809	1.715
13	1.707	1.764	1.758	1.835	1.874	1.982	2.016	2.066	1.979	1.867	1.778	1.741
14	1.734	1.756	1.765	1.827	1.888	2.020	2.037	2.085	1.934	1.914	1.788	1.740
15	1.744	1.772	1.795	1.846	1.879	2.015	2.066	2.061	1.971	1.919	1.816	1.754
16	1.757	1.792	1.792	1.885	1.886	2.008	2.008	2.048	1.982	1.914	1.826	1.770
17	1.734	1.763	1.755	1.823	1.863	2.025	2.048	2.037	1.917	1.869	1.800	1.752
18	1.640	1.698	1.716	1.797	1.826	1.983	1.991	2.059	1.923	1.810	1.747	1.690
19	1.693	1.724	1.716	1.792	1.836	1.999	2.034	2.041	1.968	1.896	1.757	1.706
20	1.701	1.722	1.765	1.806	1.856	1.993	2.009	2.008	1.953	1.872	1.775	1.724
21	1.732	1.735	1.721	1.783	1.915	2.057	2.078	2.055	1.980	1.883	1.786	1.716
22	1.686	1.724	1.754	1.840	1.895	2.013	2.043	2.005	1.932	1.877	1.778	1.703
23	1.651	1.671	1.692	1.782	1.849	1.937	2.010	1.948	1.916	1.858	1.758	1.679
24	1.699	1.736	1.737	1.798	1.838	1.966	2.054	2.005	1.918	1.855	1.764	1.694
25	1.720	1.746	1.765	1.841	1.836	1.962	2.046	2.039	1.950	1.943	1.809	1.756
26	1.741	1.779	1.787	1.850	1.932	2.064	2.036	2.068	1.990	1.895	1.844	1.768
27	1.714	1.746	1.737	1.813	1.826	1.986	2.051	2.061	1.937	1.855	1.766	1.728
28	1.635	1.681	1.697	1.753	1.806	1.941	1.964	1.961	1.862	1.791	1.721	1.639
29	1.702	1.721	1.738	1.819	1.837	2.032	2.052	2.035	1.978	1.872	1.751	1.701
30	1.701	1.737	1.730	1.836	1.851	2.041	2.044	2.054	1.974	1.897	1.772	1.716
31	1.712	1.718	1.728	1.779	1.850	2.032	2.038	2.094	1.965	1.891	1.758	1.702
32	1.718	1.742	1.755	1.812	1.882	2.007	2.023	2.034	1.950	1.880	1.776	1.747
33	1.702	1.710	1.729	1.782	1.849	1.942	1.916	1.928	1.923	1.864	1.762	1.687
34	1.752	1.789	1.775	1.814	1.872	2.019	2.054	2.100	1.978	1.887	1.848	1.769
35	1.712	1.746	1.754	1.832	1.867	2.012	2.030	2.052	1.962	1.863	1.823	1.726
36	1.704	1.722	1.753	1.788	1.839	1.972	1.994	2.021	1.934	1.868	1.768	1.701
37	1.719	1.763	1.771	1.842	1.851	2.011	2.038	2.043	1.972	1.888	1.822	1.740
38	1.690	1.743	1.743	1.791	1.840	1.980	2.024	2.054	1.958	1.866	1.774	1.730
39	1.701	1.743	1.740	1.780	1.839	1.978	2.031	2.045	1.968	1.875	1.785	1.704
40	1.709	1.719	1.763	1.833	1.830	1.992	2.038	2.063	1.952	1.865	1.777	1.702
41	1.697	1.722	1.712	1.797	1.835	1.969	2.015	2.063	1.970	1.855	1.755	1.704
42	1.681	1.703	1.711	1.791	1.859	2.009	2.011	2.022	1.958	1.870	1.785	1.682
43	1.767	1.766	1.799	1.849	1.885	2.033	2.089	2.060	1.977	1.916	1.820	1.765
44	1.720	1.730	1.746	1.821	1.843	1.983	2.011	2.060	1.921	1.883	1.784	1.718
45	1.730	1.756	1.798	1.861	1.874	2.022	2.043	2.053	1.955	1.889	1.817	1.744
46	1.723	1.718	1.756	1.797	1.907	2.038	2.063	2.060	1.997	1.919	1.835	1.747
47	1.715	1.724	1.738	1.766	1.858	1.993	2.009	2.060	1.941	1.858	1.760	1.718
48	1.745	1.770	1.788	1.839	1.887	2.005	2.043	2.042	1.971	1.937	1.804	1.771
49	1.695	1.731	1.729	1.825	1.892	2.011	2.049	2.055	1.949	1.892	1.781	1.720
50	1.739	1.736	1.741	1.819	1.883	2.033	2.080	2.079	1.979	1.940	1.813	1.741
51	1.731	1.756	1.781	1.837	1.853	2.029	2.055	2.070	1.958	1.915	1.781	1.740
52	1.676	1.697	1.715	1.761	1.817	1.902	1.982	2.037	1.877	1.889	1.768	1.700
53	1.724	1.747	1.756	1.818	1.853	2.004	2.032	2.035	1.949	1.874	1.798	1.735
54	1.626	1.682	1.670	1.717	1.803	1.944	2.033	2.035	1.899	1.792	1.708	1.683

Table A3. α and β values used in Equation (6) for each period and each station to calculate mean annual erosivity (R_p) in Model C.

ID Station	Period 1	2	3	ID Station	Period 1	2	3
1	0.098	0.116	0.148	1	1.714	1.830	2.009
2	0.104	0.121	0.150	2	1.728	1.859	2.017
3	0.100	0.116	0.141	3	1.785	1.821	1.996
4	0.092	0.110	0.144	4	1.767	1.861	1.995
5	0.095	0.114	0.147	5	1.785	1.868	2.015
6	0.094	0.109	0.143	6	1.777	1.873	2.016
7	0.100	0.117	0.146	7	1.717	1.814	1.977
8	0.094	0.110	0.143	8	1.778	1.868	2.011
9	0.094	0.111	0.141	9	1.715	1.806	1.980
10	0.088	0.103	0.132	10	1.773	1.853	2.011
11	0.096	0.116	0.149	11	1.770	1.874	2.021
12	0.099	0.119	0.145	12	1.710	1.838	1.985
13	0.087	0.111	0.144	13	1.741	1.831	2.007
14	0.087	0.108	0.142	14	1.745	1.845	2.003
15	0.090	0.106	0.133	15	1.764	1.862	2.025
16	0.092	0.107	0.135	16	1.776	1.872	2.009
17	0.083	0.096	0.127	17	1.752	1.837	1.997
18	0.092	0.109	0.135	18	1.688	1.789	1.981
19	0.091	0.111	0.137	19	1.710	1.813	2.007
20	0.096	0.119	0.147	20	1.726	1.823	1.980
21	0.097	0.117	0.141	21	1.726	1.832	2.035
22	0.097	0.118	0.142	22	1.715	1.839	1.991
23	0.095	0.115	0.140	23	1.673	1.802	1.942
24	0.079	0.093	0.125	24	1.718	1.810	1.971
25	0.087	0.108	0.135	25	1.748	1.855	1.991
26	0.095	0.115	0.153	26	1.767	1.876	2.036
27	0.082	0.104	0.136	27	1.731	1.808	1.996
28	0.091	0.108	0.139	28	1.664	1.768	1.923
29	0.099	0.119	0.143	29	1.713	1.809	2.015
30	0.098	0.119	0.142	30	1.720	1.830	2.021
31	0.097	0.119	0.131	31	1.712	1.813	2.029
32	0.089	0.110	0.146	32	1.741	1.832	1.998
33	0.094	0.111	0.130	33	1.704	1.809	1.927
34	0.086	0.104	0.130	34	1.772	1.854	2.031
35	0.095	0.115	0.142	35	1.735	1.842	2.009
36	0.099	0.118	0.141	36	1.720	1.815	1.972
37	0.088	0.104	0.136	37	1.749	1.844	2.009
38	0.097	0.116	0.142	38	1.726	1.812	2.000
39	0.098	0.119	0.146	39	1.721	1.817	2.003
40	0.099	0.117	0.145	40	1.722	1.821	2.008
41	0.099	0.116	0.152	41	1.707	1.802	1.998
42	0.098	0.117	0.152	42	1.698	1.819	1.993
43	0.090	0.112	0.142	43	1.775	1.861	2.030
44	0.085	0.105	0.137	44	1.728	1.828	1.985
45	0.091	0.113	0.144	45	1.758	1.855	2.013
46	0.101	0.121	0.152	46	1.737	1.861	2.039
47	0.096	0.113	0.130	47	1.723	1.803	2.001
48	0.095	0.114	0.148	48	1.768	1.861	2.011
49	0.099	0.114	0.140	49	1.720	1.840	2.011
50	0.101	0.119	0.146	50	1.739	1.862	2.033
51	0.091	0.108	0.143	51	1.753	1.841	2.020
52	0.096	0.116	0.135	52	1.697	1.807	1.935
53	0.090	0.109	0.138	53	1.741	1.832	1.998
54	0.084	0.108	0.135	54	1.668	1.752	1.964

References

1. Hewitt, R.; Escobar, F. The Territorial Dynamics of Fast-Growing Regions: Unsustainable Land Use Change and Future Policy Challenges in Madrid, Spain. *Appl. Geogr.* **2011**, *31*, 650–667. [CrossRef]
2. Li, Z.; Fang, H. Impacts of Climate Change on Water Erosion: A Review. *Earth-Sci. Rev.* **2016**, *163*, 94–117. [CrossRef]
3. Dvořák, J.; Novák, L.; Antal, J. (Eds.) *Soil Conservation and Silviculture*; Developments in Soil Science; Elsevier: Amsterdam, The Netherlands, 1994; ISBN 978-0-08-086991-9.
4. Uber, M.; Haller, M.; Brendel, C.; Hillebrand, G.; Hoffmann, T. Past, Present and Future Rainfall Erosivity in Central Europe Based on Convection-Permitting Climate Simulations. *Hydrol. Earth Syst. Sci.* **2024**, *28*, 87–102. [CrossRef]
5. Kosmas, C.; Danalatos, N.; Cammeraat, L.H.; Chabart, M.; Diamantopoulos, J.; Farand, R.; Gutierrez, L.; Jacob, A.; Marques, H.; Martinez-Fernandez, J.; et al. The Effect of Land Use on Runoff and Soil Erosion Rates under Mediterranean Conditions. *CATENA* **1997**, *29*, 45–59. [CrossRef]
6. Boardman, J.; Poesen, J. (Eds.) *Soil Erosion in Europe*, 1st ed.; Wiley: Chichester, UK, 2006; ISBN 978-0-470-85910-0.
7. Peña-Angulo, D.; Nadal-Romero, E.; González-Hidalgo, J.C.; Albaladejo, J.; Andreu, V.; Bahri, H.; Bernal, S.; Biddoccu, M.; Bienes, R.; Campo, J.; et al. Relationship of Weather Types on the Seasonal and Spatial Variability of Rainfall, Runoff, and Sediment Yield in the Western Mediterranean Basin. *Atmosphere* **2020**, *11*, 609. [CrossRef]
8. Wischmeier, W.H.; Smith, D.D. *Predicting Rainfall Erosion Losses: A Guide to Conservation Planning*; Agriculture Handbook No. 537; USDA, Science and Education Administration: Washington, DC, USA, 1978.
9. Todisco, F.; Vergni, L.; Ortenzi, S.; Di Matteo, L. Soil Loss Estimation Coupling a Modified USLE Model with a Runoff Correction Factor Based on Rainfall and Satellite Soil Moisture Data. *Water* **2022**, *14*, 2081. [CrossRef]
10. Moss, A.; Green, P. Movement of Solids in Air and Water by Raindrop Impact. Effects of Drop-Size and Water-Depth Variations. *Soil Res.* **1983**, *21*, 257. [CrossRef]
11. Bryan, R.B. Soil Erodibility and Processes of Water Erosion on Hillslope. *Geomorphology* **2000**, *32*, 385–415. [CrossRef]
12. Renard, K.G. (Ed.) *Predicting Soil Erosion by Water: A Guide to Conservation Planning with the Revised Universal Soil Loss Equation (RUSLE)*; Agriculture Handbook: Washington, DC, USA, 1997; ISBN 978-0-16-048938-9.
13. Diodato, N. Estimating RUSLE's Rainfall Factor in the Part of Italy with a Mediterranean Rainfall Regime. *Hydrol. Earth Syst. Sci.* **2004**, *8*, 103–107. [CrossRef]
14. Capolongo, D.; Diodato, N.; Mannaerts, C.M.; Piccarreta, M.; Strobl, R.O. Analyzing Temporal Changes in Climate Erosivity Using a Simplified Rainfall Erosivity Model in Basilicata (Southern Italy). *J. Hydrol.* **2008**, *356*, 119–130. [CrossRef]
15. Nazzareno, D.; Luigi, G.; Paola, R.; Gerardo, G.; Maria, G.F. Spatial Pattern of Hydrological Predictability of Landslide-Prone Areas. In *Engineering Geology for Society and Territory—Volume 2*; Lollino, G., Giordan, D., Crosta, G.B., Corominas, J., Azzam, R., Wasowski, J., Sciarra, N., Eds.; Springer International Publishing: Cham, Switzerland, 2015; pp. 1611–1613. ISBN 978-3-319-09056-6.
16. Bagarello, V.; Ferro, V.; Pampalone, V.; Porto, P.; Todisco, F.; Vergni, L. Predicting Soil Loss in Central and South Italy with a Single USLE-MM Model. *J. Soils Sediments* **2018**, *18*, 3365–3377. [CrossRef]
17. Marques, M.J.; Bienes, R.; Jiménez, L.; Pérez-Rodríguez, R. Effect of Vegetal Cover on Runoff and Soil Erosion under Light Intensity Events. Rainfall Simulation over USLE Plots. *Sci. Total Environ.* **2007**, *378*, 161–165. [CrossRef]
18. Bonilla, C.A.; Vidal, K.L. Rainfall Erosivity in Central Chile. *J. Hydrol.* **2011**, *410*, 126–133. [CrossRef]
19. Di Lena, B.; Curci, G.; Vergni, L. Analysis of Rainfall Erosivity Trends 1980–2018 in a Complex Terrain Region (Abruzzo, Central Italy) from Rain Gauges and Gridded Datasets. *Atmosphere* **2021**, *12*, 657. [CrossRef]
20. Petroselli, A.; Apollonio, C.; De Luca, D.L.; Salvaneschi, P.; Pecci, M.; Marras, T.; Schirone, B. Comparative Evaluation of the Rainfall Erosivity in the Rieti Province, Central Italy, Using Empirical Formulas and a Stochastic Rainfall Generator. *Hydrology* **2021**, *8*, 171. [CrossRef]
21. Mikoš, M.; Jošt, D.; Petkovšek, G. Rainfall and Runoff Erosivity in the Alpine Climate of North Slovenia: A Comparison of Different Estimation Methods. *Hydrol. Sci. J.* **2006**, *51*, 115–126. [CrossRef]
22. Angulo-Martínez, M.; Beguería, S. Estimating Rainfall Erosivity from Daily Precipitation Records: A Comparison among Methods Using Data from the Ebro Basin (NE Spain). *J. Hydrol.* **2009**, *379*, 111–121. [CrossRef]
23. Micić Ponjiger, T.; Lukić, T.; Wilby, R.L.; Marković, S.B.; Valjarević, A.; Dragićević, S.; Gavrilov, M.B.; Ponjiger, I.; Durlević, U.; Milanović, M.M.; et al. Evaluation of Rainfall Erosivity in the Western Balkans by Mapping and Clustering ERA5 Reanalysis Data. *Atmosphere* **2023**, *14*, 104. [CrossRef]
24. Meusburger, K.; Steel, A.; Panagos, P.; Montanarella, L.; Alewell, C. Spatial and Temporal Variability of Rainfall Erosivity Factor for Switzerland. *Hydrol. Earth Syst. Sci.* **2012**, *16*, 167–177. [CrossRef]
25. Vrieling, A.; Hoedjes, J.C.B.; Van Der Velde, M. Towards Large-Scale Monitoring of Soil Erosion in Africa: Accounting for the Dynamics of Rainfall Erosivity. *Glob. Planet. Change* **2014**, *115*, 33–43. [CrossRef]

26. Panagos, P.; Ballabio, C.; Borrelli, P.; Meusburger, K. Spatio-Temporal Analysis of Rainfall Erosivity and Erosivity Density in Greece. *CATENA* **2016**, *137*, 161–172. [CrossRef]
27. Goovaerts, P. Using Elevation to Aid the Geostatistical Mapping of Rainfall Erosivity. *CATENA* **1999**, *34*, 227–242. [CrossRef]
28. Borrelli, P.; Diodato, N.; Panagos, P. Rainfall Erosivity in Italy: A National Scale Spatio-Temporal Assessment. *Int. J. Digit. Earth* **2016**, *9*, 835–850. [CrossRef]
29. Panagos, P.; Ballabio, C.; Borrelli, P.; Meusburger, K.; Klik, A.; Rousseva, S.; Tadić, M.P.; Michaelides, S.; Hrabalíková, M.; Olsen, P.; et al. Rainfall Erosivity in Europe. *Sci. Total Environ.* **2015**, *511*, 801–814. [CrossRef]
30. Todisco, F.; Vergni, L.; Vinci, A.; Pampalone, V. Practical Thresholds to Distinguish Erosive and Rill Rainfall Events. *J. Hydrol.* **2019**, *579*, 124173. [CrossRef]
31. Foster, G.R. *RUSLE2 Science Documentation*; USDA-Agricultural Research Service: Washington, DC, USA, 2005.
32. Brown, L.C.; Foster, G.R. Storm Erosivity Using Idealized Intensity Distributions. *Trans. ASAE* **1987**, *30*, 0379–0386. [CrossRef]
33. Richardson, C.W.; Foster, G.R.; Wright, D.A. Estimation of Erosion Index from Daily Rainfall Amount. *Trans. ASAE* **1983**, *26*, 0153–0156. [CrossRef]
34. Bagarello, V.; D'Asaro, F. Estimating Single Storm Erosion Index. *Trans. ASAE* **1994**, *37*, 785–791. [CrossRef]
35. D'Asaro, F.; D'Agostino, L.; Bagarello, V. Assessing Changes in Rainfall Erosivity in Sicily during the Twentieth Century. *Hydrol. Process.* **2007**, *21*, 2862–2871. [CrossRef]
36. Mann, H.B. Nonparametric Tests Against Trend. *Econometrica* **1945**, *13*, 245. [CrossRef]
37. Kendall, M.G. *Rank Correlation Measures*; Charles Griffin: London, UK, 1975.
38. Sen, P.K. Estimates of the Regression Coefficient Based on Kendall's Tau. *J. Am. Stat. Assoc.* **1968**, *63*, 1379–1389. [CrossRef]
39. Bronaugh, D.; Schoeneberg, A. Zyp: Zhang + Yue-Pilon Trends Package; R Package Version 0.11-1, 2012. Available online: https://CRAN.R-project.org/package=zyp (accessed on 9 March 2025).
40. McLeod, A.I. McLeod Kendall: Kendall Rank Correlation and Mann-Kendall Trend Test; R Package Version 2.2.1, 2005. Available online: https://CRAN.R-project.org/package=Kendall (accessed on 9 March 2025).
41. Todisco, F.; Brocca, L.; Termite, L.F.; Wagner, W. Use of Satellite and Modeled Soil Moisture Data for Predicting Event Soil Loss at Plot Scale. *Hydrol. Earth Syst. Sci.* **2015**, *19*, 3845–3856. [CrossRef]
42. Todisco, F.; Vergni, L.; Bomba, C. Definizione di modelli semplificati per la stima dell'indice di erosività della pioggia e curve di isoerosività per la regione Umbria. *Quad. Idronomia Mont.* **2010**, *29*, 331–347.
43. Nearing, M.A.; Yin, S.; Borrelli, P.; Polyakov, V.O. Rainfall Erosivity: An Historical Review. *CATENA* **2017**, *157*, 357–362. [CrossRef]

water

MDPI

Article

^{137}Cs-Based Assessment of Soil Erosion Rates in a Morphologically Diverse Catchment with Varying Soil Types and Vegetation Cover: Relationship with Soil Properties and RUSLE Model Predictions

Aleksandar Čupić [1,*], Ivana Smičiklas [1], Miloš Manić [2], Mrđan Đokić [3], Ranko Dragović [3], Milan Đorđević [3], Milena Gocić [3], Mihajlo Jović [1], Dušan Topalović [1], Boško Gajić [4] and Snežana Dragović [1]

1 "VINČA" Institute of Nuclear Sciences, National Institute of the Republic of Serbia, University of Belgrade, Mike Petrovića Alasa 12-14, 11351 Belgrade, Serbia; ivanat@vin.bg.ac.rs (I.S.); mjovic@vin.bg.ac.rs (M.J.); dusan.topalovic@vin.bg.ac.rs (D.T.); sdragovic@vin.bg.ac.rs (S.D.)
2 Faculty of Geography, University of Belgrade, Studentski Trg 3, 11000 Belgrade, Serbia; milos.manic@pmf.edu.rs
3 Faculty of Sciences and Mathematics, Department of Geography, University of Niš, Višegradska 33, 18000 Niš, Serbia; mrdjan.djokic@pmf.edu.rs (M.Đ.); ranko.dragovic@pmf.edu.rs (R.D.); milan.djordjevic@pmf.edu.rs (M.Đ.); milena.j.gocic@gmail.com (M.G.)
4 Faculty of Agriculture, University of Belgrade, Nemanjina 6, 11080 Belgrade, Serbia; bonna@agrif.bg.ac.rs
* Correspondence: aleksandar.cupic@vin.bg.ac.rs; Tel.: +38-1631166540

Abstract: This study assessed soil erosion intensity and soil properties across the Crveni Potok catchment in Serbia, a region of diverse morphology, geology, pedology, and vegetation. Soil samples were collected using a regular grid approach to identify the underlying factors contributing to erosion and the most vulnerable areas. Based on ^{137}Cs activities and the profile distribution (PD) model, severe erosion (>10 t ha^{-1} y^{-1}) was predicted at nearly 60% of the studied locations. The highest mean erosion rates were detected for the lowest altitude range (300–450 m), Rendzic Leptosol soil, and grass-covered areas. A significant negative correlation was found between the erosion rates, soil organic matter, and indicators of soil structural stability (OC/clay ratio and S_t), indicating that the PD model successfully identifies vulnerable sites. The PD and RUSLE (revised universal soil loss equation) models provide relatively similar mean erosion rates (14.7 t ha^{-1} y^{-1} vs. 12.7 t ha^{-1} y^{-1}) but significantly different median values (13.1 t ha^{-1} y^{-1} vs. 5.5 t ha^{-1} y^{-1}). The model comparison revealed a positive trend. The observed inconsistencies were interpreted by the models' spatiotemporal frameworks and RUSLE's sensitivity to input data quality. Land use stands out as a significant factor modifying the variance of erosion rate, highlighting the importance of land management practices in mitigating erosion.

Keywords: soil loss; Chernobyl fallout; profile distribution model; physiographic factors; soil texture; organic matter; soil structural stability indicators

1. Introduction

Erosion is a natural process caused by water, wind, ice, and gravity, which gradually remove surface soil and parent rock material. Various factors, including topography, land cover, soil properties, rainfall, runoff, and agricultural practices, influence the vulnerability to erosion [1,2]. Soil erosion is a major global environmental issue that impacts water resources, biodiversity, carbon storage, and ecosystems [3,4]. Removal of the fertile soil layer rich in organic matter disrupts biogeochemical cycles of carbon, nitrogen, and phosphorus [5]. It affects sediment transport and off-site accumulation, harming crop yields, water quality, and ecosystem sustainability [6].

The consequences of erosion extend beyond the loss of soil fertility; they also lead to changes in soil properties through the selective mobilization and redistribution of its components [7–9] and can be a factor driving variability in these properties [10–14]. Therefore, establishing a relationship between erosion rates, soil properties, and soil quality indicators is essential for understanding the impact of soil erosion.

Currently, the global rate of soil loss exceeds the rate of soil formation, and as a result, many soils are now categorized as fair, poor, or very poor [15]. Soil degradation, primarily caused by deforestation and the expansion of arable land, has increased by 2.5% worldwide in the last two decades [16,17]. It is estimated that almost one-third of the soil resources have degraded globally, and the number could go up to 90% in the next 25 years [15]. At the European Union (EU) level, a quarter of the total soil resources is degraded above the proposed sustainable threshold (2 t ha^{-1} y^{-1}), and over 6% of the agricultural land has undergone severe erosion (11 t ha^{-1} y^{-1}), which in turn has caused a loss of over 1 billion EUR due to diminished productivity [18]. Water erosion is particularly prevalent, affecting 24% of EU land [19].

To address ongoing challenges [20], researchers and policymakers struggle to improve soil redistribution assessment and control tools. Various direct and indirect techniques are used to quantify erosion rates. Direct field-based methods like plot measurements [21], erosion pins [22], sediment traps [23–25], and tracer techniques [26–28] provide valuable data but can be labor-intensive and time-consuming and may lack resolution and variability, leading to uncertainty in data. Semi-quantitative methods, which assess erosion magnitude [29–32], may also be insufficient due to a lack of precision. On the other hand, indirect methods relying on remote sensing and geographic information systems (GISs) cover large areas and monitor changes over time [33,34], but their accuracy depends on proper calibration. Commonly used erosion models, such as the universal soil loss equation (USLE) [35] and its revised version (RUSLE) [36], estimate erosion rates based on factors like rainfall and soil type but rely on data that may be unavailable or have insufficient resolution. These models are cost-effective for large areas but may lack precision and fail to capture local variations. Given that each method has its trade-offs in terms of accuracy, scale (spatial and temporal), cost, data requirements, and expertise, choosing the most suitable one is contingent on specific objectives and available resources.

Fallout radionuclides, particularly ^{137}Cs from nuclear tests and disasters, are helpful in quantifying soil redistribution rates across different soil types [37]. Conversion models for cultivated and uncultivated soils are used to estimate the average annual erosion and deposition rates at the sampling location based on the difference in ^{137}Cs inventory between the sampled site and a nearby stable, undisturbed reference site that has not been affected by erosion or deposition. Measurements of ^{137}Cs offer advantages over traditional methods by providing spatially distributed estimates and reflecting medium-term erosion history (past 50 years). This retrospective erosion data can be gathered from a single visit without significant disruption to the landscape, thus avoiding costly and time-consuming monitoring surveys [38]. Additionally, once historical insight into overall past erosion is established, short-term erosion processes can be monitored through ^{137}Cs concentrations by periodically or occasionally sampling soil at the exact locations. Despite challenges stemming from fallout variability, such as deposition differences due to vegetation, land cover, and microrelief [39,40], careful site selection can mitigate these issues, allowing ^{137}Cs to remain valuable for developing management practices.

The simultaneous use of various methods and techniques helps to address their limitations and drawbacks, thereby reducing the risk of significant errors in assessing erosion and sedimentation rates. For instance, measurements of ^{137}Cs in soil have been employed to validate different methods for soil erosion assessment, calibrate and validate soil erosion models [41,42], and evaluate the relative contributions of various erosion processes within the study area [43].

The present study is set to evaluate soil erosion rates across the challenging landscape of the Crveni Potok catchment in Serbia. Despite the high vulnerability of soil to erosion, Serbia has yet to establish an adequate system for collecting and processing erosion data at the national level. As a result, information on soil erosion in Serbia is missing from relevant European databases, such as the European Soil Data Centre, that monitor European soil conditions. Erosion caused by wind and water affects approximately 86% of Serbia's territory [44]. The country's southeastern part, where the study area is located, has been identified as most severely impacted by water erosion. The Crveni Potok catchment is selected based on the following essential features: large production of sediments due to frequent destructive torrential floods, significant spatial and temporal variability of eroded sediments in water, significant variability of factors that control erosion on the small area, diverse land use and land cover, and heterogeneous geological, soil, and geomorphological properties. Previous research, which utilized remote sensing and nuclear techniques, identified two types of soil erosion in the catchment: linear rill/gully and sheet erosion. Additionally, the ^{137}Cs inventories revealed that sheet erosion was more intense than rill erosion [45,46].

This study's objective was to deepen the understanding of the combined effects of multiple factors on soil erosion in the Crveni Potok catchment. To achieve this, soil samples were collected throughout the catchment using a regular grid approach. The following methodological steps were employed: (i) assessing soil erosion rates using measurements of fallout radionuclide ^{137}Cs, (ii) analyzing the relationships between the ^{137}Cs-derived erosion rates and factors such as physiographic characteristics, soil physico-chemical properties, and indicators of soil structure stability, and (iii) comparing the ^{137}Cs-based soil erosion estimates with those derived from the RUSLE model. By exploring the sources of discrepancies between estimates obtained by these two models and aligning the predicted erosion intensities with field observations and actual soil conditions, the research aims to contribute to a better interpretation and understanding of erosion modeling outputs in future applications across complex landscapes.

2. Materials and Methods

2.1. Study Area and Sampling Approach

The studied terrains in the Crveni Potok catchment are located in the foothills between the western slopes of the Svrljiške mountains in the east and Debeli Del in the west. The Crveni Potok catchment is part of the Malčanska River Basin, a right tributary of the Nišava River, the largest right tributary of the Južna Morava (Figure 1). In geotectonic and morphostructural terms, the studied area belongs to the Carpatho-Balkanides [47].

In the Crveni Potok catchment and the Malčanska River Basin, the dominant initial forms of relief are hilly and undulating structures of different morphology and height. The Crveni Potok catchment covers an area of about 8 km^2. It is elevated at an altitude of over 700 m and was formed due to long-term geotectonic and sedimentary processes. The height difference between the highest and lowest points of the Crveni Potok catchment is 445 m. The lowest point is at the confluence of the Crveni Potok and the Malčanska River at 284 m above sea level, and the highest point is located on the northern catchment at 729 m above sea level [45].

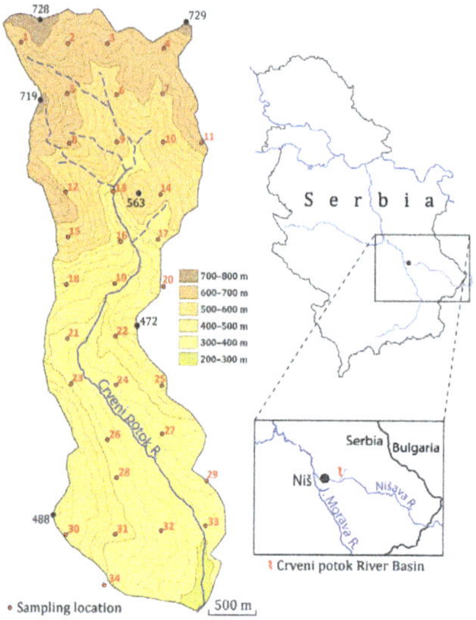

Figure 1. Geographical location of the Crveni Potok catchment with sampling sites indicated by numbers.

Sedimentation of materials, interruptions in sedimentation, or phases of transgression, regression, and lithification dominantly influenced a stratigraphic column in the northern part where red sandstones and residual clays dominate. In the northernmost part of the Crveni Potok catchment, red sandstones of Permian age (Upper Paleozoic) are widespread, made of granular quartz and mica with iron oxides (limonite), which give it a reddish color, while the binding material is clay. The central part of the catchment has the most widespread Mesozoic sandstones and siltstones, and in the lower parts, there are Miocene lake sediments and Quaternary deposits (alluvium) [48]. Pliocene and Quaternary deposits of sand, gravel, clay, gravel–sandy sediment, and conglomerates are present along the main course of Crveni Potok. Quaternary alluvial deposits have accumulated at its confluence.

In the lower parts of the Crveni Potok catchment, redeposited/diluvial/colluvial soils are dominant, and residual soils dominate its upper and higher parts. The dominant soil types are acidic and humus-silicate brown to brown-red soils on red sandstones, lithosol on sandstone, and Haplic Cambisol formed on Permian sandstones. These soils retain some physicochemical characteristics of the parent substrate on which they were formed. Soils formed on red sandstones are susceptible to erosion processes. In the middle and lower parts of the Crveni Potok catchment, carbonate brownish Vertisols on lake sediments, Distric Cambisols on smaller areas, and Calcomelansol on limestone dominate [49]. Leptosol is characterized by a shallow depth of soil for the development of the root system of dendroflora. Due to insufficient soil depth and nutrient accessibility, Leptosol prevents the development of the root system of most cultivated plants. The depth of the humus horizon of the soil in the Crveni Potok catchment varies from 1 to 3 cm, depending on the slope of the terrain, vegetation, and exposure to pluvial processes. Systematic sampling has determined that, at higher altitudes, the soil cover is shallower and in an earlier stage of formation. The soils have a powdery to fine-grained loamy texture with small skeletal fragments.

The complex geological structure of the Crveni Potok catchment generates various geomorphological forms. This catchment's physiographic predisposition and biogeography have caused the formation of gullies, and recent morphological forms arose due to the anthropogenic impact on the indigenous forest cover and grass associations. The formation of gullies is indirectly related to agrotechnical measures and overgrazing in the period after World War II until the end of the 1980s. Agricultural land use has exceeded the capacity for natural regeneration of pastures and arable land, resulting in torrential flows during periods of increased rainfall intensity and, in the following phase, the formation of gullies. Today, slower soil erosion processes occur in forest gullies. Factors that have influenced the slowdown of soil erosion are as follows: (i) the regressive phase of erosion; (ii) the erosion process has already removed the surface layer of weakly bound soil; (iii) soils now affected by erosion contain a significant percentage of mechanical clay fraction; (iv) cut forest vegetation (barren) has been partially restored with pseudo-forest vegetation; and (v) rainy periods are shorter due to regional climate changes.

The dominant vegetation types in the Crveni Potok catchment are grass formations and sapling forests of low ecological value. Grass associations are mainly found in areas where deforestation has been significant in the recent past. This type of vegetation has a limited effect on preventing water erosion. Today, these areas consist of meadows interspersed with deciduous trees and shrubs. The edification species is the Balkan sessile oak (*Quercus dalechampii*), but the white hornbeam (*Carpinus betulus*) and the black pine (*Pinus nigra*) are also present in mosaic form. The distribution of pseudo-forest vegetation inherent in trees grown from seeds is significant. In locations where several oak trees have been identified, the roots are branched and firmly attached to the parent soil. In locations with several trees with branched roots, soil erosion and the development of gullies is slowed down.

The nearest meteorological station is Niš, about 11 km from the Crveni Potok catchment. Deviations of meteorological elements in the Crveni Potok catchment from the values recorded at the meteorological station in Niš are minimal. The most important meteorological element for the erosion process is precipitation. According to Đokić, the predicted average annual precipitation is between 632 mm in the lower parts of the river catchment and 861 mm in its highest parts [50]. Although the amount of precipitation is not large, other factors also affect the occurrence of erosion and the formation of gullies in the Crveni Potok catchment. In addition to the amount of precipitation, its annual distribution, intensity, slope of the topographic surface, length of the slope, type of soil, grain size composition, water-holding capacity of the soil, water permeability and capillarity, water capacity and evaporation capacity, soil vulnerability, plant cover, saturation of the ground layer with water vapor, and air aridity are also important for the studied area.

The method for collecting data, whether through individual or multiple transects or a regular grid, depends on the study's objectives. While ^{137}Cs measurements along transects are useful for steep, relatively short, and regular slopes, the grid approach is preferred for landscapes with more complex topography [51]. Determining a reliable reference inventory is a critical step in the applying the ^{137}Cs technique. To achieve this, a targeted sampling strategy was developed to identify a suitable reference site. The reference site was selected based on its proximity to the study area, absence of slope, lack of soil erosion or deposition, undisturbed soil conditions, and sparse vegetation cover. A cylinder corer with a 10 cm diameter was utilized to collect soil samples. The depth distribution of ^{137}Cs was analyzed by dividing the soil cores into 5 cm intervals extending to a maximum depth of 20 cm. In this study the ^{137}Cs inventories from 12 sampling points at the reference site in the northern part of the basin were used to account for the microscale spatial variability of this radionuclide, as reported by Đokić et al. [45].

In total, soil samples were collected at 34 locations following a 500 m grid as a general guideline, each extracted using a cylinder corer with a 10 cm diameter to a depth of up to 20 cm. However, the sampling locations in the grid were slightly adjusted to account for key factors influencing erosion intensity, including terrain topography, slope, soil type, geology, climatic conditions, and vegetation (Figure 1). These adjustments ensured that the selected points more accurately represented the variability and dynamics of erosion processes in the study area.

2.2. Soil Analysis

Topsoil samples, up to 20 cm in depth, were air-dried at room temperature. Stones, larger organic matter, and other non-soil materials were handpicked and disposed of. Such samples were subsequently sieved to obtain a fraction up to 2 mm appropriate for gamma-ray spectrometry. Activity concentrations of ^{137}Cs in the soil samples were measured using a gamma-ray spectrometer (Baltic Scientific Instruments, Riga, Latvia) with a high-purity germanium (HPGe) detector of 65% relative efficiency and 1.74 keV energy resolution for ^{60}Co at 1.33 MeV. The efficiency calibration was conducted before measuring soil samples, utilizing a mixed calibration source containing licensed reference materials. During this process the same geometry was maintained, ensuring quality assurance throughout. The measurements were taken over a live time of 7200 s, which resulted in a counting error of up to 10% for the radionuclide of interest, ^{137}Cs. Data processing was performed using the SpectraLine (LSRM, Moscow, Russia) 1.6 software [52]. The activity concentrations of ^{137}Cs were determined from the gamma-ray line at 661.6 keV and were then converted into inventories (Bq m^{-2}) based on the soil bulk density.

Standard procedures were followed to conduct physicochemical analyses, including determining particle size distribution (coarse sand: 2.00–0.20 mm; fine sand: 0.20–0.05 mm; silt: 0.05–0.002 mm; and clay: <0.002 mm) (ISO 11277:1998(E)) [53], soil pH in water (ISO 10390:2021) [54], and electrical conductivity (ISO 11265:1994(E)) [55]. The water content (moisture) in air-dried soil samples was determined at 105 °C, following the ISO 11465:1993 [56]. The organic carbon (OC) and total nitrogen (N) content were determined by ISO 10694:2005 [57] and ISO 11261:2005 [58], respectively. The soil samples were analyzed for Al, Fe, Ca, Mg, K, and Mn content. Given the diversity of soil types, the metals included are among the most abundant in the soil and are linked to soil mineral composition. Thus, establishing the relationships between their content, other soil properties, and erosion rates helps understand soil type's effects on erosion vulnerability. Samples were digested in aqua regia (SRPS ISO 11466:2004) [59], and metal concentration measurements were conducted using inductively coupled plasma-optical emission spectrometry (ICP-OES) on the Avio 200 instrument (Perkin Elmer, Waltham, MA, USA) according to ISO 22036:2024 [60].

The indicators of soil structure stability (OC/clay ratio and S_t) were calculated based on the results from the investigated locations. The physical properties of soil, such as water retention capacity and clay dispersibility, are better understood through the interaction between OC and clay content rather than by analyzing these characteristics individually [61]. Additionally, the susceptibility of soil to structural disintegration, referred to as S^t, was evaluated based on organomineral complexes, which are defined as soil organic matter (SOM) bound to the fine fractions of both silt and clay, calculated as follows [62]:

$$S_t(\%) = \frac{\mathrm{SOM}}{(\mathrm{Clay} + \mathrm{Silt})} \times 100 \tag{1}$$

In Equation (1), a Van Bemmelen factor of 1.724 is used to convert the OC content into the SOM content.

2.3. Erosion Assessment Methods

Two models are used to estimate soil erosion rates in the investigated area: the profile distribution (PD) model and the revised universal soil loss equation (RUSLE) model. The details of the models and input parameters used in this study are given below.

2.3.1. The Profile Distribution (PD) Model

The PD model developed by Walling and He is widely applied to obtain assessments of soil redistribution (both erosion and deposition) rates based on the ^{137}Cs inventories in uncultivated soils [63]. In the study area, a notable part of the ^{137}Cs fallout input occurred after the 1986 Chernobyl disaster. Namely, it was found that only around one-fifth of the measured soils' ^{137}Cs inventories do not account for the Chernobyl-induced fallout activity [64–66]. On the other hand, there was no notable contribution to the total ^{137}Cs inventory in Serbia due to the Fukushima accident at the beginning of 2011 [67].

In this study, the modified PD model was applied, which includes the abovementioned factors, namely the year 1986 as a substitute for 1963, which is associated with thermonuclear bomb tests' maximum fallout (Equation (2)), written as follows:

$$Y = \frac{10}{t - 1986} \ln\left(1 - \frac{X}{100}\right) h_0 \tag{2}$$

where Y represents the annual soil loss (t ha^{-1} y^{-1}), i.e., the erosion rate (with a negative value for an eroding point, having a lesser total ^{137}Cs inventory compared to the reference value); t indicates the sampling year (yr); X represents the relative reduction in the ^{137}Cs inventory against the local ^{137}Cs reference value in %; and h_0 is a coefficient named the profile shape factor, which reflects the depth of penetration of ^{137}Cs into the soil [63,68], and its value increases with the penetration depth.

The percentage reduction in the ^{137}Cs inventory X is calculated as in Equation (3):

$$X = \left[\frac{A_{ref} - A_u}{A_{ref}}\right] 100 \tag{3}$$

where A_{ref} and A_u represent the total ^{137}Cs inventory measured at the reference and the sampling point (Bq m^{-2}), respectively.

The profile shape factor h_0 is calculated by fitting the simple undisturbed soil's exponential function below (Equation (4)) to the obtained vertical distribution profile of ^{137}Cs at the time of sampling at the reference site [68,69]. In many cases, for the undisturbed stable soil, this profile exhibits a simple exponential decline with depth (Equation (4)), written as follows [70,71]:

$$A(x) = A(0) e^{-x/h_0} \tag{4}$$

where A(x) represents the ^{137}Cs activity concentration at mass depth x (Bq kg^{-1}); A(0) is the ^{137}Cs activity concentration of the surface soil (Bq kg^{-1}); while x represents the mass depth relative to the soil surface (kg m^{-2}).

The deposition rate R' may be estimated using the following relationship (Equation (5)):

$$R' = \frac{A_{ex}}{\int_{t_0}^{t} Cd(t') e^{-\lambda(t-t')} dt'} = \frac{A_u - A_{ref}}{\int_{t_0}^{t} Cd(t') e^{-\lambda(t-t')} dt'} \tag{5}$$

where A_{ex} is defined as the excess ^{137}Cs inventory and Cd (t′) is the ^{137}Cs concentration of deposited sediment.

2.3.2. The Revised Universal Soil Loss Equation (RUSLE) Model

The RUSLE [35] is an empirical model used in studies worldwide to estimate erosion rates [72–74]. Due to its applicability in different environments and efficiency in data provision and model parameterization costs, the RUSLE is the most widely used model in soil conservation and management. The RUSLE model can be described as follows (Equation (6)):

$$A = R \times K \times LS \times C \times P \quad (6)$$

where A is the average annual soil loss rate (t ha^{-1} y^{-1}); R is the rainfall erosivity factor (MJ mm ha^{-1} h^{-1} $year^{-1}$); K is the soil erodibility factor (t ha h ha^{-1} MJ^{-1} mm^{-1}); LS is the topographic factor (dimensionless); C is the cover management factor (dimensionless); and P is the support practice factor (dimensionless).

The numerical value assigned to the R factor quantifies the raindrop impact and accounts for the potential runoff volume and rate resulting from the rainfall [36,75]. Traditional approaches for calculating the R factor typically rely on extensive rainfall data, frequently unavailable in many areas. In the study area, the following equation (Equation (7)) provided by Đokić was used to estimate annual precipitation (P) based on altitude (H) [50]:

$$P = 243.06 \cdot \ln(H) - 741.19 \quad (7)$$

For precipitation calculations, the advanced land observing satellite digital elevation model (ALOS DEM) was used, initially at 12.5 m resolution, resampled to 10 m. Digital elevation model (DEM) was converted from ellipsoidal heights to the EGM96 vertical datum to ensure our elevation data accurately reflects the mean sea level. The R factor was calculated according to Van et al. model (Equation (8)), written as follows [76]:

$$R = a \cdot P \quad (8)$$

where *a* factor is 1.3 and P is the annual precipitation (mm).

The K factor indicates the soil's ability to resist erosion caused by the impact of raindrops, and the rate and volume of runoff generated for that rainfall under standardized conditions [77]. In this study, the K factor is calculated using a model developed by Williams as follows (Equations (9)–(13)) [78]:

$$K_{USLE} = f_{csand} \cdot f_{cl-si} \cdot f_{orgc} \cdot f_{hisand} \quad (9)$$

$$f_{csand} = \left(0.2 + 0.3 \cdot \exp\left[-0.256 \cdot m_s \cdot \left(1 - \frac{m_{silt}}{100}\right)\right]\right) \quad (10)$$

$$f_{cl-si} = \left(\frac{m_{silt}}{m_c + m_{silt}}\right)^{0.3} \quad (11)$$

$$f_{orgc} = \left(1 - \frac{0.25 \cdot orgC}{orgC + \exp[3.72 - 2.95 \cdot orgC]}\right) \quad (12)$$

$$f_{hisand} = \left(1 - \frac{0.7 \cdot \left(1 - \frac{m_s}{100}\right)}{\left(1 - \frac{m_s}{100}\right) + \exp\left[-5.51 + 22.9 \cdot \left(1 - \frac{m_s}{100}\right)\right]}\right) \quad (13)$$

where K_{USLE} is the soil erodibility factor; f_{csand} (coarse sand factor) is function of the high coarse sand content of the soil; $f_{cl\text{-}si}$ (clay-to-silt ratio factor) is a function of the clay and silt of the soil; f_{orgc} (organic carbon factor) is a function of the organic carbon content of the soil; f_{hisand} (high sand content factor) is the function of the high sand fraction content (0.05–2.00 mm diameter) in the soil m_s (%, Table S2); m_{silt} is the silt fraction content (0.002–0.05 mm diameter) (%); m_c is the clay fraction content (<0.002 mm diameter) (%); and orgC is the organic carbon content (%, Table S2).

The LS factor includes the slope length factor (L) and the slope steepness factor (S). The LS factor was calculated using ALOS DEM with a resampled resolution of 10 × 10 m through the methodology proposed by Moore et al. using SAGA 7.8.2. [79]. A sink fill algorithm based on Wang and Liu was used to remove surface depressions in the digital elevation model before LS calculation [80].

The C factor reflects how different types of land cover influence the soil's protection against erosion. The value of the C factor is directly influenced by vegetation type, its growth stage, and the percentage of ground covered by the vegetation [77]. In this study, the C factor values are assigned to all land use types based on the values cited in the literature [81]. The ESA WorldCover dataset with a 10 m resolution based on Sentinel-1 and Sentinel-2 data was used for land cover input [82]. Field observations were conducted at every sample site for cases where the classifications according to the ESA WorldCover dataset and our sampling locations differed. The C factor for sampling locations was modified based on direct observations and accurate terrain data to reflect current conditions.

The P factor is the ratio of soil loss after soil conservation measures to the amount of soil loss without any soil conservation measures [83,84].

2.4. Data Analysis

The basic statistics included the analysis of central tendency, variability, and normality. As skewness and kurtosis values indicated the data's deviation from a normal distribution, the correlation analysis was conducted using log-transformed data. The Pearson correlation was used to examine the strength of relationships between the analyzed variables. To investigate how variations in physiographic factors (altitude, slope, geology, soil type, and land cover) impact erosion rates, a one-way Welch's analysis of variance (ANOVA) was performed. This was followed by Games–Howell pairwise comparison tests to evaluate the differences among the means of the various groups. All statistical analyses were performed using R version 4.3.1 [85].

3. Results

3.1. Physiographic Properties of the Sampling Locations

The detailed physiographic data for each sampling location can be found in Table S1. The distribution of sampling points based on geology is as follows: 15 sites are situated on clastic Permian sediments, 11 on Miocene lacustrine sediments, 5 on Pliocene and Quaternary lacustrine sediments, and 3 on limestone. In terms of soil types, Haplic Cambisols are the most dominant, found at 15 locations, followed by Vertisols at 10 locations, Regic Anthrosols at 4 locations, Mollic Leptosols at 3 locations, and Rendzic Leptosols at 2 locations.

Vegetation cover at the sampling sites varies, with grasslands of different densities found at 16 locations, forests at 15 locations, and shrubs at 3 locations.

The slopes of the sites range from 3° to 23°, with an average of 12°. Additionally, the altitude ranges from 303 to 673 m above sea level, with an average altitude of 485 m (Table 1).

Table 1. Descriptive statistics of altitude, slope, soil physicochemical properties, soil structure stability indicators, and soil erosion rates estimated by the PD and RUSLE models (N = 34).

Variable	Mean	SD *	CV **	Skewness	Kurtosis	Min.	Median	Max.
Topography								
Altitude (m)	485	107.4	22.16	0.27	−1.25	303	464	673
Slope (°)	12	5.2	41.89	0.31	−0.52	3	11	23

Table 1. *Cont.*

Variable	Mean	SD *	CV **	Skewness	Kurtosis	Min.	Median	Max.
Soil physicochemical properties								
Coarse sand (%)	14.4	6.31	43.78	0.59	1.78	2.30	14.75	34.50
Fine sand (%)	15.0	7.29	48.66	1.43	2.91	3.7	14.0	39.7
Silt (%)	43.3	10.50	24.23	0.41	−0.61	24.8	40.5	66.4
Clay (%)	27.3	13.45	49.32	0.08	−1.48	9.2	29.0	52.6
Water content (%)	3.3	1.71	51.93	0.14	−1.42	0.9	3.1	6.3
pH (H_2O)	6.9	1.27	18.59	−0.10	−1.24	4.6	6.6	8.6
EC (mS m^{-1})	7	3.14	47.61	0.80	−0.48	3	6	13
OC (%)	3.090	1.12	36.20	0.04	−0.93	0.900	3.033	5.027
N (%)	0.259	0.09	36.93	0.95	1.12	0.110	0.232	0.541
OC/N	11.9	2.23	19.16	0.83	0.80	8.2	11.8	18.3
Ca (mg kg^{-1})	23,480	41,898	178.5	2.54	6.21	453	6770	178,000
Al (mg kg^{-1})	23,310	10,164	43.61	0.18	−0.24	5110	22,700	48,300
Fe (mg kg^{-1})	30,780	8388	27.25	0.08	−0.39	13,200	30,300	46,300
K (mg kg^{-1})	3780	1831	48.47	0.14	−0.96	737	3465	7510
Mg (mg kg^{-1})	6460	4831	74.75	2.06	5.81	608	5275	24,700
Mn (mg kg^{-1})	773	4096	52.98	0.83	−0.44	288	604	1800
^{137}Cs (Bq kg^{-1})	15.0	6.95	46.47	0.17	−1.22	3.4	14.5	27.6
Indicators of soil structure stability								
OC/Clay	0.156	0.109	70.16	0.91	−0.28	0.022	0.117	0.399
S_t (%)	7.6	2.71	35.4	−0.14	−0.95	2.3	7.8	11.9
Erosion rates (t ha^{-1} y^{-1})								
PD model	14.7	11.15	75.75	0.75	0.07	0.4	13.1	43.3
RUSLE	12.7	16.63	130.3	1.43	1.34	0.1	5.5	62.7

Note(s): * Standard deviation; ** coefficient of variation.

3.2. Physicochemical Properties of the Soil Samples

Table S2 encapsulates the physical and chemical properties of the soil. The soil across different locations within the Crveni Potok catchment demonstrates significant variations in the percentage composition of individual particle size fractions. The Coefficient of Variation (CV), calculated as the ratio of the standard deviation to the mean, indicates that the most notable variations occur in the clay (CV = 49.32%) and fine sand (CV = 48.66%) fractions. In the lower part of the study area, soils predominantly exhibit finer textures, such as clay and clay loam (Table S2). In contrast, the upper section is mainly characterized by silty loam, loam, and occasionally sandy loam textures. The ability of these soils to retain moisture varies widely, as evidenced by the water content measured at 105 °C, which ranges from 0.9% to 6.3%, resulting in a CV of 51.93%.

The soil pH values ranged from 4.6 to 8.6, with a CV of 18.59%. Based on the general interpretation of soil pH measured in a 1:5 soil/water ratio [86], the soil pH values in the study area can be categorized as very strongly acidic (pH 4.5–5.0; five samples), moderately acidic (pH 5.6–6.0; three samples), slightly acidic (6.1–6.5; nine samples), neutral (6.6–7.3; four samples), mildly alkaline (pH 7.4–7.8; one sample), moderately alkaline (pH 7.9–8.4; seven samples), and strongly alkaline (pH 8.5–9.0; five samples). Acidic soils, specifically Haplic Cambisol and Regic Anthrosol, were commonly found in forested areas.

Electrical conductivity (EC) values varied from 3 mS m^{-1} to 13 mS m^{-1} across different soil samples, resulting in a CV of 47.61%. Salinity rating based on the EC measurements taken at a 1:5 soil-to-water ratio and soil texture related to clay content [86] indicates that soils have very low (28 samples) and low salinity (6 samples).

The soils also showed varying levels of organic carbon (OC) and total nitrogen (N), with CVs of 36.20% for OC and 36.93% for N. The recorded OC values ranged from 0.900% to 5.027%, while N values ranged from 0.110% to 0.541%. According to the classification of OC levels in the topsoil horizon [87], most samples (30) exhibited medium levels of OC

(2.0% to 6%), while three samples had low levels (1.0% to 2%), and one sample had a very low level (less than 1% OC). The OC/N ratio, which indicates organic matter turnover and nitrogen availability, fluctuated between 8.2 and 18.3 (CV = 19.16%).

In the study area, the mean concentrations of elements in the surface soil (Table 1) displayed that the most abundant ones were Fe, Ca, and Al, with mean concentrations of 30,780 mg kg^{-1}, 23,480 mg kg^{-1}, and 23,310 mg kg^{-1}, respectively. Next, Mg was found at a mean concentration of 6460 mg kg^{-1}, K at 3780 mg kg^{-1}, and Mn at 773 mg kg^{-1}. Iron exhibited the least variability among the selected elements (CV = 27.25%), while Ca showed the most significant variability (CV = 178.5%).

3.3. Soil Indices for Assessing Structural Quality and Susceptibility to Erosion

Aggregate stability is a relevant indicator of soil vulnerability to runoff and erosion [88]. Improved aggregation of soil particles rich in OC enhances soil structure, making it less prone to compaction and more stable in water [61]. By examining the relationship between soil organic matter and the content of fine-sized particles (clay or clay + silt) in soils with a known structural condition, including poorly structured and/or eroded ones, soils with adequate organic matter were distinguished from those that have lost their structure and are highly susceptible to erosion [62,89].

The calculated values of the indicators by location are given in Table S3. The mean OC/Clay ratio in the study area was 0.156 (Table 1). This ratio varied from 0.022 to 0.399, highlighting the complexity and diversity of soil composition and stability in the catchment, as indicated by a high coefficient of variation (CV = 70.16%).

Research has demonstrated that the OC/Clay ratio can effectively differentiate soils with varying structural quality. Values of 1/8, 1/10, and 1/13 are proposed thresholds for classifying soil structural conditions as very good, good, moderate, and degraded [61,89]. While the validity of these thresholds was challenged in a study by Johannes et al. [90], which primarily examined agricultural soils, it was confirmed that a decline in soil structure correlates with a decrease in the OC/Clay ratio, with the 1/8 ratio representing an average for soils considered to have very good structure.

Out of 34 samples from the Crveni Potok catchment, the soil at 17 locations exhibited an OC/Clay ratio greater than 1/8, indicating a very good structure. Most of these soils (13 samples) were collected in forested areas (Tables S1 and S3). In contrast, the OC/Clay ratio was lower than 1/13 at 10 grassland and one shrubland location. The results indicate the significant influence of land use on the OC/Clay ratio. Among various factors, such as land use, average annual precipitation, major soil groups, pH levels, flood risk, and topsoil depth, land use was identified as the primary factor influencing the variance of the OC/Clay ratio in the soils of England and Wales [89].

The S_t varied between 2.3% and 11.9% (CV = 35.4%) with a mean of 7.6. The S_t level of 5% was set as an empirical critical limit, below which a loss of structure leads to a great susceptibility to erosion [62]. Furthermore, S_t values between 5% and 7% are recommended as indicators of soil instability and heightened risk of structural decline. Conversely, soils with a S_t greater than 9% are not considered at immediate risk. Based on this S_t classification, the soil at six locations was physically degraded, 16 samples could be classified as at risk, and 12 as soils with stable structure at no immediate risk. Notably, all areas where soil has lost its structural integrity are covered with grass.

3.4. Soil Erosion Rates Estimated by the PD Model and Relationships with Physiographic Factors

The activity concentrations of ^{137}Cs and the soil erosion rates calculated using the PD model are presented in Table S3. The mean activity of ^{137}Cs across the study area was 15.0 Bq kg^{-1} (Table 1). The CV is 46.47%, indicating high variability among locations within the Crveni Potok catchment, with values ranging from 3.4 Bq kg^{-1} to 27.6 Bq kg^{-1}. No signs of soil deposition were observed, as the ^{137}Cs inventories at the investigated grid points were lower than those at the reference site [45].

Based on the measured ^{137}Cs activities and using Equation (2), the PD model estimated erosion intensities from 0.4 to 43.3 t ha^{-1} y^{-1}, with a CV of 75.75%. The mean erosion rate was found to be 14.7 t ha^{-1} y^{-1}, which is slightly lower than the mean erosion rate of 16 t ha^{-1} y^{-1} reported by the same methodology in gullies located in the upper part of the study area [45,46].

In terms of erosion rate classifications used by the European Union (low: <5 t ha^{-1} y^{-1}; moderate: 5–10 t ha^{-1} y^{-1}; and severe: >10 t ha^{-1} y^{-1}) [91], the study area included 9 locations characterized by low erosion intensity, 5 with moderate, and 20 with severe erosion intensity.

The impact of variations in physiographic factors on the erosion rates derived from the PD model was analyzed using ANOVA (Figure 2). Although erosion rates show high variability within individual classes of factors, the ANOVA test followed by a pairwise comparison test revealed some significant differences between the classes. In each panel of Figure 2, different letters (i.e., A and B) indicate groups with significantly different means. Conversely, groups that share at least one letter (e.g., A and AB and AB and B) are not significantly different from each other at the $p < 0.05$ significance level.

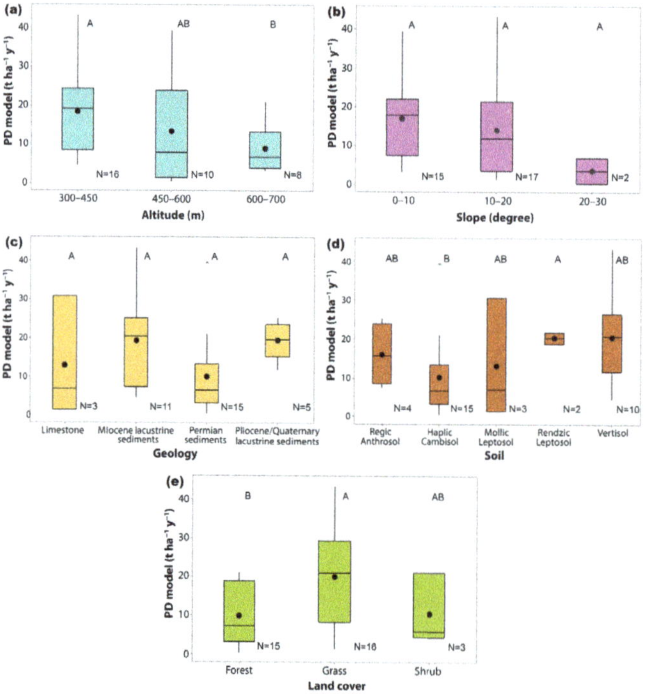

Figure 2. Box plots showing the distribution of erosion rates estimated by the PD model based on physiographic factors: (**a**) altitude, (**b**) slope, (**c**) geology, (**d**) soil, and (**e**) land cover. The box frames the interquartile range; the whisker extends from the lowest to the highest value; a star symbol indicates the outliers; the horizontal line indicates the data median; the circle indicates the mean value for the group, and N is the number of samples. Groups that do not share the same letter (i.e., A and B) are significantly different ($p < 0.05$) based on the Games–Howell pairwise comparison tests.

The mean values of erosion rates decreased with elevation (Figure 2a), demonstrating a significant difference between the lowest elevation range (300–450 m) and the highest range (600–700 m). Similarly, the mean erosion rate was lower for steeper slopes (20–30°) than those on slopes up to 20° (Figure 2b). However, the difference across these slope categories was not statistically significant, likely due to the limited number of data points, which makes reliable conclusions difficult.

No significant differences were observed when comparing groups with different geological compositions (Figure 2c). The lowest mean erosion rate was associated with Permian sediments. The distribution of erosion rates among different soil groups (Figure 2d) revealed that Haplic Cambisol, formed on consolidated Permian sediments, had a lower mean erosion rate than soils developed on unconsolidated sediments. The difference between the Haplic Cambisol and Rendzic Leptosol groups was significant.

Finally, the vegetation cover type at sampling locations influenced the estimated erosion rates (Figure 2e), revealing a significant difference between forested areas and grasslands. The lowest erosion rates were measured in forested areas, followed by shrublands, with the highest rates observed in locations with grassland cover. Moreover, the variability of erosion rates in forested areas was much lower than in grasslands.

3.5. Soil Erosion Rates Estimated by the RUSLE Model

Figure 3 shows raster maps for RUSLE model factors, including rainfall erosivity (R-factor), soil erodibility (K-factor), slope length and steepness (LS-factor), cover management (C-factor), and support practices (P-factor). The P factor was set to a value of 1.0 across the entire study area based on the observation that no specific soil conservation practices are currently implemented. The resultant soil loss map, derived from integrating these factors, provides a detailed visualization of the potential soil erosion across the area. This map highlights the spatial distribution of soil loss, identifying areas with varying erosion intensities, from low to severe, and offering valuable insights for land management and soil conservation planning.

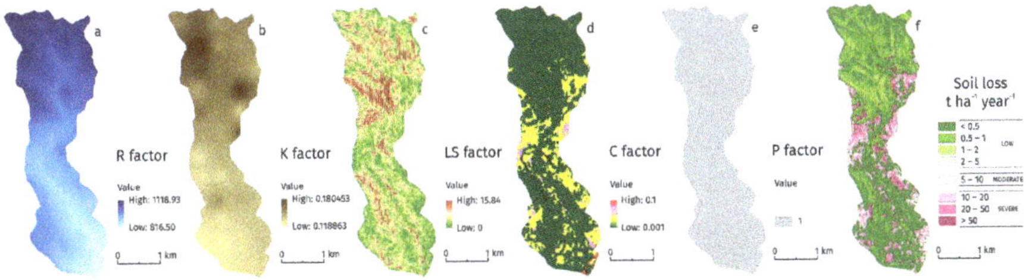

Figure 3. RUSLE Factor maps (**a–e**) and RUSLE model based soil loss map (**f**).

The RUSLE model predicted erosion rates at investigated locations (Table S3) ranged from 0.1 to 62.7 t ha^{-1} y^{-1}, scoring a CV of 130.3% (Table 1). The mean value of 12.7 t ha^{-1} y^{-1} was close to the one predicted by the PD model (14.7 t ha^{-1} y^{-1}), but the median value was significantly lower (5.5 t ha^{-1} y^{-1} vs. 13.1 t ha^{-1} y^{-1}, respectively (Table 1)). Following the classification of erosion intensities [91], low-intensity erosion was anticipated at 17 locations, moderate at 4, and severe at 13 locations.

3.6. Correlation of Estimated Soil Erosion Rates with Soil Physicochemical Properties and Indicators of Soil Structural Stability

The associations between soil properties, indices of soil stability, and soil erosion rates, tested using Pearson's correlation coefficients, are presented in Figure 4.

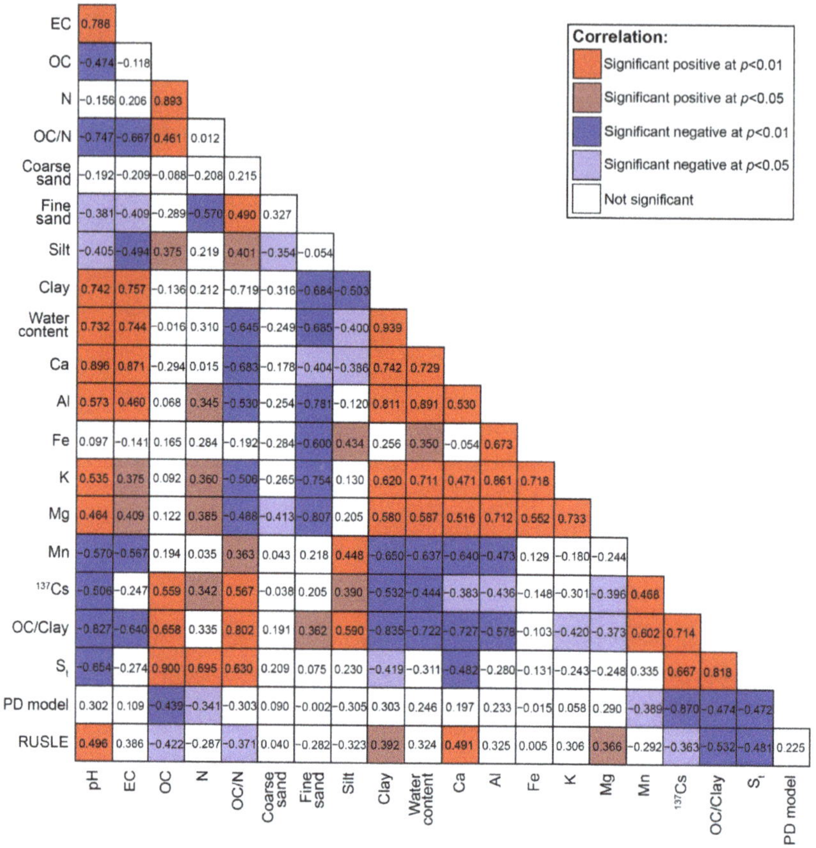

Figure 4. Pearson correlation coefficient (r) of soil erosion rates estimated by PD and RUSLE models, soil physicochemical properties, and indicators of soil structural stability (N = 34).

In the study area, the mass activities of ^{137}Cs showed significant positive correlations with OC, N, the OC/N ratio, and the silt fraction. Additionally, there was a positive correlation between ^{137}Cs and Mn concentration, which aligns well with the observed association between Mn-rich minerals and the silt fraction. Conversely, ^{137}Cs activity had a significant negative correlation with clay content. This finding was reinforced by a negative correlation of ^{137}Cs with water content and, specifically, the Al, a common element in the crystal lattice of clay minerals. Additionally, the clay fraction increased significantly with a higher Ca and Mg content and soil pH, demonstrating a high presence of clay-sized particles in carbonate-rich soil. Meanwhile, ^{137}Cs activity concentrations were in a strong negative correlation with these parameters.

Overall, soil texture and organic matter are the primary factors influencing the concentration of ^{137}Cs in the topsoil of the Crveni Potok catchment. The interaction between ^{137}Cs and soil occurs primarily through cation exchange sites with varying specificity and reversibility, mainly involving organic matter and clay minerals [92]. The strong covariance of OC and ^{137}Cs arises from similar transportation pathways during soil redistribution along the sloping terrains [93]. Positive correlations between ^{137}Cs and OC have been detected in eroding soils across different types and land uses, including forests [94–96], uncultivated areas [97], abandoned farming sites [96], and certain cultivated soils [98–101].

The relationship between the activity of ^{137}Cs and different particle size fractions is complex and influenced by specific local conditions such as soil type and structure and the extent of erosion. Studies have shown both positive correlations with clay [99] and negative correlations [38], while other research indicates a positive correlation with both silt and clay fractions [102].

Erosion rates estimated using the PD model, based on ^{137}Cs activities, demonstrated a significant negative correlation with OC, N, and Mn content in the soil (Figure 4). Areas with high erosion rates are at risk of not only a further decline in organic matter content but also changes in the composition of OC fractions in the remaining soil. Water erosion tends to selectively increase the proportion of mineral-associated organic matter while decreasing the proportion of lighter particulate organic matter [103,104]. Moreover, a recent analysis examining how soil nitrogen concentration and composition respond to varying erosion intensities confirmed that erosion diminishes total nitrogen content in eroded areas and alters the composition of dissolved nitrogen, ultimately reducing its availability [105]. Erosion rates tended to increase with higher clay and lower silt content in the soil, but these trends were not statistically significant. Furthermore, the correlation between PD-derived erosion rates and the investigated elements' content was not significant, except for Mn. A notable negative correlation with Mn-rich minerals seems to arise from Mn's association with the silt fraction. Haplic Cambisol with silt and silt loam texture was characterized by notably higher Mn concentrations than soils with finer textures (Table S2). This underlines the impact of soil type on erosion intensity and supports the findings from Figure 2d.

Soil structure stability indicators displayed a significant ($p < 0.01$) positive correlation with ^{137}Cs activity concentrations and, consequently, a negative correlation ($p < 0.01$) with the PD model estimates (Figure 4). This shows that the predictions of the PD model and the values of physicochemical indicators of soil structural quality align in identifying endangered sites. Likewise, a significant correlation has been reported between soil redistribution rates estimated by ^{137}Cs and S_t in croplands within a semi-humid region [106].

In our study area, the OC/Clay ratio and the S_t were negatively correlated with the soil pH. Such a relationship results from higher OC and lower clay content in the upper part of the catchment, predominated by acidic forest soil (Tables S1 and S2). An increase in the OC/Clay ratio in acidic soils, particularly those with a pH below 5, has been previously observed in permanent grass and woodland areas [89].

Erosion rates estimated by RUSLE were significantly positively correlated with clay fraction and clay fraction-associated parameters pH, Ca, and Mg, while presenting negative correlations with ^{137}Cs, OC, OC/N, OC/Clay, and S_t. The OC/N ratio serves as a meaningful proxy of organic matter quality and the composition of the soil microbial community [107]. Unlike particulate organic matter, mainly composed of plant residues with low N content, mineral-associated organic matter consists of low molecular weight organic compounds with higher N content [108]. The association of this fraction with minerals makes it less vulnerable to disturbance. Therefore, the significant reduction in the OC/N ratio with increased erosion intensity supports the dominance of the more stable mineral-associated organic matter over the particulate organic fraction in areas experiencing severe erosion by water.

The linear relationships observed among all examined parameters and predictive models are consistent in direction, although they vary in strength (Figure 4). The RUSLE estimates demonstrate a stronger correlation with the OC and clay content since these data were incorporated within the model. Finally, the correlation of log-normalized data for the PD and RUSLE model estimates shows a positive trend; however, it is not statistically significant at $p < 0.05$.

4. Discussion

Our findings are consistent with the observed intensity of erosion in semi-arid environments. The erosion rates estimated using the ^{137}Cs method reach 39 t ha^{-1} yr^{-1} on a Mediterranean coast [109] and even 116 t ha^{-1} yr^{-1} in an agricultural area of a watershed in Turkey [110]. Considering different methods, numerous studies [109–112] have shown that higher soil erosion rates can be observed when using the ^{137}Cs method compared to the RUSLE model. Using both methods, the highest values of erosion rates were obtained for croplands and the lowest for forested areas. However, other local factors can affect large variability in forest erosion rates [113]. In mountainous regions of Europe, the results obtained from the universal soil loss equation (USLE) are similar in magnitude to those based on ^{137}Cs across land use types, except for hayfields and pastures lacking dwarf shrubs, where the USLE underestimated erosion intensity [114]. In contrast, erosion and sedimentation rates estimated by the USLE, RUSLE, and ^{137}Cs were completely different in a watershed in Brazil [115].

A comparison of the PD and RUSLE model results can provide essential insights into soil erosion dynamics in our study area. Although the PD model, by definition, accounts for all types of erosion, its estimates could be compared with those from the RUSLE, focusing solely on water erosion, as water is the only erosion agent in the study area. After conducting a correlation analysis on the log-normalized data (Figure 4), the original data were analyzed to compare erosion rates obtained from both approaches in meaningful real-world units. The correlation was weakly positive (r = 0.372) and statistically significant at $p < 0.03$, indicating that both models identify similar erosion trends, although their quantitative outputs do not always align (Figure 5).

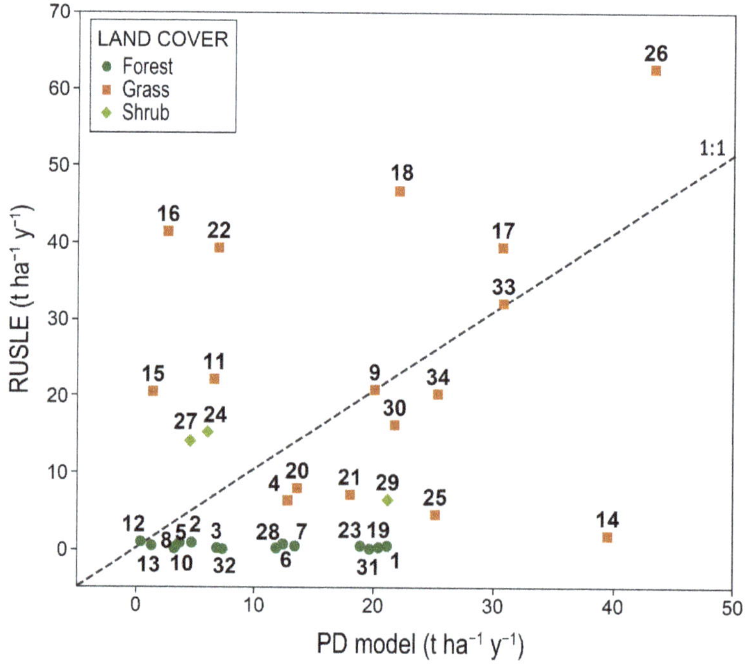

Figure 5. Scatterplot of RUSLE vs. PD model estimates (original data). The numbers refer to the sampling sites.

The comparison reveals a discrepancy in the upper area of the Crveni Potok catchment covered with forest vegetation, with the nuclear method showing predominantly higher erosion rate values. This difference arises from the two methods' distinct temporal and spatial perspectives. Namely, the ^{137}Cs method offers an insight into the long-term erosion rate since the Chernobyl nuclear accident in 1986, encompassing decades of varying land use and vegetation cover. Historically, the area was characterized by sparse grass and shrub vegetation, which provided limited protection against erosion and facilitated intense gully erosion, exacerbated by heavy grazing, primarily by goats [45]. Over time, natural vegetation succession led to the establishment of dense forests and thick grasslands, stabilizing the soil and significantly reducing active erosion. The gully systems, once a major contributor to sediment transport, have largely transitioned into a dormant phase. Consequently, the ^{137}Cs method reflects the historical dynamics of soil redistribution and captures changes over time.

In contrast, the RUSLE model estimates erosion rates based on present-day land cover and vegetation (low C-factor, indicating lower vulnerability to erosion hazards), which now effectively protects the soil in most parts of the area. Despite the area's steep slopes and substantial rainfall, the dense vegetation mitigates surface runoff and limits the detachment and transport of soil particles. This results in much lower erosion rates predicted by RUSLE, as it does not account for the historical phases of land use and erosion intensity that significantly influenced the long-term sediment budget. This comparison underscores the importance of considering historical and contemporary factors to comprehensively assess erosion dynamics and the interplay between land use, vegetation changes, and soil stability.

The limitations of the RUSLE model are evident, as it does not account for gully erosion processes [116–118]. This is particularly significant in the forested areas of the upper part of the Crveni Potok catchment, where gully erosion is a major contributor to overall topsoil degradation by water [46]. Due to its inability to model soil loss caused by gravity flow [119], the RUSLE model may significantly underestimate actual soil loss rates.

An exception to this common trend of RUSLE estimates is observed at two sampling locations characterized by grass vegetation rather than forest cover. At sampling location 9, the erosion rates are nearly the same (20.1 by PD and 20.9 t ha^{-1} y^{-1} by RUSLE), and at sampling location 11, the erosion rate predicted by the RUSLE method is higher (22.2 t ha^{-1} y^{-1}) than that obtained from the ^{137}Cs method (6.6 t ha^{-1} y^{-1}).

In the lower part of the catchment, the ^{137}Cs method indicates higher erosion rates at several locations (19, 23, 28, 31, and 32) covered by forest, compared to the RUSLE model. This can be explained by the proximity to human settlements, where, in the past, vegetation was sparse and often controlled by human activity. With fewer people in the area today, vegetation is gradually regrowing, including denser forest cover. At sampling locations 22 and 15, the ^{137}Cs method indicates a significantly lower erosion rate (6.9 t ha^{-1} y^{-1} and 1.4 t ha^{-1} y^{-1}) compared to the value estimated by the RUSLE model (39.3 t ha^{-1} y^{-1} and 20.6 t ha^{-1} y^{-1}). The RUSLE method, considering the steep terrain and sparse vegetation, predicts a high erosion rate, which seems unrealistic due to the shallow depth, slow formation of soil on limestone, and the limited amount of soil that can be eroded. In these cases, the RUSLE method probably overestimates the erosion rate.

In grassland areas, the calculation showed mean erosion rates of 20.03 t ha^{-1} y^{-1} and 24.41 t ha^{-1} y^{-1} and the median values of 19.52 t ha^{-1} y^{-1} and 20.74 t ha^{-1} y^{-1} for the PD and the RUSLE models, respectively. Despite the agreement of mean and median values, the outputs of the two models still differed in absolute values (Figure 5).

The extreme values in erosion predictions derived from the RUSLE model can primarily be attributed to its strong association with the LS factor. The RUSLE model, which directly incorporates topographic factors into its calculations, is very sensitive to changes in slope characteristics [120], and the LS factor can overestimate soil erosion in complex topographies [121]. Similarly, there is often a lack of detailed, site-specific rainfall amount and intensity data, which are critical for calculating the rainfall erosivity (R) factor. The absence of such precise information introduces uncertainties in the model's predictions. The observed differences in predictions obtained by PD and RUSLE models also arise from their distinct conceptual frameworks (relying on the vertical distribution of ^{137}Cs within a soil profile related to the soil characteristics vs. an empirical model based on several external parameters that are not always field-based and/or reliable) and the estimates they provide (gross erosion rates vs. net erosion rates not considering the sediment yield).

The comparison of estimated erosion rates with soil conditions is a practical way for model verification and agreement checking [115]. Soil structure is a key factor affecting water retention and infiltration, gas exchange, soil organic matter and nutrient dynamics, root penetration, and erosion susceptibility [122]. Various mineral and organic agents contribute to soil aggregation and structure preservation. The rearrangement of particles, flocculation, and cementation are the main drivers for soil particle aggregation, facilitated by SOM, biota, clay, Fe- and Al-(hydr)oxides, and carbonates [123,124].

In our study area, both investigated models were negatively correlated with structural stability indicators (Figure 4). Lower S_t and OC/clay values are generally found in the lower part of the study area, which is characterized by higher erosion rates. The high clay and silt and low SOM content in this area contribute to the soil's increased susceptibility to sealing [125]. Surface sealing and the resulting crusting are recognized as significant forms of soil degradation, primarily driving runoff and promoting inter-rill soil erosion. The formation of a thin layer at the soil surface reduces porosity and increases resistance to penetration, for which the S_t value serves as a valuable indicator [126]. The S_t shows a significant positive correlation with saturated hydraulic conductivity and a negative correlation with bulk density and penetration resistance [127].

Conversely, higher SOM levels contribute to soil stability in the upper parts. The interaction of OC with the surface of mineral particles through sorption on pedogenic oxides, clay minerals, and by co-precipitation with polyvalent cations is an effective mechanism of soil stabilization [128]. In addition, the increase in SOM content reduces susceptibility to erosion by altering pore size distribution, providing higher air-filled porosity and water-holding capacity [129]. High SOM content protects the soil aggregates by creating a hydrophobic coating and inhibits slaking and swelling, which destroys the soil structure.

Notably, both models and indicators of soil structure stability agree that location 26 is the most severely affected by erosion, where measures for soil quality and erosion control should be prioritized.

The study underscores the importance of high-resolution and precise terrain data for the accuracy of RUSLE outputs, which can be obtained through advanced non-terrestrial methods such as drone surveys or light detection and ranging (LiDAR) technology. Further study limitations stem from the inherent assumptions of the methods themselves, such as the PD model's inability to account for the time-dependent nature of ^{137}Cs fallout input and its post-fallout redistribution within the soil profile or the RUSLE method's exclusion of sediment yield. However, the combined use of these methods enhances the reliability of their outputs and provides valuable information on the most vulnerable areas. Evaluating the association of PD and RUSLE outputs with soil stability indicators across different geological, hydrological, and climatic conditions in future research will enhance the reliability of the study's initial findings.

5. Conclusions

Predictions from the PD model indicate that severe erosion impacts 20 out of 34 locations in the study area. Significant influences from elevation, soil type, and land cover have been identified. Erosion rates were negatively correlated with the content of organic matter and the soil aggregate stability indicators. This supports the model's findings by providing evidence of actual soil conditions and vulnerability to erosion. Increased susceptibility to erosion detected in the lower part of the basin was affiliated with predominant grass and sparse shrub vegetation, clay and clay loam soils, and low organic matter content failing to effectively stabilize soil aggregates. Adopting enhanced forest management practices is essential for effectively mitigating soil erosion in this area. This includes implementing measures to prevent deforestation and actively promoting reforestation to increase organic matter levels and enhance soil stability.

The research contributes to promoting the value of the ^{137}Cs method even over very complex terrains and the usefulness of model predictions for further assessing the combined influence of various factors on the erosion process. Comparing the results of the PD and RUSLE models proved helpful for understanding the erosion process from the distinct assumptions of the models. Furthermore, assessing soil conditions by structural stability indicators can help users and decision-makers adjust their practices and effectively direct efforts to reduce erosion intensity. Once these indicators are established, they can facilitate tracking changes in soil vulnerability over time, particularly concerning soil erosion and the effectiveness of erosion control measures.

Supplementary Materials: The following supporting information can be downloaded at https://www.mdpi.com/article/10.3390/w17040526/s1: Table S1: Coordinates of sampling locations with information on corresponding altitude, slope, geology, soil type, and land cover; Table S2: Physicochemical properties of the soil at sampling locations, including particle size distribution, water content, pH, electrical conductivity (EC), organic carbon (OC), total nitrogen (N), and concentrations of selected metals (Ca, Al, Fe, K, Mg, and Mn); Table S3: Activity concentrations of ^{137}Cs, soil erosion rates estimated using profile distribution (PD) and revised universal soil loss equation (RUSLE) model, and soil structure stability indicators (OC/Clay and St) for each location.

Author Contributions: Conceptualization, A.Č., I.S. and S.D.; methodology, A.Č., I.S., M.M., B.G. and S.D.; validation, I.S., M.M. and S.D.; formal analysis, A.Č., I.S., M.M., M.J., B.G. and S.D.; investigation, A.Č., I.S., M.M., M.Đ. (Mrđan Đokić), M.M. and D.T.; resources, I.S., B.G. and S.D.; writing—original draft preparation, A.Č., I.S., M.M., M.Đ. (Mrđan Đokić), R.D., M.G. and S.D.; writing—review and editing, A.Č., I.S., B.G. and S.D.; visualization, I.S., M.M., M.Đ. (Mrđan Đokić), M.Đ. (Milan Đorđević) and M.J.; supervision, I.S. and S.D.; project administration, S.D.; funding acquisition, A.Č., I.S., M.Đ. (Mrđan Đokić), R.D., M.Đ. (Milan Đorđević), M.G., M.J., D.T., B.G. and S.D. All authors have read and agreed to the published version of the manuscript.

Funding: This research was supported by the Science Fund of the Republic of Serbia (grant No. 7047), Development of erosion prediction tool for sustainable soil management-Predict-Er. This paper was also supported by the Ministry of Science, Technological Development and Innovation of the Republic of Serbia (contracts 451-03-136/2025-03/200017, 451-03-137/2025-03/200124, and 451-03-137/2025-03/200116).

Data Availability Statement: The datasets generated during the current study are available upon request.

Conflicts of Interest: The authors declare no conflicts of interest. The funders had no role in the design of the study; in the collection, analyses, or interpretation of data; in the writing of the manuscript; or in the decision to publish the results.

Abbreviations

The following abbreviations are used in this manuscript:

PD	Profile Distribution
USLE	Universal Soil Loss Equation
RUSLE	Revised Universal Soil Loss Equation
ALOS DEM	Advanced Land Observing Satellite Digital Elevation Model
CV	Coefficient of Variation
SOM	Soil Organic Matter
OC	Organic Carbon

References

1. Cao, Y.; Ouyang, Z.Y.; Zheng, H.; Huang, Z.G.; Wang, X.K.; Miao, H. Effects of Forest Plantations on Rainfall Redistribution and Erosion in the Red Soil Region of Southern China. *Land Degrad. Dev.* **2008**, *19*, 321–330. [CrossRef]
2. Kayet, N.; Pathak, K.; Chakrabarty, A.; Sahoo, S. Evaluation of Soil Loss Estimation Using the RUSLE Model and SCS-CN Method in Hillslope Mining Areas. *Int. Soil Water Conserv. Res.* **2018**, *6*, 31–42. [CrossRef]
3. Hou, D.; Bolan, N.S.; Tsang, D.C.W.; Kirkham, M.B.; O'Connor, D. Sustainable Soil Use and Management: An Interdisciplinary and Systematic Approach. *Sci. Total Environ.* **2020**, *729*, 138961. [CrossRef] [PubMed]
4. Quinton, J.N.; Fiener, P. Soil Erosion on Arable Land: An Unresolved Global Environmental Threat. *Prog. Phys. Geogr.* **2024**, *48*, 136–161. [CrossRef]
5. Berhe, A.A.; Barnes, R.T.; Six, J.; Marín-Spiotta, E. Role of Soil Erosion in Biogeochemical Cycling of Essential Elements: Carbon, Nitrogen, and Phosphorus. *Annu. Rev. Earth Planet. Sci.* **2018**, *46*, 521–548. [CrossRef]
6. Karlen, D.L.; Mausbach, M.J.; Doran, J.W.; Cline, R.G.; Harris, R.F.; Schuman, G.E. Soil Quality: A Concept, Definition, and Framework for Evaluation (A Guest Editorial). *Soil Sci. Soc. Am. J.* **1997**, *61*, 4–10. [CrossRef]
7. Fullen, M.A. Soil Organic Matter and Erosion Processes on Arable Loamy Sand Soils in the West Midlands of England. *Soil Technol.* **1991**, *4*, 19–31. [CrossRef]
8. Fullen, M.A.; Brandsma, R.T. Property Changes by Erosion of Loamy Sand Soils in East Shropshire, UK. *Soil Technol.* **1995**, *8*, 1–15. [CrossRef]
9. Fullen, M.A.; Yi, Z.; Brandsma, R.T. Comparison of Soil and Sediment Properties of a Loamy Sand Soil. *Soil Technol.* **1997**, *10*, 35–45. [CrossRef]
10. Kreznor, W.R.; Olson, K.R.; Banwart, W.L.; Johnson, D.L. Soil, Landscape, and Erosion Relationships in a Northwest Illinois Watershed. *Soil Sci. Soc. Am. J.* **1989**, *53*, 1763–1771. [CrossRef]
11. Pennock, D.J. Chapter 7 Effects of Soil Redistribution on Soil Quality: Pedon, Landscape, and Regional Scales. *Dev. Soil Sci.* **1997**, *25*, 167–185. [CrossRef]
12. Van Oost, K.; Govers, G.; Van Muysen, W.; Quine, T.A. Modeling Translocation and Dispersion of Soil Constituents by Tillage on Sloping Land. *Soil Sci. Soc. Am. J.* **2000**, *64*, 1733–1739. [CrossRef]
13. Ampontuah, E.O.; Robinson, J.S.; Nortcliff, S. Assessment of Soil Particle Redistribution on Two Contrasting Cultivated Hillslopes. *Geoderma* **2006**, *132*, 324–343. [CrossRef]
14. Pennock, D.J. *New Perspectives on the Soil Erosion-Soil Quality Relationship*; IAEA: Vienna, Austria, 1998.
15. FAO and ITPS. *Status of the World's Soil Resources (SWSR)—Main Report*; Food and Agriculture Organization of the United Nations and Intergovernmental Technical Panel on Soils: Rome, Italy, 2015.
16. Borrelli, P.; Robinson, D.A.; Fleischer, L.R.; Lugato, E.; Ballabio, C.; Alewell, C.; Meusburger, K.; Modugno, S.; Schütt, B.; Ferro, V.; et al. An Assessment of the Global Impact of 21st Century Land Use Change on Soil Erosion. *Nat. Commun.* **2017**, *8*, 2013. [CrossRef] [PubMed]
17. Pereira, P.; Barceló, D.; Panagos, P. Soil and Water Threats in a Changing Environment. *Environ. Res.* **2020**, *186*, 109501. [CrossRef]
18. Panagos, P.; Standardi, G.; Borrelli, P.; Lugato, E.; Montanarella, L.; Bosello, F. Cost of Agricultural Productivity Loss Due to Soil Erosion in the European Union: From Direct Cost Evaluation Approaches to the Use of Macroeconomic Models. *Land. Degrad. Dev.* **2018**, *29*, 471–484. [CrossRef]

19. Arias Navarro, C.; Baritz, R.; Jones, A.; Akca, E.; Aldrian, U.; Alewell, C.; Anzalone, E.; Arcidiacono, A.; Auclerc, A.; Aydinsakir, K.; et al. *The State of Soils in Europe*; Publications Office of the European Union: Luxembourg, 2024.
20. Transforming Our World: The 2030 Agenda for Sustainable Development | Department of Economic and Social Affairs. Available online: https://sdgs.un.org/2030agenda (accessed on 29 December 2024).
21. Boix-Fayos, C.; Martínez-Mena, M.; Arnau-Rosalén, E.; Calvo-Cases, A.; Castillo, V.; Albaladejo, J. Measuring Soil Erosion by Field Plots: Understanding the Sources of Variation. *Earth Sci. Rev.* **2006**, *78*, 267–285. [CrossRef]
22. Kearney, S.P.; Fonte, S.J.; García, E.; Smukler, S.M. Improving the Utility of Erosion Pins: Absolute Value of Pin Height Change as an Indicator of Relative Erosion. *Catena* **2018**, *163*, 427–432. [CrossRef]
23. Orr, B. A Rapid Method of Estimating Soil Erosion Using Trapped Sediment: A Case Study from a Forest Road on the Cumberland Plateau of Tennessee. *Mt. Res. Dev.* **1996**, *16*, 419. [CrossRef]
24. Albert, J.; Lang, W.; Michael Aust, M.; Bolding, C.; Kevin, J.; McGuire, E.B.S. Comparing Sediment Trap Data with Erosion Models for Evaluation of Forest Haul Road Stream Crossing Approaches. *Trans. ASABE* **2017**, *60*, 393–408. [CrossRef]
25. Wade, C.R.; Bolding, M.C.; Aust, W.M.; Iii, W.A.L.; Schilling, E.B. Schilling Comparing Sediment Trap Data with the USLE-Forest, RUSLE2, and WEPP-Road Erosion Models for Evaluation of Bladed Skid Trail BMPs. *Trans. ASABE* **2012**, *55*, 403–414. [CrossRef]
26. Zapata, F. The Use of Environmental Radionuclides as Tracers in Soil Erosion and Sedimentation Investigations: Recent Advances and Future Developments. *Soil Tillage Res.* **2003**, *69*, 3–13. [CrossRef]
27. Parwada, C.; Chipomho, J.; Tibugari, H. Comparison of Conventional and Artificial Fallout Radionuclide (FRNs) Methods in Assessing Soil Erosion. *Sustain. Environ.* **2023**, *9*, 2236406. [CrossRef]
28. Guzmán, G.; Quinton, J.N.; Nearing, M.A.; Mabit, L.; Gómez, J.A. Sediment Tracers in Water Erosion Studies: Current Approaches and Challenges. *J. Soils Sediments* **2013**, *13*, 816–833. [CrossRef]
29. Low, A.J. The Effect of Cultivation on the Structure and Other Physical Characteristics of Grassland and Arable Soils (1945–1970). *J. Soil Sci.* **1972**, *23*, 363–380. [CrossRef]
30. Ebeid, M.M.; Lal, R.; Hall, G.F.; Miller, E. Erosion Effects on Soil Properties and Soybean Yield of a Miamian Soil in Western Ohio in a Season with below Normal Rainfall. *Soil Technol.* **1995**, *8*, 97–108. [CrossRef]
31. Fenton, T.E.; Kazemi, M.; Lauterbach-Barrett, M.A. Erosional Impact on Organic Matter Content and Productivity of Selected Iowa Soils. *Soil. Tillage Res.* **2005**, *81*, 163–171. [CrossRef]
32. Malo, D.D.; Schumacher, T.E.; Doolittle, J.J. Long-Term Cultivation Impacts on Selected Soil Properties in the Northern Great Plains. *Soil Tillage Res.* **2005**, *81*, 277–291. [CrossRef]
33. Karydas, C.G.; Panagos, P.; Gitas, I.Z. A Classification of Water Erosion Models According to Their Geospatial Characteristics. *Int. J. Digit. Earth* **2014**, *7*, 229–250. [CrossRef]
34. Panagos, P.; Borrelli, P.; Poesen, J.; Ballabio, C.; Lugato, E.; Meusburger, K.; Montanarella, L.; Alewell, C. The New Assessment of Soil Loss by Water Erosion in Europe. *Environ. Sci. Policy* **2015**, *54*, 438–447. [CrossRef]
35. Wischmeier, W.H.; Smith, D.D. *Predicting Rainfall Erosion Losses, A Guide to Conservation Planning*; The USDA Agricultural Handbook No. 537; U.S. Department of Agriculture: Washington, DC, USA, 1978.
36. Renard, K.; Foster, G.; Weesies, G.; McCool, D.; Yoder, D. *Predicting Soil Erosion by Water: A Guide to Conservation Planning with the Revised Universal Soil Loss Equation (RUSLE)*; US Department of Agriculture, Agriculture Handbook No.703USDA; USDA: Washington, DC, USA, 1997.
37. Mabit, L.; Benmansour, M.; Walling, D.E. Comparative Advantages and Limitations of the Fallout Radionuclides 137Cs, 210Pbex and 7Be for Assessing Soil Erosion and Sedimentation. *J. Environ. Radioact.* **2008**, *99*, 1799–1807. [CrossRef] [PubMed]
38. Du, P.; Walling, D.E. Using 137Cs Measurements to Investigate the Influence of Erosion and Soil Redistribution on Soil Properties. *Appl. Radiat. Isot.* **2011**, *69*, 717–726. [CrossRef]
39. Parsons, A.J.; Foster, I.D.L. What Can We Learn about Soil Erosion from the Use of 137Cs? *Earth Sci. Rev.* **2011**, *108*, 101–113. [CrossRef]
40. Meusburger, K.; Evrard, O.; Alewell, C.; Borrelli, P.; Cinelli, G.; Ketterer, M.; Mabit, L.; Panagos, P.; van Oost, K.; Ballabio, C. Plutonium Aided Reconstruction of Caesium Atmospheric Fallout in European Topsoils. *Sci. Rep.* **2020**, *10*, 11858. [CrossRef] [PubMed]
41. Komissarov, M.; Golosov, V.; Zhidkin, A.; Fomicheva, D.; Konoplev, A. The Cross-Verification of Different Methods for Soil Erosion Assessment of Natural and Agricultural Low Slopes in the Southern Cis-Ural Region of Russia. *Land* **2024**, *13*, 1767. [CrossRef]
42. Walling, D.E.; He, Q.; Whelan, P.A. Using 137Cs Measurements to Validate the Application of the AGNPS and ANSWERS Erosion and Sediment Yield Models in Two Small Devon Catchments. *Soil Tillage Res.* **2003**, *69*, 27–43. [CrossRef]
43. Konz, N.; Baenninger, D.; Konz, M.; Nearing, M.; Alewell, C. Process Identification of Soil Erosion in Steep Mountain Regions. *Hydrol. Earth Syst. Sci.* **2010**, *14*, 675–686. [CrossRef]

44. Lazarević, R. *Map Erosion SR Serbia 1: 500000. Institute of Forestry and Wood Industry of Serbia;* Department of Erosion and Melioration: Belgrade, Serbia, 1983.
45. Đokić, M.; Manić, M.; Đorđević, M.; Gocić, M.; Čupić, A.; Jović, M.; Dragović, R.; Gajić, B.; Smičiklas, I.; Dragović, S. Remote Sensing and Nuclear Techniques for High-Resolution Mapping and Quantification of Gully Erosion in the Highly Erodible Area of the Malčanska River Basin, Eastern Serbia. *Environ. Res.* **2023**, *235*, 116679. [CrossRef]
46. Manić, M.; Đorđević, M.; Đokić, M.; Dragović, R.; Kićović, D.; Đorđević, D.; Jović, M.; Smičiklas, I.; Dragović, S. Remote Sensing and Nuclear Techniques for Soil Erosion Research in Forest Areas: Case Study of the Crveni Potok Catchment. *Front. Environ. Sci.* **2022**, *10*, 897248. [CrossRef]
47. Menkovic, L.; Koscal, M.; Milivojevic, M.; Djokic, M. Morphostructure Relations on the Territory of the Republic of Serbia. *Glas. Srp. Geogr. Drus.* **2018**, *98*, 1–28. [CrossRef]
48. Dimitrijević, M.; Dragić, D.; Karamata, S.; Petrović, B.; Sikošek, B.; Veselinović, D. *Osnovna geološka karta 1:100 000, Tumač za list Aleksinac, K 34-20;* Zavod za Geološka i Geollzička Istraživanja: Belgrade, Yugoslavia, 1980.
49. Nikodijević, V. *Soil Map of SFRY, Niš-1, 1:50,000;* Secretariat for Agriculture, Forestry, and Water Management SRS: Republic Science Association SRS: Belgrade, Yugoslavia, 1979.
50. Đokić, M.M. Nišava—Potamology Study. Ph.D. Thesis, University of Niš, Niš, Serbia, 2015.
51. Fulajtar, E.; Mabit, L.; Renschler, C.S.; Lee, A.; Yi, Z. *Use of 137 Cs for Soil Erosion Assessment;* FAO: Rome, Italy, 2017.
52. Laboratory of Spectrometry and Radiometry (LSRM), Moscow, 2019. SpectraLine (1.6). [Software]. Available online: http://lsrm.ru/en/products/detail.php?ELEMEN%20T_CODE=spectraline_1.6 (accessed on 4 January 2025).
53. *SRPS EN ISO 17892-4:2017;* Geotechnical Investigation and Testing—Laboratory Testing of Soil—Part 4: Determination of Particle Size Distribution. Institute for Standardization of Republic of Serbia: Belgrade, Serbia, 2017. Available online: https://iss.rs/en/project/show/iss:proj:54602 (accessed on 1 July 2024).
54. *ISO 10390:2021;* Soil, Treated Biowaste and Sludge—Determination of PH. International Organization for Standardization: Geneva, Switzerland, 2021. Available online: https://www.iso.org/standard/75243.html (accessed on 29 December 2024).
55. *ISO 11265:1994;* Soil Quality—Determination of the Specific Electrical Conductivity. International Organization for Standardization: Geneva, Switzerland, 1994. Available online: https://www.iso.org/standard/19243.html (accessed on 29 December 2024).
56. *ISO 11465:1993;* Soil Quality—Determination of Dry Matter and Water Content on a Mass Basis—Gravimetric Method. International Organization for Standardization: Geneva, Switzerland, 1993. Available online: https://www.iso.org/standard/20886.html (accessed on 29 December 2024).
57. *ISO 10694:1995;* Soil Quality—Determination of Organic and Total Carbon after Dry Combustion (Elementary Analysis). International Organization for Standardization: Geneva, Switzerland, 1995. Available online: https://www.iso.org/standard/18782.html (accessed on 29 December 2024).
58. *ISO 11261:1995;* Soil Quality—Determination of Total Nitrogen—Modified Kjeldahl Method. International Organization for Standardization: Geneva, Switzerland, 1995. Available online: https://www.iso.org/standard/19239.html (accessed on 29 December 2024).
59. *SRPS ISO 11466:2004;* Soil Quality—Extraction of Trace Elements Soluble in Aqua Regia. SRPS: Institute for Standardization of Republic of Serbia: Belgrade, Serbia, 2004. Available online: https://iss.rs/en/project/show/iss:proj:16775 (accessed on 1 July 2024).
60. *ISO 22036:2024;* Environmental Solid Matrices—Determination of Elements Using Inductively Coupled Plasma Optical Emission Spectrometry (ICP-OES). International Organization for Standardization: Geneva, Switzerland, 2024. Available online: https://www.iso.org/standard/82632.html (accessed on 29 December 2024).
61. Dexter, A.R.; Richard, G.; Arrouays, D.; Czyż, E.A.; Jolivet, C.; Duval, O. Complexed Organic Matter Controls Soil Physical Properties. *Geoderma* **2008**, *144*, 620–627. [CrossRef]
62. Pieri, C.J.M.G. *Fertility of Soils;* Springer Series in Physical Environment; Springer Berlin Heidelberg: Berlin/Heidelberg, Germany, 1992; Volume 10, ISBN 978-3-642-84322-8.
63. Walling, D.E.; He, Q. Improved Models for Estimating Soil Erosion Rates from Cesium-137 Measurements. *J. Environ. Qual.* **1999**, *28*, 611–622. [CrossRef]
64. Byrne, A.R. Radioactivity in Fungi in Slovenia, Yugoslavia, Following the Chernobyl Accident. *J. Environ. Radioact.* **1988**, *6*, 177–183. [CrossRef]
65. Popović, D.; Spasić-Jokić, V. Consequences of the Chernobyl Disaster in the Region of the Republic of Serbia. *Vojnosanit. Pregl.* **2006**, *63*, 481–487. [CrossRef]
66. Federal Committee for Labour, Health and Social Policy. *The Level of Radioactive Contamination of Human Environment and Exposure of Population in Yugoslavia in 1986 Due to Nuclear Power Plant Disaster in Chernobyl;* IAEA: Belgrade, Yugoslavia, 1987.

67. Bikit, I.; Mrda, D.; Todorovic, N.; Nikolov, J.; Krmar, M.; Veskovic, M.; Slivka, J.; Hansman, J.; Forkapic, S.; Jovancevic, N. Airborne Radioiodine in Northern Serbia from Fukushima. *J. Environ. Radioact.* **2012**, *114*, 89–93. [CrossRef]
68. Porto, P.; Walling, D.E.; Ferro, V. Validating the Use of Caesium-137 Measurements to Estimate Soil Erosion Rates in a Small Drainage Basin in Calabria, Southern Italy. *J. Hydrol.* **2001**, *248*, 93–108. [CrossRef]
69. Porto, P.; Walling, D.E.; Ferro, V.; Di Stefano, C. Validating Erosion Rate Estimates Provided by Caesium-137 Measurements for Two Small Forested Catchments in Calabria, Southern Italy. *Land. Degrad. Dev.* **2003**, *14*, 389–408. [CrossRef]
70. Xinbao, Z.; Higgitt, D.L.; Walling, D.E. A Preliminary Assessment of the Potential for Using Caesium-137 to Estimate Rates of Soil Erosion in the Loess Plateau of China. *Hydrol. Sci. J.* **1990**, *35*, 243–252. [CrossRef]
71. Walling, D.E.; Quine, T.A. *Use of Caesium-137 as a Tracer of Erosion and Sedimentation: Handbook for the Application of the Caesium-137 Technique*; UK Overseas Development Administration Research Scheme R4579; IAEA: Vienna, Austria, 1993.
72. Kumar, R.; Deshmukh, B.; Kumar, A. Using Google Earth Engine and GIS for Basin Scale Soil Erosion Risk Assessment: A Case Study of Chambal River Basin, Central India. *J. Earth Syst. Sci.* **2022**, *131*, 228. [CrossRef]
73. Wang, H.; Zhao, H. Dynamic Changes of Soil Erosion in the Taohe River Basin Using the RUSLE Model and Google Earth Engine. *Water* **2020**, *12*, 1293. [CrossRef]
74. Perović, V.; Životić, L.; Kadović, R.; Dordević, A.; Jaramaz, D.; Mrvić, V.; Todorović, M. Spatial Modelling of Soil Erosion Potential in a Mountainous Watershed of South-Eastern Serbia. *Environ. Earth Sci.* **2013**, *68*, 115–128. [CrossRef]
75. Xu, Y.Q.; Shao, X.M.; Kong, X.B.; Jian, P.; Cai, Y.L. Adapting the RUSLE and GIS to Model Soil Erosion Risk in a Mountains Karst Watershed, Guizhou Province, China. *Environ. Monit. Assess.* **2008**, *141*, 275–286. [CrossRef]
76. Van Der Knijff, J.; Jones, R.; Montanarella, L. *Soil Erosion Risk Assessment in Italy*; European Soil Bureau, European Commission: Brussels, Belgium, 1999.
77. Ghosal, K.; Das Bhattacharya, S. A Review of RUSLE Model. *J. Indian Soc. Remote Sens.* **2020**, *48*, 689–707. [CrossRef]
78. Williams, J.R. The EPIC Model, Chapter 25. In *Computer Models of Watershed Hydrology;* Singh, V.P., Ed.; Water Resources Publications: Highlands Ranch, CO, USA, 1995.
79. Moore, I.D.; Grayson, R.B.; Ladson, A.R. Digital Terrain Modelling: A Review of Hydrological, Geomorphological, and Biological Applications. *Hydrol. Process* **1991**, *5*, 3–30. [CrossRef]
80. Wang, L.; Liu, H. An Efficient Method for Identifying and Filling Surface Depressions in Digital Elevation Models for Hydrologic Analysis and Modelling. *Int. J. Geogr. Inf. Sci.* **2006**, *20*, 193–213. [CrossRef]
81. Rozos, D.; Skilodimou, H.D.; Loupasakis, C.; Bathrellos, G.D. Application of the Revised Universal Soil Loss Equation Model on Landslide Prevention. An Example from N. Euboea (Evia) Island, Greece. *Environ. Earth Sci.* **2013**, *70*, 3255–3266. [CrossRef]
82. Zanaga, D.; Van De Kerchove, R.; De Keersmaecker, W.; Souverijns, N.; Brockmann, C.; Quast, R.; Wevers, J.; Grosu, A.; Paccini, A.; Vergnaud, S.; et al. ESA WorldCover 10 m 2020 V100. 2021. Available online: https://zenodo.org/records/5571936 (accessed on 9 January 2025).
83. Chen, P.; Feng, Z.; Mannan, A.; Chen, S.; Ullah, T. Assessment of Soil Loss from Land Use/Land Cover Change and Disasters in the Longmen Shan Mountains, China. *Appl. Ecol. Environ. Res.* **2019**, *17*, 11233–11247. [CrossRef]
84. Kuok, K.K.K.; Mah, D.Y.S.; Chiu, P.C. Evaluation of C and P Factors in Universal Soil Loss Equation on Trapping Sediment: Case Study of Santubong River. *J. Water Resour. Prot.* **2013**, *5*, 1149–1154. [CrossRef]
85. R Core Team. *R A Language and Environment for Statistical Computing*; R Foundation for Statistical Computing: Vienna, Austria, 2023. Available online: https://www.scirp.org/reference/referencespapers?referenceid=3582659 (accessed on 29 December 2024).
86. Hazelton, P.; Murphy, B. *Interpreting Soil Test Results, What Do All the Numbers Mean*; CSIRO Publishing: Collingwood, Australia, 2007.
87. Jones, R.J.A.; Hiederer, R.; Rusco, E.; Loveland, P.J.; Montanarella, L. *The Map of Organic Carbon in Topsoils in Europe*; Office for Official Publications of the European Communities: Luxembourg, 2004.
88. Barthès, B.; Roose, E. Aggregate Stability as an Indicator of Soil Susceptibility to Runoff and Erosion; Validation at Several Levels. *Catena* **2002**, *47*, 133–149. [CrossRef]
89. Prout, J.M.; Shepherd, K.D.; McGrath, S.P.; Kirk, G.J.D.; Haefele, S.M. What Is a Good Level of Soil Organic Matter? An Index Based on Organic Carbon to Clay Ratio. *Eur. J. Soil. Sci.* **2021**, *72*, 2493–2503. [CrossRef]
90. Johannes, A.; Matter, A.; Schulin, R.; Weisskopf, P.; Baveye, P.C.; Boivin, P. Optimal Organic Carbon Values for Soil Structure Quality of Arable Soils. Does Clay Content Matter? *Geoderma* **2017**, *302*, 14–21. [CrossRef]
91. Eurostat Agri-Environmental Indicator—Soil Erosion. Available online: https://ec.europa.eu/eurostat/statistics-explained/index.php?title=Agri-environmental_indicator_-_soil_erosion (accessed on 9 January 2025).
92. Rigol, A.; Vidal, M.; Rauret, G. An Overview of the Effect of Organic Matter on Soil–Radiocaesium Interaction: Implications in Root Uptake. *J. Environ. Radioact.* **2002**, *58*, 191–216. [CrossRef]

93. Xiaojun, N.; Xiaodan, W.; Suzhen, L.; Shixian, G.; Haijun, L. 137Cs Tracing Dynamics of Soil Erosion, Organic Carbon and Nitrogen in Sloping Farmland Converted from Original Grassland in Tibetan Plateau. *Appl. Radiat. Isot.* **2010**, *68*, 1650–1655. [CrossRef] [PubMed]
94. Federico, R.; Paolo, P.; Katharina, K.; Axel, M.; Adele, M. Soil Biological Indicators and Caesium-137 to Estimate Soil Erosion in Areas with Different Forest System Management. *Eur. J. For. Res.* **2020**, *139*, 67–81. [CrossRef]
95. Romeo, F.; Porto, P.; Mallamaci, C.; Muscolo, A. The Relationships between Selected Soil Properties and Caesium-137 Identify Organic Carbon, Nitrogen and Water Soluble Phenols as Indicators of Soil Erosion Processes in Different Forest Stands. *J. For. Res.* **2021**, *32*, 2589–2598. [CrossRef]
96. Ayoubi, S.; Zhao, S.; Yousefifard, M.; Amiri, F.; Abdi, M.R.; Abbaszadeh Afshar, F. Quantification of Mid-Term Soil Redistribution in Hilly Regions under Abandoned Rainfed Farming and Oak Forest Using 137Cs Radionuclide and Magnetic Susceptibility Techniques. *Catena* **2024**, *244*, 108270. [CrossRef]
97. Meliho, M.; Nouira, A.; Benmansour, M.; Boulmane, M.; Khattabi, A.; Mhammdi, N.; Benkdad, A. Assessment of Soil Erosion Rates in a Mediterranean Cultivated and Uncultivated Soils Using Fallout 137Cs. *J. Environ. Radioact.* **2019**, *208–209*, 106021. [CrossRef] [PubMed]
98. Mabit, L.; Bernard, C. Relationship between Soil 137Cs Inventories and Chemical Properties in a Small Intensively Cropped Watershed. *Comptes Rendus Académie Sci. Ser. IIA-Earth Planet. Sci.* **1998**, *327*, 527–532. [CrossRef]
99. Theocharopoulos, S.P.; Florou, H.; Walling, D.E.; Kalantzakos, H.; Christou, M.; Tountas, P.; Nikolaou, T. Soil Erosion and Deposition Rates in a Cultivated Catchment Area in Central Greece, Estimated Using the 137Cs Technique. *Soil Tillage Res.* **2003**, *69*, 153–162. [CrossRef]
100. Mabit, L.; Chhem-Kieth, S.; Toloza, A.; Vanwalleghem, T.; Bernard, C.; Amate, J.I.; González de Molina, M.; Gómez, J.A. Radioisotopic and Physicochemical Background Indicators to Assess Soil Degradation Affecting Olive Orchards in Southern Spain. *Agric. Ecosyst. Environ.* **2012**, *159*, 70–80. [CrossRef]
101. Li, T.; Zhang, H.; Wang, X.; Cheng, S.; Fang, H.; Liu, G.; Yuan, W. Soil Erosion Affects Variations of Soil Organic Carbon and Soil Respiration along a Slope in Northeast China. *Ecol. Process* **2019**, *8*, 28. [CrossRef]
102. Mesrar, H.; Sadiki, A.; Faleh, A.; Quijano, L.; Gaspar, L.; Navas, A. Vertical and Lateral Distribution of Fallout 137 Cs and Soil Properties along Representative Toposequences of Central Rif, Morocco. *J. Environ. Radioact.* **2017**, *169–170*, 27–39. [CrossRef] [PubMed]
103. Wairiu, M. Soil Organic Carbon in Relation to Cultivation and Topsoil Removal on Sloping Lands of Kolombangara, Solomon Islands. *Soil Tillage Res.* **2003**, *70*, 19–27. [CrossRef]
104. He, Y.; Zhang, F.; Yang, M.; Li, X.; Wang, Z. Insights from Size Fractions to Interpret the Erosion-Driven Variations in Soil Organic Carbon on Black Soil Sloping Farmland, Northeast China. *Agric. Ecosyst. Environ.* **2023**, *343*, 108283. [CrossRef]
105. Shi, J.; Zhang, Z.; Wang, Z.; Peng, Y.; Wang, X. Soil Erosion Alters the Composition of Soil Nitrogen and Induces Nitrogen Immobilization along a Sloping Agricultural Landscape. *Soil Use Manag.* **2024**, *40*, e13067. [CrossRef]
106. Khodadadi, M.; Meusburger, K.; Mirzaei, M.; Strauss, P.; Blake, W.H.; Moghaseh, E.; Alewell, C. Spatial Cross-Correlation of Surface Soil Physicochemical Properties with Soil Erosion Estimated by Fallout Radionuclides in Croplands in a Semi-Humid Region of Iran. *Catena* **2024**, *237*, 107836. [CrossRef]
107. Ballabio, C.; Lugato, E.; Fernández-Ugalde, O.; Orgiazzi, A.; Jones, A.; Borrelli, P.; Montanarella, L.; Panagos, P. Mapping LUCAS Topsoil Chemical Properties at European Scale Using Gaussian Process Regression. *Geoderma* **2019**, *355*, 113912. [CrossRef] [PubMed]
108. Francesca Cotrufo, M.; Lavallee, J.M.; Zhang, Y.; Hansen, P.M.; Paustian, K.H.; Schipanski, M.; Wallenstein, M.D. In-N-Out: A Hierarchical Framework to Understand and Predict Soil Carbon Storage and Nitrogen Recycling. *Glob. Chang. Biol.* **2021**, *27*, 4465–4468. [CrossRef]
109. Azbouche, A.; Kessaissia, A.; Hamoudi, A.; Morsli, B.; Moulla, A.S. Land Degradation Assessment Using Fallout Cesium-137 in a Coastal Mediterranean Environment (Ténès, Algeria). *EuroMediterr J. Environ. Integr.* **2023**, *8*, 55–65. [CrossRef]
110. Saç, M.M.; Uğur, A.; Yener, G.; Özden, B. Estimates of Soil Erosion Using Cesium-137 Tracer Models. *Environ. Monit. Assess.* **2007**, *136*, 461–467. [CrossRef]
111. Soileau, J.M.; Hajek, B.F.; Touchton, J.T. Soil Erosion and Deposition Evidence in a Small Watershed Using Fallout Cesium-137. *Soil Sci. Soc. Am. J.* **1990**, *54*, 1712–1719. [CrossRef]
112. Turnage, K.M.; Lee, S.Y.; Foss, J.E.; Kim, K.H.; Larsen, I.L. Comparison of Soil Erosion and Deposition Rates Using Radiocesium, RUSLE, and Buried Soils in Dolines in East Tennessee. *Environ. Geol.* **1997**, *29*, 1–10. [CrossRef]
113. Naghdi, R.; Zahedi, S.S.; Gharibreza, M.; Gholami, V.; Mirzaei, M. Measuring Soil Erosion Rates in Hyrcanian Forests: An Application of the 137Cs Method. *J. Earth Syst. Sci.* **2024**, *133*, 28. [CrossRef]

114. Konz, N.; Schaub, M.; Prasuhn, V.; Baenninger, D.; Alewell, C. Cesium-137-Based Erosion-Rate Determination of a Steep Mountainous Region. *J. Plant Nutr. Soil Sci.* **2009**, *172*, 615–622. [CrossRef]
115. Bacchi, O.O.S.; Reichardt, K.; Sparovek, G. Sediment Spatial Distribution Evaluated by Three Methods and Its Relation to Some Soil Properties. *Soil Tillage Res.* **2003**, *69*, 117–125. [CrossRef]
116. Bircher, P.; Liniger, H.P.; Prasuhn, V. Comparison of Long-Term Field-Measured and RUSLE-Based Modelled Soil Loss in Switzerland. *Geoderma Reg.* **2022**, *31*, e00595. [CrossRef]
117. Alewell, C.; Borrelli, P.; Meusburger, K.; Panagos, P. Using the USLE: Chances, Challenges and Limitations of Soil Erosion Modelling. *Int. Soil Water Conserv. Res.* **2019**, *7*, 203–225. [CrossRef]
118. Mekonnen, E.; Kebede, A.; Asfaw, S.; Feyissa, S. Optimizing Soil Erosion Estimates of RUSLE Model by Analyzing Land Use/Cover Dynamics in Upper Awash River Basin, Central Ethiopia. *Geomatics, Nat. Hazards Risk* **2023**, *14*, 2257363. [CrossRef]
119. Yesuph, A.Y.; Dagnew, A.B. Soil Erosion Mapping and Severity Analysis Based on RUSLE Model and Local Perception in the Beshillo Catchment of the Blue Nile Basin, Ethiopia. *Environ. Syst. Res.* **2019**, *8*, 17. [CrossRef]
120. Perovic, V.; Jaramaz, D.; Zivotic, L.; Cakmak, D.; Mrvic, V.; Milanovic, M.; Saljnikov, E. Design and Implementation of WebGIS Technologies in Evaluation of Erosion Intensity in the Municipality of NIS (Serbia). *Environ. Earth Sci.* **2016**, *75*, 211. [CrossRef]
121. Mejía-Parada, C.; Mora-Ruiz, V.; Vallejo-Borda, J.A.; Arrieta-Baldovino, J. Influence of LS Factor Overestimation Soil Loss on RUSLE Model for Complex Topographies. *J. Indian Soc. Remote Sens.* **2024**, *52*, 1661–1674. [CrossRef]
122. Rabot, E.; Wiesmeier, M.; Schlüter, S.; Vogel, H.-J. Soil Structure as an Indicator of Soil Functions: A Review. *Geoderma* **2018**, *314*, 122–137. [CrossRef]
123. Bronick, C.J.; Lal, R. Soil Structure and Management: A Review. *Geoderma* **2005**, *124*, 3–22. [CrossRef]
124. Totsche, K.U.; Amelung, W.; Gerzabek, M.H.; Guggenberger, G.; Klumpp, E.; Knief, C.; Lehndorff, E.; Mikutta, R.; Peth, S.; Prechtel, A.; et al. Microaggregates in Soils. *J. Plant Nutr. Soil Sci.* **2018**, *181*, 104–136. [CrossRef]
125. Udom, B.E.; Kamalu, O.J. Sealing Index, Air-Filled Porosity and Hydrological Behaviour of a Tropical Ultisol as Affected by Incidental Flooding and Soil Disturbance. *Int. J. Soil Sci.* **2016**, *11*, 79–86. [CrossRef]
126. Valentin, C.; Bresson, L.M. Soil Crusting. In *Methods of Assessment of Soil Degradation*; CRC Press: Boca Raton, FL, USA, 1997; pp. 89–107.
127. Bassouny, M. Soil Shrinkage, Sealing Index and Hydro-Physical Properties of Vertisols as Influenced by Long-Term Cultivation Systems in Northern Egypt. *J. Soil Sci. Agric. Eng.* **2018**, *9*, 173–181. [CrossRef]
128. Wang, X.; Cammeraat, E.L.H.; Cerli, C.; Kalbitz, K. Soil Aggregation and the Stabilization of Organic Carbon as Affected by Erosion and Deposition. *Soil. Biol. Biochem.* **2014**, *72*, 55–65. [CrossRef]
129. Carter, M.R.; Stewart, B.A. *Structure and Organic Matter Storage in Agricultural Soils*; Carter, M.R., Stewart, B.A., Eds.; CRC Press: Boca Raton, FL, USA, 2020; ISBN 978-100-307-556-1.

water

MDPI

Article

Temporal Changes in the Average Contribution of Land Uses in Sediment Yield Using the ^{137}Cs Method and Geochemical Tracers

Negin Ghaderi Dehkordi [1], Abdulvahed Khaledi Darvishan [1,*], Mohamad Reza Zare [2] and Paolo Porto [3,4]

1 Department of Watershed Management, Faculty of Natural Resources, Tarbiat Modares University, Noor P.O. Box 4641776489, Iran
2 Department of Physics, Faculty of Sciences, University of Isfahan, Isfahan P.O. Box 8174673441, Iran
3 Department of Agraria, University Mediterranea of Reggio Calabria, 89124 Calabria, Italy; paolo.porto@unirc.it
4 Faculty of Geographical Sciences, Kazimierz Wielki University, 85-064 Bydgoszcz, Poland
* Correspondence: a.khaledi@modares.ac.ir

Abstract: The study highlights the increasing significance of understanding sediment sources and their contributions within a watershed, particularly in relation to different land use types. As the demand for effective source apportionment grows, this research aims to quantify how various land uses—specifically rangeland, rainfed agriculture, irrigated agriculture, and orchards—contribute to sediment yield over time. To achieve this, the researchers employed geochemical tracers and ^{137}Cs to assess sediment contributions in a small sub-basin located in western Iran. The methodology involved creating a working unit map by overlaying land use maps from 1967 and 2021 with a slope map of the region. A total of 75 and 31 soil samples were systematically collected across different land uses to ensure a representative analysis of ^{137}Cs and geochemical methods, respectively. The study utilized specific models to calculate the average contributions of each land use type. For non-agricultural lands, a diffusion and migration model was applied, while agricultural lands were analyzed using a mass balance type II model. The FingerPro program in R software 4.2.2 facilitated the selection of suitable tracers and allowed for the determination of sediment source contributions through a multivariate mixed model algorithm. The findings revealed significant changes in sediment yield contributions over the past 60 years. In 2021, rainfed agriculture accounted for 72.26% of sediment yield, down from 85.49% six decades earlier. Conversely, irrigated agriculture showed an increase from 1.80% to 15.06%. Rangeland and orchard contributions remained relatively stable but low, at approximately 8% and 4%, respectively. The total erosion rate for the sub-basin was estimated at 526.87 t y^{-1}, with rainfed agriculture being responsible for the majority at 450.43 t y^{-1}.

Keywords: FingerPro; sediment fingerprinting; sediment source; soil disturbance; tillage effect

1. Introduction

Soil is a fundamental element of ecosystems and a critical resource for food production, playing an essential role in human survival. Consequently, it is vital to support and protect soil while preventing erosion, as soil erosion and sediment transfer to waterways represent significant environmental challenges and concerns for sustainable development [1]. Soil erosion, the most prevalent form of land degradation, results from the interplay of human activities and natural phenomena [2]. This process has historically been a major contributor to land degradation, leading to various environmental and ecological issues, including limitations on agricultural production, hindrances to regional economic development, and deteriorating living conditions [3,4]. The intensity of soil erosion directly correlates with its impacts and is influenced by numerous natural and human factors. Among these factors, land use emerges as a crucial determinant affecting both the intensity of soil erosion and its spatial distribution [5,6]. Specifically, converting land for agricultural use on lower slopes exacerbates soil erosion, highlighting the necessity for targeted land management practices [7].

To effectively implement soil conservation measures, a precise understanding of sediment sources and their contributions to sediment yield is essential. This knowledge aids in identifying sensitive areas within watersheds. However, traditional assessment methods—such as visual inspections, field surveys using erosion plots to measure soil loss, and monitoring rills and gullies—face significant challenges related to time constraints and operational demands. These methods often require substantial financial resources [8,9] and are primarily designed for measuring erosion rather than sediment yield. Consequently, they fall short in estimating the sediments that reach the end of the basin or in linking these sediments back to their primary sources. In summary, addressing soil erosion requires innovative approaches that go beyond traditional methods to effectively relate sediment sources to rivers and sediment yield at the basin's end [8].

There are various methods for determining and identifying sediment sources and their contributions, with the source tracing or fingerprinting method gaining significant attention from researchers due to its efficiency. Source tracing involves measurement techniques based on field surveys that apportion eroded soil from multiple sources using tracer properties and composite mathematical models [10]. In recent years, this sediment source tracing method has been recognized as a reliable approach for determining the contributions of primary sediment sources and assessing their relative importance [11,12]. This technique operates by comparing the physical and chemical characteristics of sediment sources with those of sediments produced at the basin outlet, allowing for the identification of high-risk areas for sediment yield [13,14]. The effectiveness of sediment source tracing lies in its ability to demonstrate the relative importance of potential sediment sources [11]. It offers simple and cost-effective principles for collecting spatial and temporal data across various watershed scales [8,11]. In employing the sediment source tracing method, it is crucial that sediment sources are distinguishable based on tracer properties [9]. One notable tracer used in this context is ^{137}Cs, a radionuclide resulting from nuclear activities during the period extending from the 1950s to the 1970s. This radionuclide is deposited into soil and serves as a marker for assessing soil erosion across different regions. With a half-life of 30.1 years, ^{137}Cs provides valuable information about changes in soil surface conditions and erosion rates. The initial research utilizing radionuclides to estimate soil erosion was conducted by Menzel in 1960, followed by studies from [15,16], who further explored the impact of ^{137}Cs in soil erosion assessments.

Materials derived from different parent rocks exhibit distinct geochemical characteristics, which are preserved in sediments resulting from erosion. This retention of geochemical properties allows researchers to identify the relative contributions of various sediment sources [17–21]. In the early 1980s, Walling pioneered the use of geochemical tracers as an innovative method for identifying sediment sources within watersheds. This approach has proven highly effective in distinguishing contributions from various sources, particularly land uses, and has gained widespread acceptance in the field. A significant advancement stemming from these developments is the creation of FingerPro package, a specialized tool for source tracing integrated into R software 4.2.2, which is freely available for users interested in sediment source analysis [22]. Additionally, understanding soil particle displacement and redistribution processes due to erosion is crucial for effective soil conservation management at various scales, from individual plots to entire watersheds. Spectroscopy of fallout radionuclides plays a key role in this understanding, with ^{137}Cs being particularly useful for quantifying soil particle displacement [23].

This study employs the ^{137}Cs method to assess the average contributions of different land uses over the past 60 years in the Khamsan representative paired watershed located in Kurdistan Province. The contributions will be compared against current land use shares, enabling an examination of changes in land use contributions and the specific share attributed to each land use type.

2. Materials and Methods

2.1. Study Area

The control sub-watershed, located in the southwest of the Khamsan representative paired watershed, is characterized by the absence of watershed management operations. According to a digital elevation model derived from photogrammetric drone data, this sub-watershed covers an area of approximately 102.14 hectares, features an average slope of 28%, and has an average elevation of 1712 m above sea level. Its geographical coordinates range from 47°4′39″ to 47°5′31″ E longitude and 34°57′52″ to 34°58′33″ N latitude (see Figure 1). Climate data from the Khamsan climatology station indicate that the average annual temperature in this region is 12.5 °C, with average annual precipitation recorded at 428 mm. The study area encompasses four types of land use/covers: orchards, rainfed agriculture, irrigated agriculture, and rangeland. The predominant type of erosion observed in this area is surface erosion [7,24].

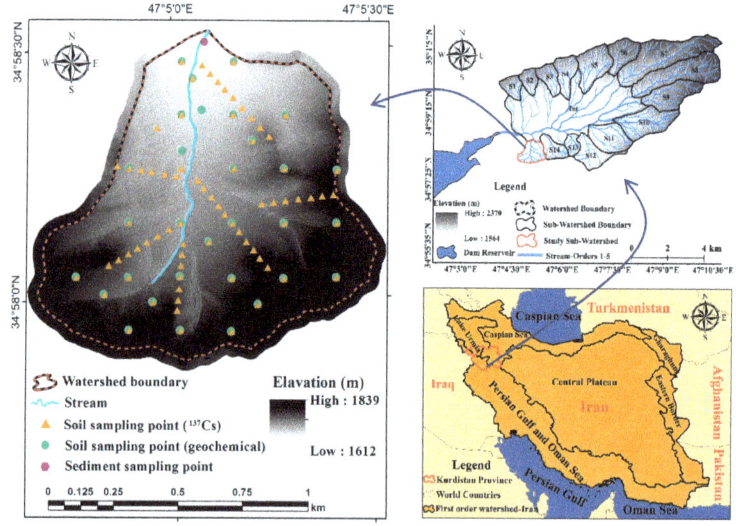

Figure 1. The location of the control sub-watershed of Khamsan representative paired watershed in Iran, along with the soil and sediment sampling points in the study area.

In areas surrounding the main stream channel near the watershed outlet, the soil depth exceeds one meter. The hydrologic group of this soil is Group B, and the soil texture is clay loam, which is suitable for agriculture. Consequently, this part of the watershed is used for orchard and irrigated agriculture. However, frequent plowing typically disrupts the natural soil layers. In the central part of the watershed, the soil depth is approximately

50–60 cm, and the hydrologic group is Group C. The soil texture here is loam, but due to unfavorable economic conditions, this area is mainly under rainfed agriculture. Plowing has led to the destruction of the natural soil layers. In the upper part of the watershed, which features rock outcrops and debris falls, the soil depth varies between 10 and 20 cm, with some limited areas exceeding 20 cm. The hydrologic group of this soil is also Group C. Due to the absence of plowing and the presence of rangeland cover, the natural soil layers are visible in this region. The O horizon has a thin layer (a few millimeters), followed by a loam-textured layer with a high percentage of gravel. The dominant vegetation types in these rangelands include species of *Astragalus* and *Festuca*.

Approximately 65% of the sub-watershed area is composed of black, dark gray, and yellow shales, silty shale, and phyllitic shale with minor sandstone and micritic limestone intercalations. Medium-level alluvial deposits cover 15% of the watershed area around the main channel. Additionally, tuff, which is light blue and thinly to medium-bedded with abundant calcite veins, is observed in the highlands of the sub-watershed, covering 14% of the area. In the western part of the sub-watershed, pink to purple, thin-bedded pelagic limestone with calcite veins constitutes 4% of the area, while 2% consists of light green to gray calcareous shale interlayered with limestone and sandstone.

To create the land use map related to 60 years ago, aerial photographs from 1967 were utilized. For the current land use map, satellite images were prepared using Google Earth Pro software 7.3.6 (see Figure 2).

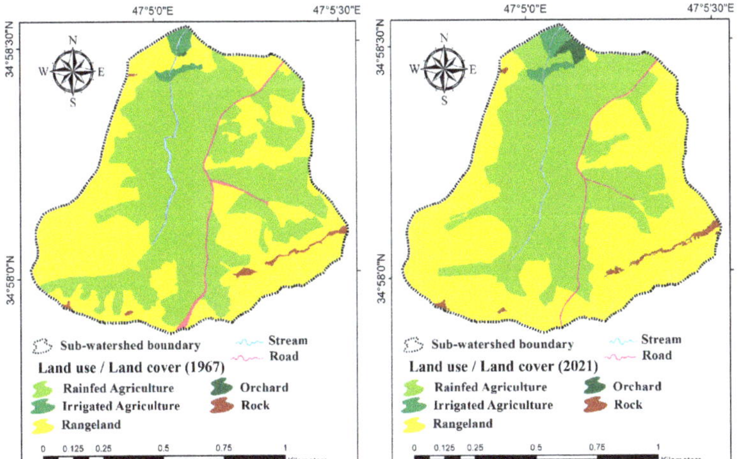

Figure 2. Land use/land cover maps of the study area based on the 1967 aerial photos (**a**) and satellite images of the 2021 (**b**).

2.2. Estimating the Average Contribution of Land Uses in Sediment Yield Using ^{137}Cs

In measuring soil erosion using ^{137}Cs, the first step involves identifying reference points that have remained undisturbed by human activity since the radioactive elements were deposited into the soil. These reference points should have undergone minimal cultivation and movement, and they must not have experienced significant erosion or deposition processes, ensuring that only the natural decay of ^{137}Cs has occurred. Typically, reference sites include protected areas such as cemeteries and parks [25]. For this study, a total of 12 undisturbed samples were collected from the reference area, ensuring a coefficient of variation in radioactive element activity of less than 20%. Four samples were obtained through incremental sampling (layered), while the remaining samples were collected using depth sampling (bulk or integrated) to analyze variations in radioactive element activity across different soil depths. The average of these 12 points was calculated within a specified confidence level to establish an acceptable reference value. Results from previously selected reference areas in this watershed were utilized [7,24]. Following this, soil sampling in the control sub-watershed was conducted using the radioactive core method, with samples proportionately distributed across all land uses/covers and varying slopes. To enhance the accuracy of erosion and deposition measurements while accounting for the combined effects of land use/cover and slope aspect, systematic-random sampling was employed [26]. A regular grid with 200 m intervals was implemented, supplemented by additional samples to ensure comprehensive coverage of different land uses/covers. In total, 31 soil samples were collected following a grid scheme, and 44 samples were taken along six transects aligned with the main slope direction of the hills, resulting in a total of 75 soil samples from the sub-watershed surface. For layered sampling, a sampling box measuring 40 × 20 × 30 cm was utilized, while depth samples were collected using an auger with a height of 25 cm and a diameter of 8 cm.

Each sample was placed in an aluminum container and dried in an oven at 105 °C for 24 h. After drying, the samples were weighed and passed through a 2 mm sieve, then ground in an industrial mill and passed through a 63 mm sieve [27]. Finally, 293 g of each sample was placed in commercial containers, coded, and sent to the Institute of Applied Physics laboratory for spectroscopy and measurement of ^{137}Cs activity using high-purity germanium detectors. The process of preparing soil and sediment samples in the laboratory is illustrated in Figure 3.

Figure 3. Preparation steps of soil and sediment samples in the laboratory for ^{137}Cs method.

The cumulative amount of ^{137}Cs in different soil layers was measured based on activity per unit mass (Bq kg^{-1}). To calculate erosion/deposition, the activity was converted to the inventory of ^{137}Cs in the unit area of the soil (Bq m^{-2}) using Equation (1) [28].

$$\mathrm{CPI} = \sum \mathrm{C_i B_i D_i} 10^3 \quad (1)$$

where C_i is the activity of ^{137}Cs in the soil (Bq kg^{-1}), which is the measured amount for each sample in the laboratory; B_i is the bulk density of the soil (g cm^{-3}); D_i is the sampling depth (m), equal to 0.25 m; and CPI is the inventory of ^{137}Cs in the soil (Bq m^{-2}).

This method allows for a comprehensive assessment of soil erosion and deposition by converting the activity measurements into a spatially relevant inventory metric. To analyze changes in radioactive nuclei levels in the soil and their relationship with soil displacement volumes, conversion models were employed. These models facilitate comparisons between the radioactive element inventory in field soils and that in reference areas [23,29]. The International Atomic Energy Agency has developed an Excel-based macro software package designed to convert radioactive element inventories into estimates of soil redistribution using various models. Initially created for ^{137}Cs, this software has since been adapted for other radionuclides such as ^{7}Be and $^{210}Pb_{ex}$ [26,29]. In this study, the software was utilized to convert ^{137}Cs inventory data into erosion and deposition rates. Different models have been proposed by researchers depending on land use types. Given that no material displacement was observed in the reference soil profile from previous studies and field surveys, a diffusion and migration model was applied for non-agricultural lands (undisturbed soil). For agricultural lands, where uncertainties regarding rainfall patterns, tillage timing, and soil displacement from plowing exist, the Mass Balance Model II was employed [7,24,26]. The necessary information for utilizing this software according to each of the two selected models is summarized in Table 1.

Table 1. Required information for the implementation of transformation models.

Transformation Model	Model Inputs
Mass balance model II	Year of Tillage Commencement, Bulk density of soil, Tillage depth
Diffusion and migration model	Proportion factor (Y), Relaxation depth (H), Migration rate (V), Annual fallout of radionuclide

After converting the ^{137}Cs inventory in the soil from Becquerels per square meter to erosion and deposition values [28], the output data from the conversion models, which include both negative and positive values, were analyzed using an Excel-based macro package designed by the International Atomic Energy Agency. This analysis involved determining the erosion and deposition ranges using the following equations:

$$\text{Boundary between steady state and erosion} = \text{Mean} - 95\%\ \text{confidence interval} \quad (2)$$

$$\text{Boundary between steady state and sedimentation} = \text{Mean} + 95\%\ \text{confidence interval} \quad (3)$$

In these equations, the mean represents the average ^{137}Cs inventory, while the 95% confidence interval indicates a confidence level with a 10% error margin between the ^{137}Cs inventory and $^{210}Pb_{ex}$ inventory in the reference area. The range defined by these erosion and deposition limits is considered stable, suggesting a relative balance between erosion and deposition processes over time [26].

The land use map from approximately 60 years ago was compared with the 2021 land use map, revealing significant changes primarily from rangeland to rainfed agriculture. To ascertain the approximate year of land use conversion and the initiation of plowing for these lands, interviews were conducted with residents of Khamsan village. This information was subsequently integrated into the conversion models for these specific points.

At each point of interest, any increase or decrease in ^{137}Cs activity that fell outside the Mean $\pm$ 95% confidence interval range relative to the reference area indicated whether erosion or deposition processes had dominated over the past 60 years.

2.3. Estimating the Average Contribution of Land Uses in Sediment Yield Using Geochemical Characteristics

Soil samples for fingerprinting studies utilizing geochemical characteristics were collected from the surface layer at a depth of 0 to 2.5 cm [30]. Initially, a thin layer of surface horizon, consisting of plant residues and organic materials, was carefully removed. Following this, a uniform mixture of the 2.5 cm surface soil layer was collected, totaling approximately 500 g. The sampling conditions ensured appropriate moisture levels, with no rainfall occurring five days prior to sampling. In this study, fine sediment bed sampling was performed using a plastic trowel, which minimizes the impact on heavy metals and other geochemical characteristics of the sediment [31–33]. This sampling targeted the 5 cm surface layer of the streambed.

After collection, soil and sediment samples were dried in a freeze dryer for 24 h, then ground in a porcelain mortar and passed through a 63-micron sieve. Approximately 10 g of each sample were prepared and coded for further analysis [34–36]. The laboratory preparation process for soil and sediment samples is illustrated in Figure 4.

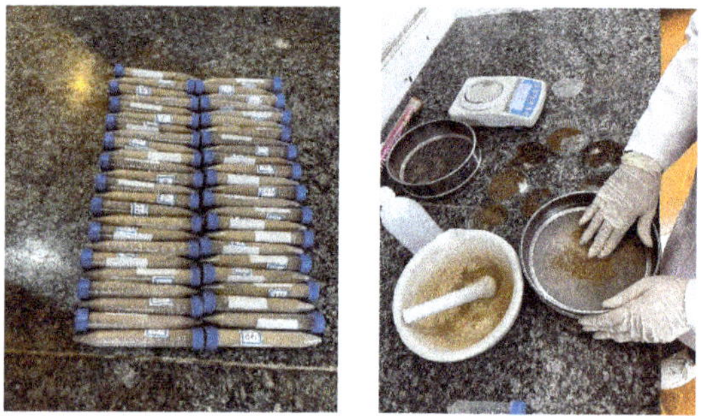

Figure 4. Preparation of soil and sediment samples in the laboratory for geochemical sourcing.

To measure geochemical elements, 31 soil samples and a target sediment sample underwent a process of acid digestion. Specifically, 3 g of dried samples, which were smaller than 63 microns, were treated with aqua regia solution—a mixture of concentrated hydrochloric acid and nitric acid in a 3:1 ratio—for two hours in a water bath. After cooling, the samples were carefully filtered using filter paper with a pore diameter of 0.2 microns. The geochemical characteristics were then measured in micrograms per gram of soil sample (ppm) using the ICP-OES Integra device (manufactured by GBC, sourced in Beijing, China), along with standard Merck samples and calibration curve plotting [37].

To assess the relative contributions of various land uses and covers, the open-source FingerPro software package, developed in the R programming language, was employed. This package is specifically tailored for sediment fingerprinting and calculates sediment source contributions by selecting suitable tracers and applying multivariate mixing model algorithms. One of the standout features of the FingerPro package is its user-friendly interface, which aids users in selecting the best tracers and visualizing results through advanced R graphical plots. This functionality allows for determining sediment source contributions without requiring extensive experience or knowledge of software [23]. The selection of tracer sets was conducted using range test (RT) methods, Kruskal–Wallis (KW) tests, and relative source contribution calculations to sediment yield through a multivariate mixing model. This systematic approach enables users to accurately identify suitable tracers and obtain precise results regarding the contributions of different land uses. This meticulous methodology underscores the importance of integrating geochemical analyses with innovative software tools to enhance our understanding of sediment dynamics within watersheds. By utilizing these techniques, we can better inform land management practices aimed at mitigating erosion and improving soil conservation efforts.

In this study, 56 elements were initially selected as potential tracers. Following this, statistical methods were employed to identify an optimal combination of tracers that could effectively distinguish between different sediment sources. The performance of source identification is influenced by the number of tracers utilized; generally, a greater number of tracers leads to reduced uncertainty in the results [38–41]. The process for selecting tracers and applying the mixing model to ascertain the relative contributions of various land uses in sediment yield was executed using coded methods within the FingerPro package in R software [22]. The stability of the geochemical characteristics of the tracers was assessed using the range test available in the FingerPro package. Tracers that exhibited significant differences between sources were subsequently selected through the Kruskal–Wallis test. After identifying suitable tracers from these two tests, the relative contributions of sediment sources to sediment yield were determined using a multivariate mixing model. To explore the various combinations of each source's contribution—ranging from 0 to 100%—Latin hypercube sampling was employed. The quality of results regarding source contributions to sediment yield was evaluated using the Goodness of Fit (GOF) index, which is based on the sum of relative error squares. The Goodness of Fit metric assesses how well a model aligns with observed data and is calculated using Equation (4) [22].

$$\mathrm{GOF} = 1 - \frac{1}{n} \times \left(\sum_{j=1}^{n} \frac{\mid b_i - \sum_{j=1}^{m} w_j a_{i,j} \mid}{\Delta i} \right) 0 \leq w_j \leq 1 \quad (4)$$

where b_i is the tracer property i (i = 1 to n) of the sediment mixture, $a_{i,j}$ represents the tracer property i in the source type j (j = 1 to m), w_j is the unknown relative contribution of the source type j, m represents the number of potential sediment sources and n is the number of tracer properties selected.

3. Results and Discussion

3.1. Results of the Average Contribution of Land Uses in Sediment Yield Using ^{137}Cs

The comparison of land use maps from 1967 and 2021 reveals significant transformations in land use patterns. Notably, the area dedicated to rainfed agriculture has decreased during this period, particularly in regions characterized by higher elevations and steeper slopes. Elevation and slope are critical factors influencing the intensity of agricultural land abandonment. The newly abandoned rainfed lands have considerable impacts on erosion and sedimentation, primarily due to declines in soil quality, limited vegetation recovery, and the effects of tillage operations along the slope. In the early years following abandonment, sediment yield tends to increase, but this trend reverses over the long term, a phenomenon confirmed by numerous researchers, e.g., [42–49]. Therefore, it is essential to plan and manage land abandonment alongside soil erosion control strategies over several years to prevent further land degradation. Table 2 presents the average contributions of various land uses to soil erosion intensity over the past 60 years, as assessed using the ^{137}Cs method. These data underscore the profound impact of agricultural activities on soil erosion dynamics. This thorough analysis highlights the need for effective land management practices that consider both current land use and historical changes.

Table 2. Average contribution of land uses in sediment yield in the last 60 years (by ^{137}Cs method).

Land Uses/Covers	Orchard	Rangeland	Irrigated Agriculture	Rainfed Agriculture	Road	Stream	Rock	Total
Area (ha)	0.04	46.17	1.29	52.32	1.03	0.54	0.75	102.14
Area (%)	0.04	45.20	1.26	51.22	1.01	0.53	0.73	100.00
Contribution (%)	3.78	8.93	1.80	85.49	-	-	-	-
land uses erosion (t y^{-1})	−19.90	−47.02	−9.51	−450.43	-	-	-	-
Specific Contribution (% per ha)	94.50	0.19	1.40	1.63	-	-	-	-
Gross erosion (t y^{-1})			−526.87		-	-	-	-

The data presented in Table 2 reveal that over 85% of soil erosion in the study area occurs within rainfed agricultural land. Interestingly, irrigated agricultural land, which comprises small, low-slope areas near the sub-watershed outlet, exhibits low erosion intensity and is considered a stable area, contributing minimally to overall sub-watershed erosion. Rangeland, covering 45% of the study area, accounts for only 8.93% of sediment yield. Although soil erosion does occur in a significant portion of this land use, the intensity remains low. According to the 95% confidence interval criterion, these areas are deemed stable and are excluded from calculations regarding sediment yield contributions. Among the four potential sediment sources analyzed, rainfed agriculture and rangeland emerged as the primary sediment sources in the study area, contributing 85.49% and 8.93% of total sediments, respectively. In contrast, the contributions from orchard and irrigated agriculture land uses were considerably lower, at 3.78% and 1.80%. It is important to note that rainfed agriculture and rangeland together comprise more than 96% of the total watershed area, which is why specific contributions from each land use were also calculated. The specific contributions per hectare for rainfed agriculture, orchards, irrigated agriculture, and rangeland were found to be 1.63%, 94.50%, 1.40%, and 0.19%, respectively. Orchards exhibited the highest specific contribution despite constituting only 0.04% of the total watershed area. This significant increase can be attributed to practices such as deep and frequent plowing, excessive use of chemical fertilizers, pit digging, soil amendment with sand, sapling planting, and other activities that lead to soil compaction and disrupt natural soil structure—all of which significantly contribute to soil erosion. Following orchards, rainfed agriculture has the next highest specific contribution with a substantial margin.

3.2. Results of the Average Contribution of Land Uses in Sediment Yield Using Geochemical Characteristics

After conducting the range test, which compared the minimum and maximum values of geochemical tracer characteristics in sediment sources using 31 soil samples, a total of 11 elements—including Al, Ba, Be, Fe, Ga, Lu, Pb, Sn, Sr, Te, and Tm—were eliminated from subsequent analyses as they fell outside the selected sources' range. The remaining elements were retained in the tracer set for further evaluation. Based on the results from the non-parametric rank-based Kruskal–Wallis test, an additional 42 elements—such as Na, K, Ca, Mg, Li, B, Sc, Ti, V, Cr, Mn, Co, Cu, Zn, Ge, As, Se, Rb, Y, Zr, Nb, Sb, Cs, La, Ce, Pr, Nd, Sm, Eu, Gd, Tb, Dy, Ho, Er, Yb, Ta, W, Hg, Tl, Bi, Th, and U—were also removed from the tracer set. Ultimately, only three elements—Ni, Mo, and Hf—remained as optimal tracers after this rigorous selection process. These results are summarized in Table 3 and include both the range and Kruskal–Wallis test outcomes for selecting the optimal combination of tracers. The characteristics of the selected tracers are visually represented using box plots for each land use in Figure 5. A box plot is an effective visual tool that illustrates data distribution through quartiles. In evaluating the performance of the remaining tracers, Ni consistently showed higher concentrations across all four land uses. The concentration order was as follows: rainfed agriculture had the highest concentration, followed by irrigated agriculture and orchard land; rangeland exhibited the lowest concentration. According to the box plots presented in Figure 5, rainfed agriculture emerged as the primary source for all three selected tracers.

The DFA plot effectively separates sediment sources by identifying a linear combination of the selected tracers and presenting the results in a 3D format. Principal Component Analysis (PCA) was employed as a complementary visual tool to aid in the correct selection of optimal tracers. Figure 6 illustrates both the DFA plot and the PCA plot, showcasing the distribution of sediment sources based on the selected optimal tracers.

Table 3. Results of the Kruskal–Wallis and Range Tests for choosing the optimal combination of tracers.

Test	Removed Trackers	Remaining Trackers
RT	Al, Ba, Be, Fe, Ga, Lu, Pb, Sn, Sr, Te, Tm	Na, K, Ca, Mg, Li, B, Sc, Ti, V, Cr, Mn, Co, Cu, Zn, Ge, As, Se, Rb, Y, Zr, Nb, Sb, Cs, La, Ce, Pr, Nd, Sm, Eu, Ni, Mo, Hf, Gd, Tb, Dy, Ho, Er, Yb, Ta, W, Hg, Tl, Bi, Th, U
KW	Na, K, Ca, Mg, Li, B, Sc, Ti, V, Cr, Mn, Co, Cu, Zn, Ge, As, Se, Rb, Y, Zr, Nb, Sb, Cs, La, Ce, Pr, Nd, Sm, Eu, Gd, Tb, Dy, Ho, Er, Yb, Ta, W, Hg, Tl, Bi, Th, U	Ni, Mo, Hf

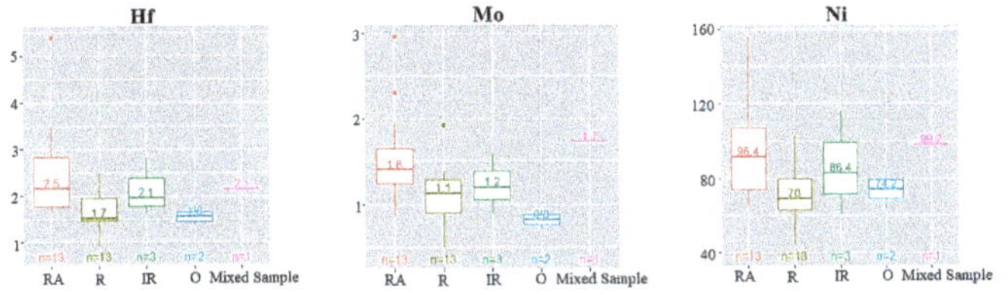

Figure 5. Box plot of changes in concentration of final tracers (mg kg^{-1}) in different sediment sources (RA: rainfed agriculture, R: rangeland, IA: irrigated agriculture, O: orchard).

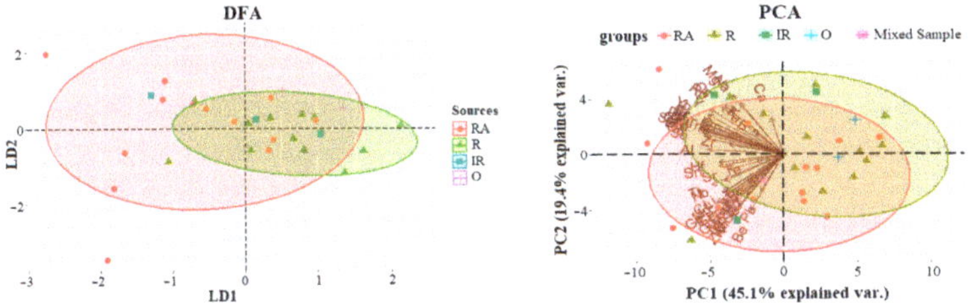

(a)

(b)

Figure 6. Discriminant Function Analysis (**a**) and Principal Component Analysis (**b**) plots (RA: rainfed agriculture, R: rangeland, IA: irrigated agriculture, O: orchard).

The results from the Principal Component Analysis (PCA) plot indicated that most of the arrows representing the optimal tracers converged, suggesting a positive correlation among the tracers. This finding aligns with previous studies by [39,50], which also demonstrated similar positive correlations between tracer elements in their analyses. Following the identification of sediment sources and their contributions using the FingerPro package in R software, the calculations and results detailing the contribution levels of various sediment sources—including rainfed agriculture, rangeland, irrigated agriculture, and orchards—are presented in Table 4. The findings reveal that rainfed agricultural lands, identified as a primary sediment source, contributed 72.26% of the total sediment yield. In addition, the relative contributions from other sources were 15.06% from irrigated agricultural lands, 7.96% from rangelands, and 4.72% from orchards. The dominance of rainfed agricultural lands in sediment yield can be attributed to improper and suboptimal land use management practices, particularly unsustainable agricultural methods such as plowing along slopes on hillsides. This observation has been corroborated by numerous studies conducted in the watershed. Moreover, the results obtained from the FingerPro package indicated that the Goodness of Fit (GOF) index for sediment source tracing was 0.78. This value is considered acceptable, especially in light of recent scientific findings suggesting that a GOF index above 0.60 is indicative of a reliable model [22,41,51–53]. The advanced capabilities of the FingerPro technique have significantly enhanced our understanding of sediment sources and enabled a more precise and comprehensive analysis of data.

Table 4. Average contribution of land use/land covers in sediment yield by geochemical method (percentage).

Land Uses/Covers	Orchard	Rangeland	Irrigated Agriculture	Rainfed Agriculture	Road	Stream	Rock	Total
Area (ha)	0.63	52.54	1.54	45.65	0.59	0.36	0.81	102.14
Area (%)	0.62	51.45	1.51	44.69	0.58	0.35	0.80	100.00
Contribution (%)	4.72	7.96	15.06	72.26	-	-	-	-
Specific Contribution (% per ha)	7.49	0.15	9.78	1.58	-	-	-	-
GOF			0.78		-	-	-	-

The specific contributions of various land uses to sediment yield were calculated and are presented in Table 4. Among the four potential sediment sources, irrigated agriculture and orchards emerged as significant contributors in the study area, with specific contributions of 9.78% and 7.49% per hectare, respectively. In contrast, the relative specific contributions from rainfed agriculture and rangeland were considerably lower, estimated at 1.58% and 0.15% per hectare, respectively. It is noteworthy that irrigated agriculture comprises only 1.51% of the total watershed area, while orchards account for 0.62%. These findings underscore the critical importance of targeted land management strategies in controlling erosion and sediment yield within the watershed. They highlight that having accurate information on soil erosion and sediment yield across different areas of a watershed is essential for planning effective soil conservation management strategies. To provide a visual comparison, Figure 7 presents a chart illustrating the percentage of relative and specific contributions of each land use to sediment yield, utilizing both the ^{137}Cs method and geochemical analyses.

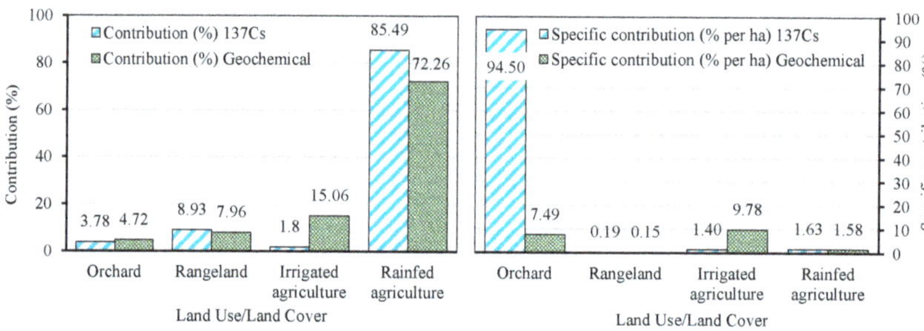

(a) (b)

Figure 7. Comparing the relative (**a**) and specific (**b**) contribution of land use/land covers in sediment yield using ^{137}Cs and geochemical methods.

Based on the results obtained from comparing contributions of rangeland, rainfed agriculture, irrigated agriculture, and orchards within the study area, it can be concluded that over the past 60 years, rangeland—occupying nearly half of the sub-watershed and characterized by steeper slopes compared to other land uses—has exhibited the lowest specific contribution to sediment yield, as clearly illustrated in Figure 7. The contribution of rainfed agriculture to sediment yield was estimated at over 85% using the ^{137}Cs method for the average of the last 60 years, while the geochemical method indicated a contribution of over 72% for the present period. This clearly demonstrates that rainfed agriculture is the primary source of sediment yield in the control sub-watershed. Furthermore, irrigated agriculture and orchards have shown the highest specific contributions during both time periods, largely due to soil displacement resulting from tillage activities, particularly practices such as pit digging and soil transfer during orchard establishment. The comparative analysis of contributions from various land uses over the past 60 years to the present period revealed a decrease of approximately one times in both rainfed agriculture and rangeland, while irrigated agriculture and orchards experienced increases of about eight and one times, respectively. Specifically, the contributions of rainfed agriculture, rangeland, orchards, and irrigated agriculture have decreased by factors of 1.03, 1.27, and 12.62, respectively, while increasing by a factor of 6.99 for irrigated agriculture. Interestingly, the relative and specific contributions of rangeland in sediment yield have remained relatively constant over the past 60 years compared to the present period. Other human-induced changes, such as soil compaction, the effects of drainage systems, infrastructures, and the length, width, and type of the roads, had not occurred significantly. During this period, a land use change (abandonment of rangeland) has been observed. Field visits to the studied sub-watershed indicated that the reduction in sediment yield from rainfed agriculture is primarily due to a decrease in the area dedicated to this land use. Additionally, the decrease in rainfall and the change in rainfall patterns across large parts of the Zagros Mountains and the associated watersheds [54], along with the barriers created, particularly plowing even within the main stream channel, have prevented runoff and sediment from rainfed agriculture from reaching the sub-watershed outlet.

Moreover, both the ^{137}Cs and geochemical methods reveal that sediment yield in rainfed agriculture is approximately 10 times higher than that in rangeland. These findings highlight significant issues regarding adherence to soil management and agricultural principles on rainfed lands, including improper plowing techniques and a lack of conservation methods. Such practices have led to reduced soil fertility and increased runoff and sedimentation. This underscores the critical importance of precise land use management and implementing effective conservation practices, particularly to prevent land use conversion in sloped areas.

4. Limitations

Source fingerprinting procedures often encounter significant challenges and limitations [3]. Although a total number of 106 soil samples were analyzed for four different land uses in a 102 h area, the distribution of the samples in each land use may still be considered as a source of uncertainty in this research. Using aqua regia digest to extract tracers from source materials and target sediment samples produced pseudo-concentrations rather than total concentrations. Tracer selection relied more on the success reported in previous global studies than on physico-chemical reasoning, resulting in statistical solutions for source discrimination and apportionment. Despite these constraints, virtual mixture tests demonstrated acceptable accuracy levels for source apportionment (Table 5).

Table 5. The known and predicted relative contributions to virtual sediment mixtures, along with the corresponding root mean square error (RMSE) and mean absolute error (MAE).

Known Sediment Source Contributions (%)				Predicted Sediment Source Contributions (%)				RMSE	MAE
RA	R	IA	O	RA	R	IA	O		
25	25	25	25	24.2	19.9	28.1	27.8	3.3	3.0
100	0	0	0	90.7	0.7	6.0	2.6	5.7	4.7
0	100	0	0	1.9	87.2	5.6	5.3	7.5	6.4

Table 5. *Cont.*

Known Sediment Source Contributions (%)				Predicted Sediment Source Contributions (%)				RMSE	MAE
RA	R	IA	O	RA	R	IA	O		
0	0	100	0	2.6	3.1	91.5	2.8	4.9	4.3
0	0	0	100	0.4	2.6	3.7	93.3	4.0	3.4
50	25	12.5	12.5	42.9	28.7	8.7	19.7	5.7	5.5
25	50	12.5	12.5	20.9	48.8	14.8	15.5	2.9	2.7
25	6.25	56.25	12.5	26.9	5.4	62.5	5.2	4.9	4.1
70	10	10	10	75.4	6.7	13.9	4.0	4.8	4.7
85	5	5	5	79.1	4.6	13.6	2.7	5.3	4.3
Average								4.9	4.3

5. Conclusions

Soil erosion is a significant challenge in watersheds, and sediment fingerprinting methods can identify sediment sources. Studies show that the contributions of land uses/covers to sediment yield have changed significantly over the past 60 years compared to the present situation. Specifically, in rainfed agriculture and rangeland, the contributions have decreased by approximately 1.2 times, while in irrigated agriculture and orchards, they have increased by about 8 times and 1.2 times, respectively. Additionally, the specific contributions of rainfed agriculture, rangeland, orchards, and irrigated agriculture over the past 60 years compared to the present period have decreased by 1.03, 12.62, and 1.27 times, respectively, and increased by 7.04 times. Field visits to the study area indicated that the reduction in sediment yield in rainfed agriculture was due to the decreased area of this land use. Additionally, over time, the study area has undergone changes such as land degradation and the creation of obstacles, especially plowing along the main watercourse and even planting crops in parts of the stream, which has prevented runoff and sediment from rainfed agriculture from reaching the sub-watershed outlet. Climate change and reduced rainfall have significantly accelerated trends in soil erosion and sediment yield, particularly in agricultural settings. The results indicate that erosion and sediment yield in irrigated agriculture have increased over the past six decades, primarily due to tillage operations that compromise soil stability and elevate sediment levels. Notably, both the ^{137}Cs and geochemical methods reveal that sediment yield in rainfed agriculture is approximately 10 times higher than that in rangeland. This finding underscores the critical need for precise land use management and the implementation of conservation practices, especially to prevent land use conversion in sloped areas, mitigating the adverse effects of plowing along slopes. The findings from this study highlight that analyzing temporal changes in land use contributions to soil erosion and sediment yield offers valuable insights into the impacts of human activities over recent decades. Such insights are essential for developing effective management plans and implementing soil and water conservation measures. However, there are undoubtedly uncertainties in our results, and it is important to acknowledge that any scientific study contains uncertainties, chiefly when we use models. Simplification of natural conditions is a key task of any model, especially when it comes to the effects of soil disturbance and the fact that the sensitivity of the model to its inputs is not well defined.

Author Contributions: Conceptualization, A.K.D. and P.P.; methodology, N.G.D., A.K.D., M.R.Z. and P.P.; software, N.G.D., A.K.D., M.R.Z. and P.P.; validation, N.G.D., A.K.D. and P.P.; formal analysis, N.G.D., A.K.D., M.R.Z. and P.P.; investigation, N.G.D., A.K.D. and P.P.; resources, N.G.D., A.K.D. and P.P.; data curation, N.G.D. and A.K.D.; writing—original draft preparation, N.G.D.; writing—review and editing, A.K.D. and P.P.; visualization, N.G.D. and A.K.D.; supervision, A.K.D., M.R.Z. and P.P.; project administration, A.K.D. All authors have read and agreed to the published version of the manuscript.

Funding: This research received no external funding.

Data Availability Statement: The raw data supporting the conclusions of this article will be made available by the authors on request.

Acknowledgments: The current research was undertaken as the doctoral dissertation of Negin Ghaderi Dehkordi, with the financial support of Tarbiat Modares University, Iran. Authors would like to thank Hamid Khodamoradi for his helps in the field sampling.

Conflicts of Interest: The authors declare no conflicts of interest.

References

1. Alatorre, L.; Beguería, S.; García-Ruiz, J. Regional scale modeling of hillslope sediment delivery: A case study in the Barasona Reservoir watershed (Spain) using WATEM/SEDEM. *J. Hydrol.* **2010**, *391*, 109–123. [CrossRef]
2. Bhattacharya, R.; Chatterjee, N.; Das, K. Estimation of erosion susceptibility and sediment yield in ephemeral channel using RUSLE and SDR model: Tropical Plateau Fringe Region, India. In *Gully Erosion Studies from India and Surrounding Regions*; Springer: Berlin/Heidelberg, Germany, 2020; pp. 163–185. [CrossRef]
3. Collin, A.L.; Pulley, S.; Foster, I.D.L.; Gellis, A.; Porto, P.; Horowitz, A.J. Sediment source fingerprinting for catchment management: A review. *J. Environ. Manag.* **2017**, *194*, 86–108. [CrossRef] [PubMed]
4. Chen, F.; Chen, S.; Zhang, X.; Chen, J.; Wang, X.; Gowan, E.J.; Qiang, M.; Dong, G.; Wang, Z.; Li, Y.; et al. Asian dust-storm activity dominated by Chinese dynasty changes since 2000 BP. *Nat. Commun.* **2020**, *11*, 992. [CrossRef] [PubMed]
5. Bastala, S.; Dialynas, Y.G.; Bras, R.L.; Noto, L.V.; Istanbulluoglu, E. The role of vegetation on gully erosion stabilization at a severely degraded landscape: A case study from Calhoun Experimental Critical Zone Observatory. *Geomorphology* **2018**, *308*, 25–39. [CrossRef]
6. Devátý, J.; Dostál, T.; Hösl, R.; Krása, J.; Strauss, P. Effects of historical land use and land pattern changes on soil erosion—Case studies from Lower Austria and Central Bohemia. *LUP* **2019**, *82*, 674–685. [CrossRef]
7. Zarei, R.; Khaledi Darvishan, A.; Porto, P.; Zare, M.R. Using radiotracers and topographic metrics for sediment budgeting at pixel and hillslope scales: A case study from western Iran. *Ecol. Indic.* **2024**, *167*, 112711. [CrossRef]
8. Collins, A.L.; Walling, D.E. Documenting catchment suspended sediment sources: Problems, approaches and prospects. *Prog. Phys. Geogr.* **2004**, *28*, 159–196. [CrossRef]
9. Franz, C.; Makeschin, F.; Weiß, H.; Lorz, C. Sediments in urban river basins: Identification of sediment sources within the Lago Paranoá catchment, Brasilia DF, Brazil—Using the fingerprint approach. *Sci. Total Environ.* **2014**, *466–467*, 513–523. [CrossRef]
10. Gruszowski, K.E.; Foster, I.D.L.; Lees, J.A.; Charlesworth, S.M. Sediment sources and transport pathways in a rural catchment, Herefordshire, UK. *Hydrol. Proc.* **2003**, *17*, 2665–2681. [CrossRef]
11. Walling, D.E. Tracing suspended sediment sources in catchments and river systems. *Sci. Total Environ.* **2005**, *344*, 159–184. [CrossRef]
12. Koiter, A.J.; Owens, P.N.; Petticrew, E.L.; Lobb, D.A. The behavioural characteristics of sediment properties and their implications for sediment fingerprinting as an approach for identifying sediment sources in river basins. *Earth-Sci. Rev.* **2013**, *125*, 24–42. [CrossRef]
13. Martínez-Carreras, N.; Krein, A.; Gallart, F.; Iffly, J.F.; Pfister, L.; Haffman, L.; Owens, P.N. Assessment of different parameter for discriminating potentioal suspende sediment sources and provenance: A multi scale study in Luxembourg. *Geomorphology* **2010**, *11*, 118–129. [CrossRef]
14. Porto, P.; Walling, D.E.; Callegari, G. Investigating sediment sources within a small catchment in southern Italy. In *Sediment*; Budgets, I., Walling, D.E., Horowitz, A.J., Eds.; IAHS Publication 291; IAHS Press: Wallingford, UK, 2005; pp. 113–122.
15. Rogowsky, A.S.; Tamura, T. Movement of ^{137}Cs by runoff, erosion and infiltration on the alluvial Captina silt loam. *Health Phys.* **1965**, *11*, 1333–1340. [CrossRef]

16. Ritchie, J.C.; McHenry, J.R. Determination of fallout ^{137}CS and naturally occurring gamma-ray emitters in sediments. *Int. J. Appl. Radiat. Isot.* **1973**, *24*, 575–578. [CrossRef]
17. Motha, J.A.; Wallbrink, P.J.; Hairsine, P.B.; Grayson, R.B. Tracer properties of eroded sediment and source material. *Hydrol. Proc.* **2002**, *16*, 1983–2000. [CrossRef]
18. Caitcheon, G.; Douglas, G.; Palmer, M. Sediment Source Tracing in the Lake Burragorang Catchment. *CSIRO Land Water Sci. Rep. Canberra* **2006**, *47*, 28.
19. Douglas, G.; Caitcheon, G.; Palmer, M. Sediment source identification and residence times in the Maroochy River estuary, southeast Queensland, Australia. *Environ. Geol.* **2009**, *57*, 629–639. [CrossRef]
20. Hughes, A.O.; Olley, J.M.; Croke, J.C.; McKergow, L.A. Sediment source changes in a dry-tropical catchment. *Geomorphology* **2009**, *104*, 262–275. [CrossRef]
21. D'Haen, K.; Verstraeten, G.; Dusar, B.; Degryse, P.; Haex, J.; Waelkens, M. Unravelling changing sediment sources in a Mediterranean mountain catchment: A Bayesian fingerprinting approach. *Hydrol. Proc.* **2013**, *27*, 896–910. [CrossRef]
22. Lizaga, I.; Latorre, B.; Gaspar, L.; Navas, A. FingerPro: An R package for tracking the provenance of sediment. *Water Resour. Manag.* **2020**, *34*, 3879–3894. [CrossRef]
23. Zapata, F. *Handbook for the Assessment of Soil Erosion and Sedimentation Using Environmental Radionuclides*; Kluwer Academic Publishers: Dordrecht, The Netherland, 2003; Volume 219, pp. 9348054–9348059. [CrossRef]
24. Sedighi, F.; Khaledi Darvishan, A.; Zare, M.R. Effect of watershed geomorphological characteristics on sediment redistribution. *Geomorphology* **2021**, *375*, 107559. [CrossRef]
25. Dercon, G.; Deckers, J.; Govers, G.; Poesen, J.; Sánchez, H.; Vanegas, R.; Ramírez, M.; Loaiza, G. Spatial variability in soil properties on slow-forming terraces in the Andes region of Ecuador. *Soil Tillage Res.* **2003**, *72*, 31–41. [CrossRef]
26. Walling, D.E.; He, Q.; Zhang, Y. Conversion models and related software. In: Guidelines for Using Fallout Radionuclides to Assess Erosion and Effectiveness of Soil Conservation Strategies. In *IAEA-TECDOC-1741*; IAEA Publisher: Vienna, Austria, 2014; Volume 46, pp. 125–148.
27. Chiu, Y.J.; Chang, K.T.; Chen, Y.C.; Chao, J.H.; Lee, H.Y. Estimation of soil erosion rates in a subtropical mountain watershed using ^{137}Cs radionuclide. *Nat. Hazards* **2011**, *59*, 271–284. [CrossRef]
28. Walling, D.E.; He, Q.; Whelan, P.A. Using ^{137}Cs measurements to validate the application of the AGNPS and ANSWERS erosion and sediment yield models in two small Devon catchments. *Soil Tillage Res.* **2003**, *69*, 27–43. [CrossRef]
29. Porto, P.; Walling, D.E.; Callegari, G. Using repeated ^{137}Cs and $^{210}Pb_{ex}$ measurements to establish sediment budgets for different time windows and explore the effect of connectivity on soil erosion rates in a small experimental catchment in Southern Italy. *Land Degrad. Dev.* **2018**, *29*, 1819–1832. [CrossRef]
30. Lamba, J.; Karthikeyan, K.G.; Thompson, A.M. Apportionment of suspended sediment sources in an agricultural watershed using sediment fingerprinting. *Geoderma* **2015**, *239–240*, 25–33. [CrossRef]
31. Smith, H.G.; Blake, W.H. Sediment fingerprinting in agricultural catchments: A critical re-examination of source discrimination and data corrections. *Geomorphology* **2014**, *204*, 177–191. [CrossRef]
32. Boudreault, M.; Koiter, A.J.; Lobb, D.A.; Liu, K.; Benoy, G.; Owens, P.N.; Li, S. Comparison of sampling designs for sediment source fingerprinting in an agricultural watershed in Atlantic Canada. *J. Soils Sediments* **2019**, *19*, 3302–3318. [CrossRef]
33. Li, T.; Sun, G.; Yang, C.; Liang, K.; Ma, S.; Huang, L.; Luo, W. Source apportionment and source-to-sink transport of major and trace elements in coastal sediments: Combining positive matrix factorization and sediment trend analysis. *Sci. Total Environ.* **2019**, *601*, 344–356. [CrossRef] [PubMed]
34. Wallbrink, P.J.; Croke, J. A combined rainfall simulator and tracer approach to assess the role of Best Management Practices in minimising sediment redistribution and loss in forests after harvesting. *For. Ecol. Manag.* **2002**, *170*, 217–232. [CrossRef]
35. Collins, A.L.; Walling, D.E.; Webb, L.; King, P. Apportioning catchment scale sediment sources using a modified composite fingerprinting technique incorporating property weightings and prior information. *Geoderma* **2010**, *155*, 249–261. [CrossRef]
36. Gholami, H.; Middleton, N.; NazariSamani, A.A.; Wasson, R. Determining contribution of sand dune potential sources using radionuclides, trace and major elements in central Iran. *Arab. J. Geosci.* **2017**, *10*, 163–171. [CrossRef]
37. Nosrati, K.; Ahmadi, F. Investigating of Seasonal Change Effects on Suspended Sediment Yield Using Geochemical Tracers. *Iran J. Geol.* **2018**, *12*, 60–73. Available online: http://geology.saminatech.ir/en/Article/9601 (accessed on 28 September 2024).
38. Smith, H.G.; Karam, D.S.; Lennard, A.T. Evaluating tracer selection for catchment sediment fingerprinting. *J. Soils Sediments* **2018**, *18*, 3005–3019. [CrossRef]
39. Vale, S.; Swales, A.; Smith, H.G.; Olsen, G.; Woodward, B. Impacts of tracer type, tracer selection, and source dominance on source apportionment with sediment fingerprinting. *Sci. Total Environ.* **2022**, *831*, 154832. [CrossRef] [PubMed]

40. Haji, K.; Khaledi Darvishan, A.; Mostafazadeh, R. Soil erosion and sediment sourcing in the Hyrcanian forests, Northern Iran: An integration approach of the G2loss model and sediment fingerprinting technique. *Model. Earth Syst. Environ.* **2024**, *10*, 1897–1914. [CrossRef]
41. Mohammadi, M.; Khaledi Darvishan, A.; Bahramifar, N.; Alavi, S.J. Spatio-temporal suspended sediment fingerprinting under different land management practices. *Int. J. Sediment. Res.* **2023**, *38*, 481–493. [CrossRef]
42. Lesschen, J.P.; Cammeraat, L.H.; Nieman, T. Erosion and terrace failure due to agricultural land abandonment in a semi-arid environment. *Earth Surf. Process. Landf.* **2008**, *33*, 1574–1584. [CrossRef]
43. Cerdà, A.; Rodrigo-Comino, J.; Novara, A.; Brevik, E.C.; Vaezi, A.R.; Pulido, M.; Giménez-Morera, A.; Keesstra, S.D. Long-term impact of rainfed agricultural land abandonment on soil erosion in the Western Mediterranean basin. *Prog. Phys. Geogr.* **2018**, *42*, 202–219. [CrossRef]
44. Rodrigo-Comino, J.; Martínez-Hernández, C.; Iserloh, T.; Cerdà, A. Contrasted impact of land abandonment on soil erosion in Mediterranean agriculture fields. *Pedosphere* **2018**, *28*, 617–631. [CrossRef]
45. Kheirfam, H.; Sadeghi, S.H.; Darki, B.Z. Soil conservation in an abandoned agricultural rain-fed land through inoculation of cyanobacteria. *Catena* **2020**, *187*, 104341. [CrossRef]
46. Lana-Renault, N.; Nadal-Romero, E.; Cammeraat, E.; Llorente, J.A. Critical environmental issues confirm the relevance of abandoned agricultural land. *Water* **2020**, *12*, 1119. [CrossRef]
47. Liu, M.; Han, G.; Li, X. Contributions of soil erosion and decomposition to SOC loss during a short-term paddy land abandonment in Northeast Thailand. *Agric. Ecosyst. Environ.* **2021**, *321*, 107629. [CrossRef]
48. Khorchani, M.; Gaspar, L.; Nadal-Romero, E.; Arnaez, J.; Lasanta, T.; Navas, A. Effects of cropland abandonment and afforestation on soil redistribution in a small Mediterranean mountain catchment. *Int. Soil Water Conserv. Res.* **2023**, *11*, 339–352. [CrossRef]
49. Sitaula, R.; Sharma, P.; Chidi, C.L. Agricultural land abandonment and its impact on soil erosion in the Madi Watershed, Gandaki Province, Nepal. *Geogr. J. Nepal.* **2024**, *19*, 53–70. [CrossRef]
50. Lizaga, I.; Gaspar, L.; Latorre, B.; Navas, A. Variations in transport of suspended sediment and associated elements induced by rainfall and agricultural cycle in a Mediterranean agroforestry catchment. *J. Environ. Manag.* **2020**, *272*, 111020. [CrossRef]
51. Chen, F.; Wang, X.; Li, X.; Wang, J.; Xie, D.; Ni, J.; Liu, Y. Using the sediment fingerprinting method to identify the sediment sources in small catchments with similar geological conditions. *Agric. Ecosyst. Environ.* **2019**, *286*, 106655. [CrossRef]
52. Gaspar, L.; Blake, W.H.; Lizaga, I.; Latorre, B.; Navas, A. Particle size effect on geochemical composition of experimental soil mixtures relevant for unmixing modelling. *Geomorphology* **2022**, *403*, 108178. [CrossRef]
53. Karimi, N.; Gholami, L.; Kavian, A.; Khaledi Darvishan, A. Determination of specific contribution of suspended sediment sources in Vaz watershed using geochemical properties. *J. Ecohydrol.* **2022**, *9*, 705–718. (In Persian) [CrossRef]
54. Eblaghian, A.; Ali, A.A.; Radmanesh, F.; Zarei, H. Trend Analysis of Temperature, Precipitation, and Relative Humidity Changes in Iran. *Irrig. Sci. Eng.* **2019**, *42*, 197–212.

water

MDPI

Article

Research on Sediment Deposition Characteristics and the Vegetation Restoration of Ecological Riverbanks in the Deep Waterway Regulation Scheme of Yangtze River

Jinpeng Li [1,*], Xuexia Wang [2] and Lei Wu [3]

[1] China Waterborne Transport Research Institute, Beijing 100088, China
[2] Institute of Plant Nutrition, Resources and Environment, Beijing Academy of Agriculture and Forestry Sciences, Beijing 100097, China; wxx0427@163.com
[3] College of Forestry, Central South University of Forestry and Technology, Changsha 410004, China; wuleiibk@163.com
* Correspondence: lijinpeng@wti.ac.cn

Abstract: In order to explore the sediment deposition characteristic of ecological riverbanks associated with vegetation restoration in the deep waterway regulation scheme of Yangtze River, two kinds of typical ecological riverbanks and a traditional riprap riverbank (TRR) in engineering areas were investigated. The vegetation community characteristics, sediment aggregate compositions, nutrient contents, total organic carbon (TOC), sediment microbial biomass carbon (MBC), sediment microbial biomass nitrogen (MBN), and sediment microbial biomass phosphorus (MBP) were determined. The results indicated that the ecological restoration effect of the lattice gabion ballasted vegetation mat riverbank (LGBVR) was best, followed by the mesh grid riverbank (MGR), and that of the TRR was relatively poor. In different ecological riverbanks, the sediment aggregated compositions were not significantly varied. The sediment contents of NH_4^+-N, available phosphorus (AP), and TOC in ecological riverbank areas were relatively higher than those of the TRR. In the LGBVR, the sediment contents of MBC were relatively higher than those of the others. The sediment deposition characteristics and ecological restoration effects in the study area should be monitored for a long time.

Keywords: sediment deposition; ecological riverbank; deep waterway; Yangtze River

1. Introduction

As an important ecotone between terrestrial and aquatic ecosystems, a riparian zone can act as an indispensable corridor for substance, energy, and biological information flows [1–3]. At the same time, riparian zones have been strongly impacted and modified by artificial projects and natural flow regime alternations [4]. Conventionally, rock and concrete materials have been widely applicated in riverbank protection projects. This emphasizes the structural stability and safety of flood control. However, these hard riverbanks and stream canalizations have negative effects on riparian ecosystems, ecological integrity, habitat connectivity, and esthetic value [5–8]. As effective methods for riparian protection and restoration, the ecological riverbank concepts have been widely accepted and applied in developed countries, such as countries in the European Union, the USA, and Japan, for years [9–11]. Ecological riverbank development is a promising tool to maintain structure stability, safety, and riparian ecological restoration [12]. Recently, ecological riverbanks have played significant roles in improving surface water quality, aquatic ecosystem health, and biodiversity protection in the lower Yangtze River basin [13], and they have become the main development trend in waterway regulation projects on ecological construction [12,14,15]. In general, for different rivers and navigation conditions, ecological riverbank development must conform to engineering stability and facilitate the growth of plants, organisms' movement, surface and groundwater flow, riparian ecosystem protection, and ecological restoration effects [16,17].

The Yangtze River waterway has unique advantages and great development potential in navigation development, and it is known as the "golden waterway", which is the main framework for building a comprehensive three-dimensional transportation corridor in the Yangtze River Economic Belt. In recent years, new riverbank structures such as concrete grids, steel wire mesh gabions, and geogrids have been adopted to the waterway regulation scheme in Yangtze River [15,18]. During the construction of the 12.5 m deep waterway regulation scheme of Yangtze River below Nanjing, a variety of ecological riverbank forms were developed and applied in some river sections. In this study, two kinds of typical ecological riverbanks and a riprap riverbank in the engineering areas were selected and surveyed in 2018. The vegetation restoration and sediment deposition characteristics associated with ecological riverbanks were investigated in this study. The objectives of this study were as follows: (1) to assess vegetation restoration and sediment deposition characteristics in different ecological riverbanks; (2) to explore the ecological effects and influence factors associated with ecological riverbank restoration; (3) to provide practices for the structural design of ecological riverbanks in the waterway regulation scheme of Yangtze River.

2. Materials and Methods

2.1. Study Area

As the national key engineer of the 12th and 13th five-year plans in China, the 12.5 m deep waterway regulation scheme of Yangtze River below Nanjing started in June 2015 and was completed in June 2017. With the river reach length of 227 km, it includes the Fujiangsha waterway, Kouanzhi waterway, Hechangzhou waterway, and Yizheng waterway [19]. Affected by tidal flow at varying degrees, the Yangtze River is located at a fluvial plain in this section near to the estuary. In this study area of Yangtze River, the hydrodynamic and sediment movements are complex, which are affected by the current to varying degrees [20].

In this study, the traditional riprap riverbank (TRR) region (Figure 1a) and two typical ecological riverbanks, including a lattice gabion ballasted vegetation-mat riverbank (LGBVR) region (Figure 1b) and a mesh grid riverbank (MGR) region (Figure 1c), were selected as study areas to explore the vegetation restoration associated with the sediment deposition characteristic in the dry season in November of 2018.

2.2. Sampling Surveys

At the spatial scale for the MGR region, 3 survey transects were set perpendicular to the river flow direction (Figure 1c). On each transect, 3 sampling sites were set at equal distances along the transverse gradient of the riparian zone. There were 12 sampling sites (E1–E12) for the MGR region. As for the LGBVR region, 2 gabions with 2 sampling sites (L1 and L2) were randomly selected (Figure 1b). As for the TRR region, 3 survey transects were also set perpendicular to the river flow direction (Figure 1a). On each transect, 3 sampling sites were set at equal distances along the transverse gradient. There were 9 sampling sites (S1–S9) for the TRR region. In addition, two control sampling sites (CSs) (D1 and D2) were randomly selected and set outside the influence range of the riverbank engineering areas (Figure 1a,c).

According to the terrestrial vegetation community survey method [21], in each sampling site, the quadrat size was 1 m × 1 m and the plants' Latin names and numbers were recorded. At the same time, the aboveground parts of plant biomasses were obtained by a collection method, and the biomasses ($g \cdot m^{-2}$) were determined in the laboratory (by drying in a 75 °C oven for 48 h). At each sampling site, surface sediment samples (0–5 cm) were collected with an inner diameter of 5 cm of soil drilling, and 3 samples were taken from each sampling site and mixed into 1 sediment sample.

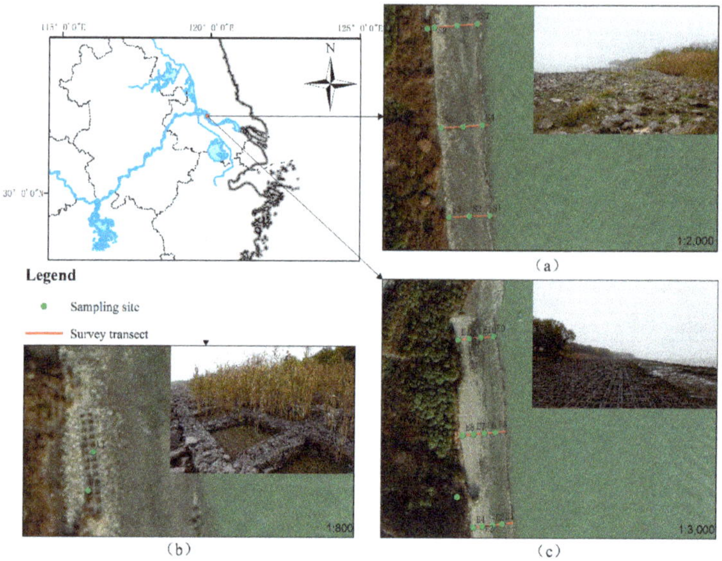

Figure 1. Sampling sites of ecological riverbanks in the deep waterway scheme of Yangtze River: (**a**) traditional riprap riverbank, TRR; (**b**) lattice gabion ballasted vegetation-mat riverbank, LGBVR; (**c**) mesh grid riverbank, MGR.

2.3. Analysis

In this study, the pH values of sediment were measured using an acidity meter (the water to sediment ratio was 2.5:1) (PHS-3C, Sanxin, China). The values of sediment electrical conductivity (EC) were measured using a conductivity meter (the water to sediment ratio was 5:1) (DDS-11A, Leici, China). The contents of ammonia nitrogen (NH_4^+-N), nitrate nitrogen (NO_3^--N), available phosphorus (AP), and available potassium (AK) in sediment were measured using flow analyzers (AA3, Bran+Luebbe, German). The contents of total nitrogen (TN) were determined via the addition of the ammonia nitrogen (NH_4^+-N) and nitrate nitrogen (NO_3^--N) contents. The contents of sediment microbial biomass carbon (MBC), microbial biomass nitrogen (MBN), and microbial biomass phosphorus (MBP) were measured by the chloroform fumigation–K_2SO_4 extraction method [22]. By using the wet sieving method [23], the sediment aggregate could be classified and calculated as large macroaggregates (>2.00 mm), small macroaggregates (2.00–0.25 mm), microaggregates (0.25–0.053 mm), and silt + clay fractions (<0.053 mm). The separated sediment aggregates were air-dried and the sediment total organic carbon (TOC) contents were measured. The sediment TOC contents were the mean values of the sediment aggregate TOC contents. The sediment organic carbon (TOC) contents were measured by the $K_2Cr_2O_7$-$FeSO_4$ oxidation method [24].

2.4. Biodiversity Index

The biodiversity index can reflect the species composition, structural stability, and complexity of a vegetation community [25]. These were calculated as follows:

Margalef index (d) [26]:

$$d = (S - 1)/lnN \tag{1}$$

Shannon–Weaver index (H') [27]:

$$H' = -\sum_{i=1}^{n} P_i ln P_i \tag{2}$$

Pielou index (J) [28]:

$$J = H'/ln(S) \quad (3)$$

where P_i is the proportion of the number of plant species (i) individuals to the total number of individuals in each sampling site, n is the total number of plant species in each sampling site, and N is the total number of all plant species individuals in each sampling site. S is the classification taxa number of plant species in each sampling site.

2.5. Data Analysis

SPSS Statistical 22.0 was performed for one-way ANOVA test and Pearson correlation analysis. To analyze the sediment characteristic differences on the ecological riverbanks, a one-way ANOVA test was employed, and the significance was tested using the LSD method ($\alpha = 0.05$). Pearson correlation analysis was employed to analyze the vegetation community and sediment characteristics of the ecological riverbanks.

3. Results

3.1. Vegetation Community Characteristics

Regarding different ecological riverbanks, Table 1 shows that the MGR region had eight plant species, the Pielou index (J) was at its highest, and the dominant species were *Kalimeris indica* and *Cardamine hirsute*. The LGBVR region had six plant species, it had the highest Margalef index (d) and Shannon-Weaver index (H'), and the dominant species were *Phragmites australis* and *Cardamine hirsute*. The TRR region had six plant species and the dominant species was *Veronica anagallis-aquatica*. The LGBVR region had the highest biomass due to large and tall phragmite communities, followed by the MGR region, and the TRR region had lowest biomass. In addition, the CSs region of the natural riparian riverbank had 12 plant species and a relatively higher biomass. The one-way ANOVA test also indicated that the biomass of the LGBVR region was significantly different compared to the MGR and TRR regions. The biomass was not significantly different between the LGBVR and CSs regions. For biodiversity indices, the Pielou index (J) of the CSs region was significantly different compared to that of the MGR, LGBVR, and TRR region.

Table 1. Vegetation community characteristics in different ecological riverbanks. Mesh grid riverbank (MGR); lattice gabion ballasted vegetation-mat riverbank (LGBVR); traditional riprap riverbank (TRR); control sampling sites (CSs). Lowercase letters in same rows indicate a significant difference at the 0.05 level by one-way ANOVA test; the same applies below.

Types of Riverbank	MGR (Mean ± SE)	LGBVR (Mean ± SE)	TRR (Mean ± SE)	CSs (Mean ± SE)
Numbers of sampling sites	12	2	9	2
Numbers of species	8	6	6	12
Margalef index (d)	0.37 ± 0.14 a	0.65 ± 0.00 a	0.47 ± 0.04 a	0.45 ± 0.01 a
Shannon–Weaver index (H')	0.66 ± 0.24 a	1.08 ± 0.17a	0.92 ± 0.07 a	0.48 ± 0.01 a
Pielou index (J)	0.82 ± 0.08 a	0.78 ± 0.12 a	0.80 ± 0.07 a	0.44 ± 0.01 b
Biomass ($g \cdot m^{-2}$)	74.53 ± 27.08 b	1225.01 ± 200.33 a	11.47 ± 4.79 b	368.34 ± 25.48 ab
Latin names of dominant species	*Kalimeris indica, Cardamine hirsute*	*Phragmites australis, Cardamine hirsute*	*Veronica anagallis-aquatica*	*Phragmites australis, Achyranthes aspera, Polygonum hydropiper*, etc.

3.2. Sediment Aggregate Compositions

Table 2 shows that the LGBVR region had relatively higher values of pH and EC in sediments. The sediment aggregate compositions of large macroaggregates (>2.00 mm) and small macroaggregates (2.00–0.25 mm) had relatively higher values in the CSs region (33.22%), followed by the LGBVR region (32.23%). Otherwise, they had relatively lower values in the MGR region (31.41%) and TRR region (30.62%). The one-way ANOVA test also indicated that the EC values of sediment were significantly different between the LGBVR and CSs regions.

Table 2. Sediment aggregate compositions in different ecological riverbanks. Lower case letters in same rows are same to Table 1.

Types of Riverbank	MGR (Mean ± SE)	LGBVR (Mean ± SE)	TRR (Mean ± SE)	CSs (Mean ± SE)
Numbers of sampling sites	12	2	9	2
pH	7.52 ± 0.08 a	7.64 ± 0.11 a	7.53 ± 0.15 a	7.29 ± 0.25 a
EC(μs·cm^{-1})	109.1 ± 8.6 ab	151.7 ± 19.9 a	111.4 ± 8.2 ab	91.4 ± 1.0 b
>2.00 mm (%)	5.4 ± 0.3 a	4.8 ± 0.1 a	4.9 ± 0.3 a	5.5 ± 1.4 a
2.00–0.25 mm (%)	26.1 ± 0.5 a	27.4 ± 0.2 a	25.7 ± 0.4 a	27.8 ± 2.8 a
0.25–0.053 mm (%)	45.9 ± 0.9 a	45.7 ± 0.5 a	46.7 ± 0.8 a	45.7 ± 2.6 a
<0.053 mm (%)	22.7 ± 0.4 a	22.1 ± 0.8 a	22.7 ± 0.7 a	21.1 ± 1.2 a

3.3. Sediment Nutrient Contents

Table 3 shows that the LGBVR region had relatively higher values of NH_4^+-N, TN, and AP content in sediments. The TRR region had relatively higher values of AK content. As for the LGBVR region, the one-way ANOVA test indicated that sediment TN contents were significantly different compared to those of the MGR (p = 0.032) and TRR (p = 0.042) regions. Its sediment NH_4^+-N contents were significantly different compared to those of the MGR (p = 0.042) and TRR (p = 0.043) regions. Furthermore, the sediment NO_3^--N, AP, and AK contents displayed no difference among the riverbanks and the CSs region.

Table 3. Sediment nutrient contents in different ecological riverbanks. Lower case letters in same rows are same to Table 1.

Types of Riverbank	MGR (Mean ± SE)	LGBVR (Mean ± SE)	TRR (Mean ± SE)	CSs (Mean ± SE)
Numbers of sampling sites	12	2	9	2
NH_4^+-N (mg·kg^{-1})	1.97 ± 0.12 b	2.71 ± 0.52 a	1.95 ± 0.15 b	2.04 ± 0.06 ab
NO_3-N(mg·kg^{-1})	2.35 ± 0.18 a	3.04 ± 0.08 a	2.42 ± 0.30 a	3.29 ± 1.06 a
TN (mg·kg^{-1})	4.32 ± 0.19 b	5.75 ± 0.43 a	4.38 ± 0.30 b	5.33 ± 1.12 ab
AP (mg·kg^{-1})	18.13 ± 1.65 a	21.97 ± 3.41 a	18.93 ± 1.54 a	21.68 ± 2.39 a
AK (mg·kg^{-1})	315.37 ± 15.15 a	339.55 ± 41.33 a	341.74 ± 21.25 a	315.88 ± 40.68 a

3.4. Sediment TOC Contents

Figure 2 shows that the MGR region had relatively higher mean values of TOC content in sediments. The LGBVR region had relatively lower mean values of TOC content. The one-way ANOVA test indicated that the sediment mean TOC content in the MGR region was significantly different compared to that in the TRR region (p = 0.026). Moreover, the sediment aggregate TOC contents were not significantly different among ecological riverbanks and the CSs region.

3.5. Sediment MBC, MBN, and MBP Contents

Table 4 shows that the values of MBC content, MBC/TOC, and MBC/MBN were highest in the LGBVR-region sediment. The sediment in the CSs region had highest MBN and MBP contents. The sediments in the TRR region had relatively higher MBN contents. The sediments in the LGBVR region had relatively higher MBP contents. As for the LGBVR region, one-way ANOVA demonstrated that the sediment MBC content was significantly different compared to that in the MGR region (p = 0.009). The sediment value of MBC/TOC in the LGBVR region was significantly different compared to that in the MGR region (p = 0.001) and TRR region (p = 0.001). The sediment value of MBC/MBN in the LGBVR region was significantly different compared to that in the MGR region (p = 0.000) and TRR region (p = 0.000).

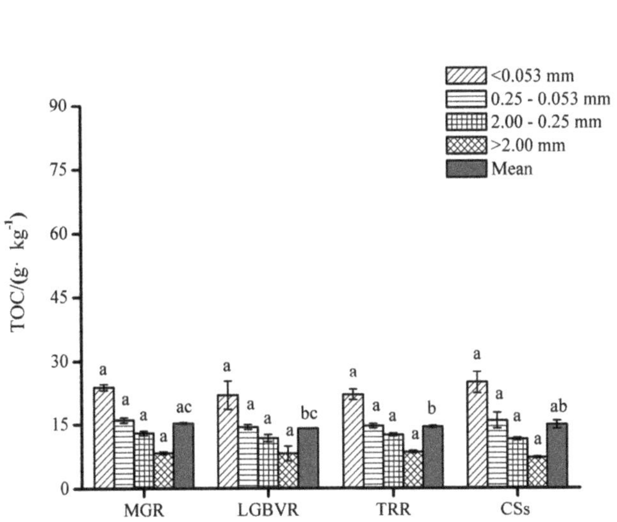

Figure 2. Sediment mean TOC content and its aggregate compositions TOC contents in different ecological riverbanks.

Table 4. Sediment MBC, MBN, and MBP contents in different ecological riverbanks. Lower case letters in same rows are same to Table 1.

Types of Riverbank	MGR (Mean ± SE)	LGBVR (Mean ± SE)	TRR (Mean ± SE)	CSs (Mean ± SE)
Numbers of sampling sites	12	2	9	2
MBC (mg·kg^{-1})	596.23 ± 27.56 b	803.67 ± 47.67 ac	671.16 ± 28.56 ab	704.90 ± 107.40 ab
MBN (mg·kg^{-1})	33.12 ± 1.53 a	32.88 ± 3.90 a	37.29 ± 1.59 a	39.16 ± 5.97 a
MBP (mg·kg^{-1})	6.58 ± 0.07 a	6.60 ± 0.20 a	6.52 ± 0.07 a	6.65 ± 0.15 a
MBC/TOC	0.010 ± 0.000 c	0.014 ± 0.001 a	0.012 ± 0.001 b	0.012 ± 0.001 abc
MBC/MBN	18.000 ± 0.000 b	24.970 ± 4.415 a	18.000 ± 0.000 b	18.000 ± 0.000 b

3.6. Correlation Analysis

Table 5 shows the Pearson correlation coefficients among the vegetation and sediment characteristics in different ecological riverbanks. The sediment contents of NO_3^--N were significantly correlated with TN ($r = 0.841$, $p < 0.01$). The sediment contents of NH_4^+-N were significantly correlated with MBC/MBN ($r = 0.674$, $p < 0.01$). The sediment contents of MBC were significantly correlated with MBN ($r = 0.966$, $p < 0.01$) and MBC/TOC ($r = 0.938$, $p < 0.01$). The sediment contents of MBN were significantly correlated with MBC/TOC ($r = 0.882$, $p < 0.01$). The biomasses were significantly correlated with MBN ($r = 0.536$, $p < 0.05$). The sediment EC values were significantly correlated with AK ($r = 0.643$, $p < 0.05$).

Table 5. Pearson correlation coefficients between vegetation and sediment characteristics in ecological riverbanks. ** means significant correlation at 0.01 level (bilateral); * means significant correlation at 0.05 level (bilateral).

	N	Biomass	pH	EC	NH_4^+-N	NO_3^--N	TN	AP	AK	TOC	MBC	MBN	MBP	MBC/TOC	MBC/MBN
N	1.000														
Biomass	0.251	1.000													
pH	0.406	−0.291	1.000												
EC	−0.481	−0.308	−0.225	1.000											
NH_4^+-N	0.067	−0.124	−0.064	−0.237	1.000										
NO_3^--N	0.219	−0.145	0.231	−0.028	−0.038	1.000									
TN	0.225	−0.192	0.165	−0.152	0.509	0.841 **	1.000								
AP	−0.143	−0.034	−0.022	−0.078	−0.075	−0.020	−0.058	1.000							
AK	−0.491	−0.383	−0.187	0.643 *	−0.273	0.047	−0.108	−0.185	1.000						
TOC	0.001	0.340	−0.145	−0.310	0.054	0.007	0.035	−0.276	−0.487	1.000					
MBC	0.401	0.480	−0.180	0.061	0.138	0.016	0.088	0.112	−0.356	−0.159	1.000				
MBN	0.360	0.536 *	−0.245	0.027	−0.039	−0.017	−0.036	0.142	−0.323	−0.100	0.966 **	1.000			
MBP	0.112	−0.203	0.372	−0.185	0.267	0.065	0.201	0.105	0.029	−0.253	−0.403	−0.509	1.000		
MBC/TOC	0.367	0.289	−0.113	0.161	0.125	0.010	0.077	0.160	−0.143	−0.487	0.938 **	0.882 **	−0.270	1.000	
MBC/MBN	0.234	−0.104	0.201	0.137	0.674 **	0.126	0.474	−0.089	−0.194	−0.251	0.333	0.079	0.301	0.402	1.000

4. Discussion

4.1. Sediment Deposition Characteristics in Ecological Riverbanks

Worldwide, rivers have been tremendously affected and modified by artificial riverbanks for navigation channels [29,30]. The accumulation of sediment deposition and organic matter in riparian zones has improved the soil conditions that are the basis for vegetation restoration and biodiversity maintenance [31]. Traditionally, in Yangtze River waterway regulation schemes, riverbank constructions were mostly made of concrete, cemented rock, riprap, and wire gabions, which are not suitable for sedimental deposition and vegetation restoration [16,32,33]. Ecological riverbanks can prevent and remove large amounts of nutrients from upland areas [30]. In recent years, ecological riverbanks have been widely used to restore the riparian ecosystem, and various forms have been developed in China [18,33]. This study demonstrated that the LGBVR region had the best effects of sedimental deposition, nutrient accumulation, and biological fixed carbon. It also had relatively higher values of NH_4^+-N, TN, AP, MBC contents, MBC/TOC, and MBC/MBN in sediments. MBC was regarded as bioactive carbon components in sediment and is sensitive to the ecological restoration effect [25]. The structure of the LGBVR was developed to facilitate sediment deposition and nutrient accumulation and does not affect the growth of plant roots. According to the survey, it had 10–15 cm of siltation in the gabion after the completion of 1 year, which made it especially suitable for the growth of perennial phragmites. In addition, the bottom layers of the MGR are geotextiles and a bamboo woven mat, and sediments are deposited among three-dimensional network components. Although it has certain advantages in promoting sedimental deposition, its structure is more suitable for the growth of annual shallow root herbaceous plants. Furthermore, the MGR region had relatively higher values of TOC content in sediments. For the TRR region, the sediments were concentrated between the stone crevices, and the sediment deposition and plant growth conditions were poor.

4.2. Vegetation Restoration and Sediment Deposition Characteristics in Ecological Riverbanks

The accumulation of sediment and organic matter improves soil conditions and was the material basis for vegetation restoration and biodiversity maintenance in riparian zones [31]. In riparian zones, ecological riverbanks can benefit sediment deposition, decrease the water flow velocity, and facilitate plant growth [34,35]. The unique structure of ecological riverbanks can promote plant colonization, growth, and sedimental deposition, but the vegetation restoration and sediment deposition effects are obviously different in different forms of ecological riverbanks. In the same climate zone, an ecological riverbank structure such as an eco-bag, an eco concrete tetrad-ball in Hang-jia-hu plain [13], or the planting of a concrete revetment in an island of the estuary of Yangtze River [8] can have a good vegetation restoration effect. The plant species composition and dominant species are important indicators for vegetation succession [36]. In this study, the LGBVR region had the highest biomass, Margalef index (d), and Shannon–Weaver index (H'), and its dominant species was *Phragmites australis.* Although the MGR has certain advantages in resisting water flow impact and promoting sedimental deposition, it is not suitable for the colonization and growth of perennial plants in the short term. In the frequently submerged area near the riverside, this study also revealed that the sediment deposition and vegetation restoration effect of the riprap riverbank were better because it was closer to the natural state. In the French and Swiss Alps region, riprap riverbanks had relatively poor vegetation restoration effects [37]. However, in sub-tropical areas, they had relatively good ecological restoration effects compared to a gabion revetment in Cuatien River, Vietnam [17]. For periodically submerged riverbank areas, the riprap riverbank's sediment deposition and vegetation restoration effects were comparatively better than those achieved using the MGR and LGBVR. Therefore, along the transverse gradient of riparian zones, ecological riverbank development should adopt the mixed use of different structure forms in the deep waterway regulation scheme of Yangtze River [38,39]. In the lower reach of the Yangtze River, the deep waterway regulation scheme is affected by both runoff and tidal currents, and the river level variation and riverbed evolution are complex [15]. In the riparian zone survey, the natural habitat conditions varied with different plant communities, which was related to the water flow magnitude and sedimental deposition characteristics [40,41]. Due to the high structural stability and safety requirements, the ecological riverbanks were only studied in parts of river sections in this project. Furthermore, the design, construction, and maintenance of ecological riverbanks needs to be further evaluated.

5. Conclusions

In this study, the sediment deposition characteristics associated with the vegetation restoration of two kinds of typical ecological riverbanks, LGBVR and MGR, and TRR were investigated in the deep waterway regulation scheme of Yangtze River. For the LGBVR region, it had the highest biomass due to large and tall phragmite communities. It also had the highest Margalef index (d) and Shannon–Weaver index (H') values, as well as the highest MBC content, MBC/TOC, and MBC/MBN in sediments. For the MGR region, it had the highest number of plant species (eight) and Pielou index (J) and a relatively higher biomass in sampling sites. It also had highest values of TOC contents in sediment. The TRR region had the lowest biomass. In general, the LGBVR region had the best vegetation restoration and sediment deposition effects, followed by the MGR region. The TRR region had a relatively poor ecological restoration effect. Furthermore, the sediment deposition and vegetation restoration effects in different forms of ecological riverbanks are still in the process of dynamic succession, and these ecological effects need to be studied for a long time.

Author Contributions: Conceptualization, J.L.; methodology, J.L., X.W. and L.W.; software, J.L.; validation, J.L.; formal analysis, J.L.; investigation, J.L., X.W. and L.W.; resources, J.L.; data curation, J.L., X.W. and L.W.; writing—original draft preparation, J.L.; writing—review and editing, J.L.; visualization, J.L.; supervision, J.L.; project administration, J.L.; funding acquisition, J.L. All authors have read and agreed to the published version of the manuscript.

Funding: This research was funded by Key science and technology projects in the transportation sector, grant number 2018-ZD3-024, and the Central public-interest scientific institution basal research fund of China, grant number WTI-62203.

Data Availability Statement: The data presented in this study are available on request from the corresponding author. The data are not publicly available due to the project was not yet completed.

Conflicts of Interest: The authors declare no conflicts of interest.

References

1. Nilsson, C.; Berggren, K. Alteration of riparian ecosystems caused by river regulation. *BioScience* **2000**, *50*, 783–792. [CrossRef]
2. Wang, Z.Y.; Pei, Y.S. Ecological risk resulting from invasive species: A lesson from riparian wetland rehabilitation. *Procedia Environ. Sci.* **2012**, *13*, 1798–1808. [CrossRef]
3. Luo, Z.L.; Zuo, Q.T.; Shao, Q.X. A new framework for assessing river ecosystem health with consideration of human service demand. *Sci. Total Environ.* **2018**, *640–641*, 442–453. [CrossRef] [PubMed]
4. Paetzold, A.; Yoshimura, C.; Tockner, K. Riparian arthropod responses to flow regulation and river channelization. *J. Appl. Ecol.* **2008**, *45*, 894–903. [CrossRef]
5. Hartig, J.H.; Zarull, M.A.; Cook, A. Soft shoreline engineering survey of ecological effectiveness. *Ecol. Eng.* **2011**, *37*, 1231–1238. [CrossRef]
6. Addy, S.; Wilkinson, M.E. Geomorphic and retention responses following the restoration of a sand-gravel bed stream. *Ecol. Eng.* **2019**, *130*, 131–146. [CrossRef]
7. Schwindt, S.; Pasternack, G.B.; Bratovich, P.M.; Rabone, G.; Simodynes, D. Hydro-morphological parameters generate lifespan maps for stream restoration management. *J. Environ. Manag.* **2019**, *232*, 475–489. [CrossRef]
8. He, Y.; Wang, P.C.; Sheng, H.; Wang, D.F.; Huang, M.S.; Cao, C.J. Sustainability of riparian zones for non-point source pollution control in Chongming Island: Status, challenges, and perspectives. *J. Clean. Prod.* **2020**, *244*, 118804. [CrossRef]
9. Die Moran, A.; El Kadi Abderrezzak, K.; Mosselman, E.; Habersack, H.; Lebert, F.; Aelbrecht, D.; Laperrousaz, E. Physical model experiments for sediment supply to the old Rhine through induced bank erosion. *Int. J. Sediment Res.* **2013**, *28*, 431–447. [CrossRef]

10. Taylor, D.L.; Bolgrien, D.W.; Angradi, T.R.; Pearson, M.S.; Hill, B.H. Habitat and hydrology condition indices for the upper Mississippi, Missouri, and Ohio rivers. *Ecol. Indic.* **2013**, *29*, 111–124. [CrossRef]
11. Chen, Q.F.; Guo, B.B.; Zhao, C.S.; Zhang, J. A comprehensive ecological management approach for northern mountain rivers in China. *Chemosphere* **2019**, *234*, 25–33. [CrossRef] [PubMed]
12. Liang, K.M.; Zhang, J.E.; Zhao, B.L.; Ye, Y.Q.; Kong, X.H. Review on the application and research of ecological riverbanks for water systems. *Trop. Geogr.* **2014**, *34*, 116–122. (In Chinese)
13. Cai, W.W.; Zhou, Z.Y.; Xia, J.H.; Wang, W.M.; Dou, C.B.; Zeng, Z. An advanced index of ecological Integrity (IEI) for assessing ecological efficiency of restauration revetments in river plain. *Ecol. Indic.* **2020**, *108*, 105762. [CrossRef]
14. Chen, Y.M.; Xu, S.D.; Jin, Y. Evaluation on ecological restoration capability of revetment in inland restricted channel. *KSCE J. Civ. Eng.* **2016**, *20*, 2548–2558. [CrossRef]
15. Liu, H.H.; Yang, S.F.; Cao, M.X. Advances in 'golden waterway' regulation technologies of Yangtze River. *Adv. Eng. Sci.* **2017**, *49*, 17–27. (In Chinese) [CrossRef]
16. Schmitt, K.; Schäffer, M.; Koop, J.; Symmank, L. River bank stabilisation by bioengineering: Potentials for ecological diversity. *J. Appl. Water Eng. Res.* **2018**, *6*, 262–273. [CrossRef]
17. Tang, V.T.; Fu, D.F.; Binh, T.N.; Rene, E.R.; Sang, T.T.T.; Singh, R.P. An investigation on performance and structure of ecological revetment in a sub-tropical area: A case study on Cuatien River, Vinh City, Vietnam. *Water* **2018**, *10*, 636. [CrossRef]
18. Xu, C.H.; Bu, H.B.; Cheng, W.H.; Chen, Z.Q. Development and application of ecological protection bank for inland waterway. *Port Waterw. Eng.* **2009**, *432*, 107–110. (In Chinese)
19. Cao, M.X.; Shen, X.; Ying, H.H. Study on ecological structures of waterway regulation in the Yangtze River below Nanjing. *Port Waterw. Eng.* **2018**, *538*, 1–11. (In Chinese)
20. Chen, X.Y. Regulation method and construction thoughts of deep-water channel of the Yangtze River below Nanjing city. *Port Waterw. Eng.* **2011**, *461*, 99–105. (In Chinese)
21. Dong, M.; Wang, Y.F.; Kong, F.Z. *Survey and Analysis Methods for Terrestrial Biocommunity*; China Standard Press: Beijing, China, 1997. (In Chinese)
22. Vance, E.D.; Brookes, P.C.J. An extraction method for measuring soil microbial biomass carbon. *Soil Biol. Biochem.* **1987**, *19*, 703–704. [CrossRef]
23. Cambardella, C.A.; Elliott, E.T. Carbon and nitrogen distribution in aggregates from cultivated and native grassland soils. *Soil Sci. Soc. Am. J.* **1993**, *57*, 1071–1076. [CrossRef]
24. Walkley, A.J.; Black, I.A. An examination of the Degtjareff method for determining soil organic matter, and a proposed modification of the chromic acid titration method. *Soil Sci.* **1934**, *37*, 29–38. [CrossRef]
25. Wang, X.X.; Zhang, L.; Liang, L.N.; Song, N.N.; Liu, D.S.; Wang, J.C. Effects of straw returning on the stability of soil organic carbon in wheat-maize rotation systems. *J. Agro-Environ. Sci.* **2020**, *39*, 1774–1782. (In Chinese)
26. Margalef, D.R. Information theory in ecology. *Soc. Gen. Syst. Res.* **1957**, *3*, 36–71.
27. Shannon, C.E.; Weaver, W. *The Mathematical Theory of Communication*; University of Illinois Press: Urbana, IL, USA, 1949.
28. Pielou, E.C. *Ecological Diversity*; John Wiley & Sons Inc: New York, NY, USA, 1975.
29. Nilsson, C.; Reidy, C.A.; Dynesius, M.; Revenga, C. Fragment and flow regulation of the world's large river system. *Science* **2005**, *308*, 405–408. [CrossRef] [PubMed]
30. Symmank, L.; Natho, S.; Scholz, M.; Schröder, U.; Raupach, K.; Schulz-Zunkel, C. The impact of bioengineering techniques for riverbank protection on ecosystem services of riparian zones. *Ecol. Eng.* **2020**, *158*, 106040. [CrossRef]
31. Zhang, J.Y.; Zhao, H.L.; Zhang, T.H.; Zhao, X.Y. Dynamics of species diversity of communities in restoration processes in Horqin sandy land. *Acta Phytoecol. Sin.* **2004**, *28*, 86–92. (In Chinese)
32. Wang, Y.; Ding, Y.R.; Fan, B.L. Study on river regime evolution and ecological revetment for Jingjiang reach after impoundment of TGP. *Resour. Environ. Yangtze Basin* **2011**, *20* (Suppl. S1), 117–122. (In Chinese)
33. Zhan, L.T.; Chen, Y.M.; Bouazza, A. *Combination of Porous Ecological Concrete and Geocell in Riverbank Protection*; Springer Singapore Pte. Limited: Singapore, 2018; pp. 280–287.
34. Nakamura, F.; Seo, J.I.; Akasaka, T.; Swanson, F.J. Large wood, sediment, and flow regimes: Their interactions and temporal changes caused by human impacts in Japan. *Geomorphology* **2017**, *279*, 176–187. [CrossRef]
35. Kiss, T.; Nagy, J.; Fehérváry, I.; Vaszkó, C. (Mis) management of floodplain vegetation: The effect of invasive species on vegetation roughness and flood levels. *Sci. Total Environ.* **2019**, *686*, 931–945. [CrossRef] [PubMed]
36. Liu, Y.Y.; Gu, B.; Wang, L. Species diversity of plant community and the niche of dominant species in Beichuan after earthquake engineering. *Chin. J. Ecol.* **2019**, *38*, 309–320. (In Chinese)
37. Cavaillé, P.; Ducasse, L.; Breton, V.; Dommanget, F.; Tabacchi, E.; Evette, A. Functional and taxonomic plant diversity for riverbank protection works: Bioengineering techniques close to natural banks and beyond hard engineering. *J. Environ. Manag.* **2015**, *151*, 65–75. [CrossRef] [PubMed]
38. Li, J.P.; Xuan, H.; Wang, X.X.; Li, T. Comprehensive evaluation on implement effect of ecological riverbank project in deep waterway regulation of Yangtze River. *Yangtze River* **2022**, *53*, 211–217. (In Chinese)
39. Li, J.P.; Wu, L.; Liu, S.; Xuan, H.; Wang, X.X. Terrestrial vegetation restoration of ecological riverbank in the deep waterway regulation scheme of Yangtze River. *J. Yangtze River Sci. Res. Inst.* **2021**, *38*, 31–37. (In Chinese)

40. Wollny, J.T.; Otte, A.; Harvolk-Schöning, S. Riparian plant species composition alternates between species from standing and flowing water bodies—Results of field studies upstream and downstream of weirs along the German rivers Lahn and Fulda. *Ecol. Eng.* **2019**, *139*, 105576. [CrossRef]
41. Wang, L.; Ge, M.T.; Chen, N.G.; Ding, J.H.; Shen, X.W. An evaluation model of riparian landscape: A case in rural Qingxi Area, Shanghai. *Land* **2022**, *11*, 1512. [CrossRef]

water

MDPI

Article

The Contradictory Issue of the Impact of Antecedent Soil Moisture to Interrill Erosion in Clay Soil: A Two-Year Field Study

Yu-Da Chen [1,*] and Chia-Chun Wu [2]

[1] Department of Civil Engineering, National Pingtung University of Science and Technology, Pingtung 912, Taiwan
[2] Department of Soil and Water Conservation, National Pingtung University of Science and Technology, Pingtung 912, Taiwan; npustccwu@gmail.com
* Correspondence: a7290963@gmail.com; Tel.: +886-987924313

Abstract: The impact of antecedent soil moisture content on soil erosion has been a contradictory issue in erosion research, as well as process-based soil loss estimation models. The objective of this study was to investigate the impact of antecedent soil moisture content on the loss of clay soil through two-year runoff plot experiments under natural rainfall. Volumetric soil moisture sensors were used to monitor soil moisture changes, and readings were used along with rainfall records to quantify the antecedent soil moisture conditions. The results of this study show that the impact of antecedent soil moisture on interrill erosion is conditional, and the impact only exists in erosion events with a low Rainfall–Runoff Erosivity Index. The coefficient of determination between antecedent soil moisture content and soil loss per the Rainfall–Runoff Erosivity Index (Soil Loss/EI_{30}) varies from 0.222 to 0.758, depending on the rainfall duration and Rainfall–Runoff Erosivity. The results of this study also suggest that accumulative rainfall within 48 h (Pp_{48}) prior to an effective erosion event is strongly correlated with Soil Loss/EI_{30}, particularly when the duration of an effective erosion event is either 3~7 h or 10~30 h. Hence, Pp_{48} can be considered as a replacement for antecedent soil moisture in process-based soil loss simulation models.

Keywords: antecedent soil moisture content; rainfall characteristics; interrill erosion; soil loss

1. Introduction

Soil erosion is one of the natural processes that shape the landscape through the action of water and wind. It causes environmental impacts such as (1) non-point source pollution resulting from eroded soil carrying pollutants from hillslopes, (2) a reduction in soil fertility that affects crop yield, and (3) sedimentation in rivers and water bodies [1–5]. Hence, the quantification of soil loss has received much research attention, and the techniques for soil loss estimation have changed from parametrized estimations such as USLE [6,7] and RUSLE [8] to processes-based simulations, for instance, CREAMS [9], AgNPS [10], WEPP [11,12], and SWAT [13]. Regardless of the simplicity or complexity of models, soil loss estimation is still based on the existing understanding of soil erosion processes.

The physical and chemical properties of soil are determined during its formation through weathering, transport, deposition, and consolidation, among other geological processes [14]. However, soil improvement and human interventions, namely, land grading, cover and crop management, and soil conservation, can alter soil erodibility [5,15]. Putting this into Universal Soil Loss Equation (USLE) terminology, these interventions are parameterized as the slope length factor (L), slope steepness factor (S), cover and management factor (C), and conservation factor (P), respectively, among which, the C factor is probably the most complicated parameter.

The C factor has gone through development and revision for almost half a century. It has advanced from the original concept of USLE, namely, C = Cc × Cs, which considers the contributions from the vegetation canopy and surface mulch, to RSUEL Version 2 (USDA-ARS, 2013), namely, C = Cc × gc × Sr × rh × Sb × Sc × Pp × am, which considers the effects of the canopy and ground cover, random and artificial surface roughness, soil biomass, soil consolidation, ponding, and antecedent soil moisture content.

The soil moisture content changes soil physical properties such as shear strength and cohesion, thereby altering soil resistance to erosion [16,17]. A positive correlation has been found between soil moisture content and surface runoff [18–21]. Fortesa et al. (2020) [22] analyzed 5-year soil moisture and flow data at the Es Fangar catchment in Mallorca, Spain (Western Mediterranean Sea), and they found that 76% of flood events occurred under wet soil conditions. Massari et al. (2023) [23] analyzed data from the Global Runoff Data Center (GRDC) and the European Space Agency's Climate Change Initiative (CCI), and they found that antecedent precipitation, soil moisture at the surface and in the root zone, pre-storm river discharge, and pre-storm total water storage all affect the scale of flood events significantly.

Jadidoleslam et al. (2019) [24] applied Soil Moisture Active/Passive (SMAP) to estimate the surface soil layer's water contents for 38 watersheds (area ranged from 80 to 1000 km^2) and correlated soil water content with runoff coefficients. They found that the runoff coefficients increased with an increase in the antecedent soil moisture content. In addition, the runoff coefficients changed significantly under humid conditions. Schoener et al. (2019) [25] observed surface runoff from watersheds at different scales (2.8 m^2 and 2.8 km^2), and the results also showed a positive correlation between the antecedent soil moisture content and runoff coefficient, regardless of the scale of the watershed. They also noted that antecedent soil moisture content affected surface runoff during small rainfall events but had less of an impact during heavy rainfall events.

While a positive correlation between antecedent soil moisture and soil erosion or runoff has been reported [26,27], there are studies that have suggested otherwise. For instance, Castillo et al. (2003) [28] conducted a field study using time-domain reflectometry (TDR) in three small semi-arid watersheds in eastern Spain with different vegetation characteristics. Numerical simulations were also carried out to study the correlation between antecedent soil moisture and surface runoff. They found that the impact of antecedent soil moisture on surface runoff depended on rainfall intensity and soil permeability. They concluded that surface runoff was not dependent on antecedent soil moisture content in the case of high-intensity rainfall or low-permeability soil; however, if low-intensity rainfall occurred on higher-permeability soil, surface runoff was controlled by the initial soil moisture content.

Truman and Bradford (1990) [29] conducted a series of experiments using a rainfall simulator to test the splash erosion rates on air-dried and prewetted soils, and the results indicated that a rise in soil moisture reduced the splash erosion rate, which might have been related to the promotion of soil sealing and aggregation stability [18,30]. Auerswald (1993) [31] indicated that increasing soil moisture from 10% to 30% reduced soil loss by 80%. Ma et al. (2014) [32] found a transition in the splash erosion rate during a rise in soil moisture content. The splash erosion rate initially decreased and then increased at a volumetric moisture threshold content of 15%.

Because of the contradictions in research results on the impact of antecedent soil moisture content on runoff and soil erosion, RUSLE2 includes the antecedent soil moisture content subfactor (a_m) in its C factor calculation as an option to reflect the impact of antecedent soil moisture content on erosion for dryland cropping in the Northwestern Wheat and Range Region (NWRR) and similar areas, where the amount of soil moisture from a prior crop affects erosion during the present crop. RUSLE2 also suggests ignoring the moisture depletion rate when implementing the model outside of the NWRR in the US.

If antecedent soil moisture content led to different opinions during the development of the RUSLE2 model and possibly other models, similar concerns will equally be raised at the users' end. Therefore, the objectives of this study are (1) to investigate the correlation and possible explanations for the contradictions between antecedent soil moisture and interrill erosion and (2) to explore the feasibility of using rainfall parameters as alternatives to replace the impact of antecedent soil moisture to interrill erosion if impact from antecedent soil moisture content deserves consideration in soil loss estimation and modeling.

2. Materials and Methods

2.1. Experiment Site and Preparation

The experiment site is in a tropical climate zone at 22°38′41″ N and 120°36′23″ E (Soil Erosion Experiment Site at the National Pingtung University of Science and Technology). The average annual rainfall is about 2500~3000 mm. The rainy season usually begins in early May and lasts until late September. The soil at the experiment site was first plowed and disked to a depth of about 35 cm at least three times earlier in the year. Residue, including roots and ground vegetation, was manually removed. The soil was also mixed along the slope, as well as between the plots, to achieve soil consistency. Due to plowing and disking, a plow sole was formed at a depth of about 35~40 cm. Soil samples were taken randomly along the slope and within the plow layer prior to plot preparation for analyses of soil physical properties. The soil at the experiment site consists of 18.84% sand, 31.73% silt, and 49.71% clay, and it is classified as clay according to the United States Department of Agriculture (USDA) soil texture classification. The average organic matter content is 1.65%.

A total of three runoff plots were prepared, and all runoff plots remained bare throughout the rainy season. Based on past experiences, rills seldom occur under natural rainfall conditions when the slope length is shorter than 10 m; therefore, each runoff plot was set up with a length of 10 m and a width of 3 m, and the plot surface was leveled to the desired gradient of 9%. Cemented brick walls were installed along the periphery of the runoff plot to a depth of 1 m to prevent additional water sources from entering the plot, as well as lateral infiltration between adjacent plots.

The field study started in early 2019 and ended in late 2020, covering two consecutive rainy seasons. Due to plot availability, only one runoff plot was used in 2020; hence, the three runoff plots that were used in 2019 for this study were considered three replications. Disking was performed at the end of the rainy season to break any surface crust, and soil was randomly exchanged and mixed between plots prior to leveling for the incoming year's experiment.

Before this study, we tested the installation of EC-5 volumetric soil moisture sensors (Onset, Bourne, MA, USA) at different depths (5 cm and 10 cm) and at various distances (2.5 m, 5 m, and 7.5 m) from the top of the runoff plot using the same field plot setup to understand how the locations of the soil moisture sensors affected the sensors' readings. After two years of testing, we found that the soil moisture data from different locations along the slope were almost synchronous. Therefore, an EC-5 volumetric soil moisture sensor was installed 10 cm below the ground surface at the mid-length of each runoff plot to monitor soil moisture content. The volumetric soil moisture content was automatically logged (Onset, Bourne, MA, USA) every 5 min throughout the entire study period. A schematic drawing of the setup of the runoff plots, rain gauge, and locations of the soil moisture sensors and collection tanks is illustrated in Figure 1.

2.2. Data Collection and Analysis

2.2.1. Rainfall

A 0.5 mm tipping-bucket rain gauge (Takeda Instruments Co., Ltd., Tokyo, Japan) attached to a data logger (HOBO Pendant Event Data Logger, Onset, Bourne, MA, USA) was used to log the rainfall records, and the records were used to extract the rainfall characteristics, including total precipitation (P, mm), maximum 30 min rain intensity (I_{30}max, mm/h), event duration (duration, h), the time interval between two adjacent effective erosion events (TILR, h), and total precipitation within the past x hours prior to an effective erosion event (Ppx, mm).

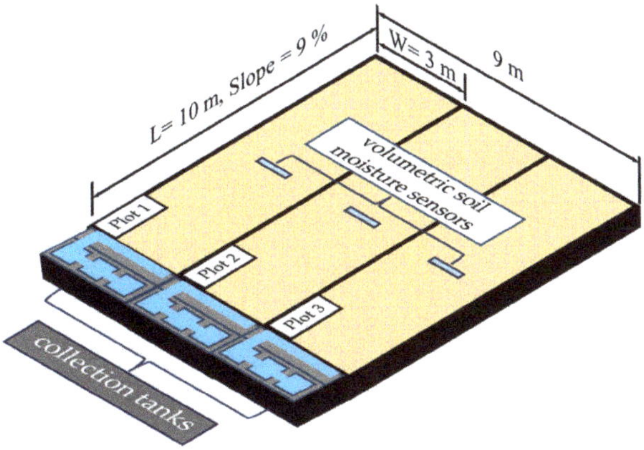

Figure 1. Schematic drawing of field experiment setup.

Effective erosion events were delineated from the rainfall records based on the effective rainfall event definition given by Wischmeier and Smith (1978) [6]. An effective rainfall event is defined as when a rainfall event satisfies the following two criteria: (1) the accumulated rainfall of an event exceeds 12.7 mm or more than 6.35 mm rain falls within 15 min and (2) the time interval between the current event and preceding event exceeds 6 h.

Based on 7 yr field observations prior to 2019, we found that surface runoff often occurred at the experiment site when the accumulated rainfall exceeded 19.0 mm. The 19 mm threshold value can result either from long-duration events or short but heavy storms, and this threshold decreases when one storm occurs after another in two to three days. Therefore, the 19 mm rainfall threshold seems to be an appropriate indicator for field erosion studies at our site. Our past field observations generally support Wischmeier and Smith's definition of an effective rainfall event. Therefore, we followed the same definition to extract the effective rainfall events (hereafter termed an effective erosion event) from our rainfall records for data analysis.

The end time (Te_a) and begin time (Tb_b) were used to extract the time interval between two adjacent effective erosion events (TILR) (Figure 2). The TILR reflects the opportunity for soil to be exposed to natural drying, a similar idea to the 6 h break between storms in Wischmeier and Smith's effective rainfall event definition. The longer the TILR, the lower the antecedent soil moisture becomes. During the process of extracting the TILR, some rain events that did not meet the criteria of an effective rainfall event were excluded.

Different climate regions produce different rainfall characteristics. We often experience downpours resulting from convectional rain, frontal rain, and typhoons during summer, but continuous drizzle seldom occurs at the experiment site. Due to the rare occurrence of continuous drizzle, we found from the volumetric soil moisture sensor data that the impact of light drizzle on surface soil moisture content was minimal, and the surface soil moisture quickly returned to the stage prior to the rain. Furthermore, from the onsite rainfall records, it was found that a total of 20 mm of discrete light drizzle occurred in 2019 and in 2020, only accounting for 0.89% of the 2239.5 mm of the total rainfall and 0.99% of the 2016 mm of the total rainfall, respectively. Therefore, we chose to exclude these ineffective events by assuming that these events had no impact on antecedent soil moisture content. The two-year field soil moisture logging and rainfall records justified our assumption.

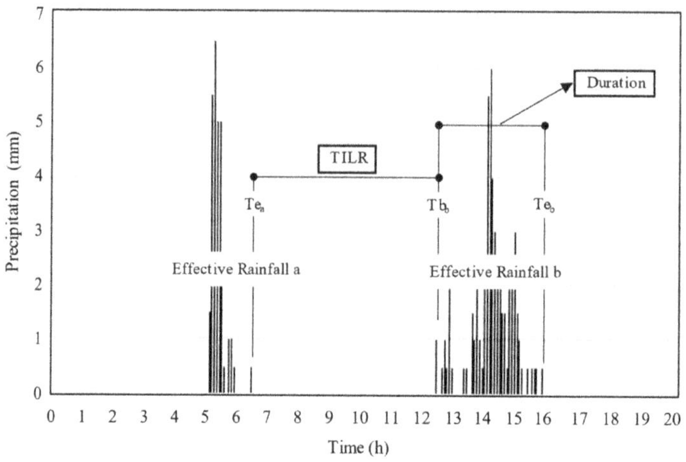

Figure 2. Delineation of effective erosion events, TILR, and duration.

"Duration" is the length of time (in hours) that an effective erosion event lasts, and it is the time span between Tb_b and Tb_e (Figure 2). Natural rainfall records may include rain-free intervals, and these intervals vary in length. Therefore, effective erosion events with a longer duration may include a higher portion of rain-free periods; conversely, events with a shorter duration typically have less chance or a lower portion of rain-free periods.

"Ppx" is defined as the accumulated rainfall (in mm) within x hours prior to the beginning of an effective erosion event. The estimation of Ppx involves tracing backward in time from T_b by a specific "x hours" time range and summing up the rainfall amounts within, as illustrated in Figure 3. Five different time ranges, namely 12, 24, 48, 72, and 120 h, were used in this study. Ppx reflects the amount of rain that soil receives before the occurrence of an effective erosion event. If the number of "x hours" selected is too low, influential antecedent rainfall may be overlooked. Conversely, if it is too high, it may include non-influential antecedent rainfall.

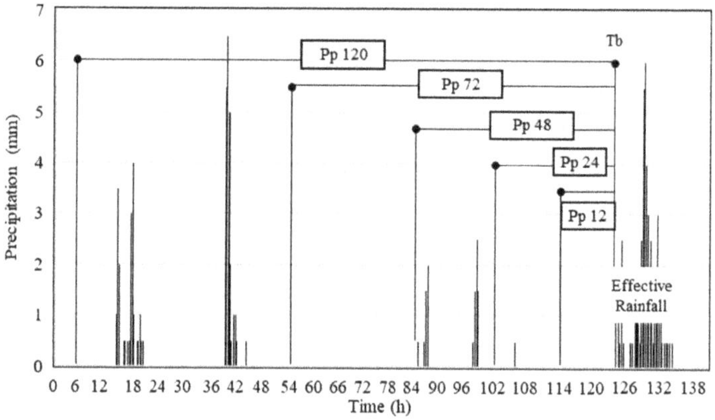

Figure 3. Determination of Ppx with different x–h time ranges.

Lastly, the Rainfall–Runoff Erosivity Index (EI_{30}) was calculated based on the USLE definition proposed by Wischmeier and Smith (1978) [6]. This index is the product of the total kinetic energy and maximum 30 min rainfall intensity ($I_{30}max$) of an effective erosion event. Wu and Wang (1996) [33] conducted a 3 yr field study on raindrop distributions and rainfall kinetic energy at the experiment site, and they found that their rainfall kinetic energy formula deviated from Wischmeier and Smith's formula when the rainfall intensity was less than 4.0 mm/h. For the sake of simplicity in the calculation, Wischmeier and Smooth's rainfall kinetic energy formula was used in this study to calculate the total kinetic energy of an effective erosion event. The I_{30}max of an effective erosion event was extracted from the rainfall records.

2.2.2. Runoff and Soil Loss

A collection tank was installed downstream of the runoff plot to collect surface runoff and soil loss. To accommodate a large runoff volume exceeding the tank design volume, two partition walls were arranged inside the collection tank, which resulted in a tank with three separate pools. A blade consisting of five 60° sharp-crested triangular weirs and a blade with single sharp-crested triangular weirs were installed at the rim of each partition wall. These two weir blades were carefully installed so that the notches were at the same elevation to ensure the evenness of runoff splitting. The excess runoff overflowing from the five-weir blade was discarded, and that from the single-weir blade overflew to the next pool. The second pool was also equipped with the same weir blades. Therefore, the second pool collected 1/6 of the runoff overflown from the first pool, and the third pool collected 1/6 of the runoff overflown from the second pool, which was 1/36 of the overflown runoff from the first pool. Hence, when runoff accumulated in both the second and third pools, the runoff volume in the second pool was multiplied by 6, and the runoff volume in the third pool was multiplied by 36; then, they were added to the runoff volume collected in the first pool, finally yielding the total runoff volume of the event. When runoff only accumulated in the first and second pools, the runoff volume in the second pool was multiplied by 6 and then added to the runoff volume in the first pool to yield the total runoff volume. Figure 4 illustrates a three-pool collection tank.

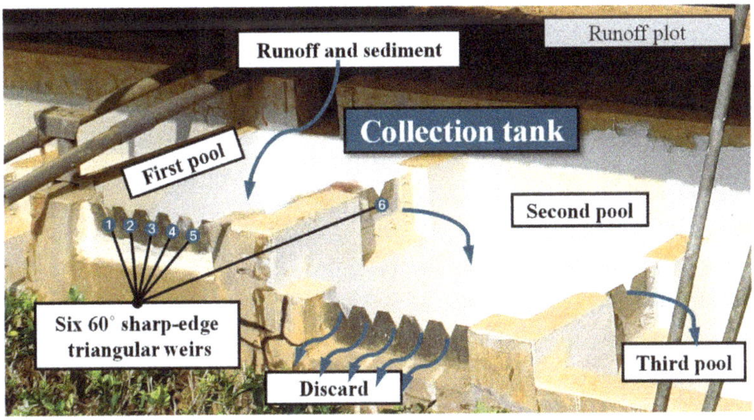

Figure 4. The collection tank.

Soil loss was calculated from grab samples of detained sediment in the first pool and added to grab samples of suspended sediment collected from each pool. Grab samples of suspended load were also taken from each pool. They were oven-dried, back-calculated using the grab sample and detained sediment volume ratios, and then added together to obtain the total soil loss. Runoff and soil loss collection was carried out on a storm-by-storm basis.

2.2.3. Antecedent Soil Moisture Content

To eliminate the effect of soil texture differences between runoff plots, the volumetric soil moisture data were first normalized as follows:

$$NVSMC_i = \frac{VSMC_i - VSMC_{min}}{VSMC_{max} - VSMC_{min}} \quad (1)$$

where $NVSMC_i$ is the normalized volumetric soil moisture content at time T_i, $VSMC_i$ is the volumetric soil moisture content at time T_i, and $VSMC_{min}$ and $VSMC_{max}$ are the minimum and maximum volumetric soil moisture contents detected during the rainy season, respectively.

T_i is the key point for analyzing the normalized antecedent volumetric soil moisture content (NAVSMC), and it must accurately represent the time right before the start of a rain event. It is slightly earlier than the beginning of a rain event (Tb). In this study, T_i is set approximately 5 to 10 min prior to the Tb based on the soil moisture contents logged throughout the study period.

3. Results and Discussion

Table 1 summarizes the 2-year field experiment data, as well as the results extracted from the rainfall, runoff, and soil loss data. Events 1~37 in Table 1 were replications; therefore, they shared the same rainfall parameters. The abbreviations in Table 1 are defined as follows: Precp. is the total precipitation of an effective erosion event, EI_{30} is the Rainfall–Runoff Erosivity Index, C is the runoff coefficient, NAVSMC is the normalized antecedent volumetric soil moisture content, Ppi is the accumulated rainfall within i hours (i = 1, 24, 48, 72, and 120) prior to the beginning of an effective erosion event, and TILR is the time interval between two adjacent effective erosion events. Unit area soil loss (kg/m^2) and unit area runoff volume (mm^3/mm^2) are the total soil loss and total runoff volume collected from an effective erosion event divided by the runoff plot area, respectively.

No rills were found in the 10 m long runoff plots during the field experiment periods. Within the scope of this study, surface runoff was mainly responsible for transporting the eroded soil that was generated by splash and sheet erosion, also known as interrill erosion. A positive correlation that fits a power function of $Y = 0.0248X^{0.8055}$ exists between the unit area runoff volume (X; mm^3/mm^2) and unit area soil loss (Y; kg/m^2), with a coefficient of determination of $R^2 = 0.82$, despite greater variabilities in natural rain settings.

A Pearson correlation analysis (Microsoft EXCEL 2021) was also conducted between the unit area soil loss; unit area runoff volume; and selected rainfall parameters, namely, NAVSMC (the normalized antecedent volumetric soil moisture content), duration, Ppx (the accumulated rainfall within x hours prior to the beginning of an erosion event), and TILR (time interval between two adjacent events). A correlation with a coefficient greater than 0.75 is classified as a strong correlation, a coefficient between 0.5 and 0.75 is classified as a moderate correlation, and a coefficient between 0.25 and 0.5 is interpreted as a weak correlation. The results of the Pearson correlation coefficients are summarized in Table 2, in which the data cells shaded with a gray background have p values greater than 0.05, and the other cells without a gray background have p values less than 0.05. The results in Table 2 indicate that the unit area soil loss is strongly positively correlated with the unit area runoff volume, precipitation, duration, and Pp_{48}. It is moderately positively correlated with Pp_{24} and Pp_{72}, and it is weakly positively correlated with Pp_{12} and Pp_{120}. However, it is negatively correlated with TILR. A weak and negative correlation was found between the unit area soil loss and NAVSMC.

Table 1. Summary of field experiment results.

Event No.	Unit Area Soil Loss	Unit Area Runoff Volume	Precp.	EI_{30}	Soil Loss/EI_{30}	C	NAVSMC	Pp_{12}	Pp_{24}	Pp_{48}	Pp_{72}	Pp_{120}	Event Duration	TILR
	kg/m^2	mm^3/mm^2	mm	MJ-mm/ha-h	(kg/m^2)/(MJ-mm/ha-h)	-	m^3/m^3	mm	mm	mm	mm	mm	h	h
1	0.080	9.60	34.0	391.75	0.020	0.28	0.15	0.0	9.0	9.0	9.0	9.5	3.17	213.00
2	0.214	9.20	34.0	391.75	0.055	0.27	0.20	0.0	9.0	9.0	9.0	9.5	3.17	213.00
3	0.313	17.02	34.0	391.75	0.080	0.50	0.51	0.0	9.0	9.0	9.0	9.5	3.17	213.00
4	0.177	19.58	39.0	451.04	0.039	0.50	0.13	0.0	0.0	0.0	34.0	43.0	3.33	68.83
5	0.346	19.43	39.0	451.04	0.077	0.50	0.18	0.0	0.0	0.0	34.0	43.0	3.33	68.83
6	0.440	18.72	39.0	451.04	0.098	0.48	0.47	0.0	0.0	0.0	34.0	43.0	3.33	68.83
7	0.972	61.85	87.5	2951.65	0.033	0.71	0.18	0.0	39.0	39.0	39.0	82.0	2.42	20.25
8	2.109	39.19	87.5	2951.65	0.071	0.45	0.22	0.0	39.0	39.0	39.0	82.0	2.42	20.25
9	0.890	77.77	87.5	2951.65	0.030	0.89	0.53	0.0	39.0	39.0	39.0	82.0	2.42	20.25
10	0.178	19.40	23.0	117.47	0.151	0.84	0.27	0.5	3.5	16.0	21.5	114.0	10.83	104.00
11	0.230	16.90	23.0	117.47	0.196	0.73	0.34	0.5	3.5	16.0	21.5	114.0	10.83	104.00
12	1.335	91.31	117.5	1731.80	0.077	0.78	0.28	0.5	3.5	16.0	21.5	114.0	61.33	11.08
13	0.679	21.39	117.5	1731.80	0.039	0.18	0.34	0.5	3.5	16.0	21.5	114.0	61.33	11.08
14	1.166	48.37	117.5	1731.80	0.067	0.41	0.56	0.5	3.5	16.0	21.5	114.0	61.33	11.08
15	2.969	93.65	402.5	7811.82	0.038	0.23	0.29	0.5	2.0	94.5	117.5	144.0	91.50	10.50
16	2.511	53.17	402.5	7811.82	0.032	0.13	0.25	0.5	2.0	94.5	117.5	144.0	91.50	10.50
17	1.689	106.28	197.5	2476.39	0.068	0.54	0.24	0.0	0.0	3.0	3.0	25.0	28.00	133.58
18	1.666	45.95	197.5	2476.39	0.067	0.23	0.38	0.0	0.0	3.0	3.0	25.0	28.00	133.58
19	1.944	114.89	197.5	2476.39	0.078	0.58	0.47	0.0	0.0	3.0	3.0	25.0	28.00	133.58
20	0.256	32.20	41.5	374.80	0.068	0.78	0.20	0.0	0.0	0.0	0.0	176.5	4.33	97.08
21	0.205	20.68	41.5	374.80	0.055	0.50	0.36	0.0	0.0	0.0	0.0	176.5	4.33	97.08
22	0.172	24.73	41.5	374.80	0.046	0.60	0.44	0.0	0.0	0.0	0.0	176.5	4.33	97.08
23	0.320	48.77	54.0	1070.56	0.030	0.90	0.28	0.0	0.0	0.0	0.0	0.0	13.08	121.42
24	0.269	21.05	54.0	1070.56	0.025	0.39	0.45	0.0	0.0	0.0	0.0	0.0	13.08	121.42
25	0.414	52.16	54.0	1070.56	0.039	0.97	0.50	0.0	0.0	0.0	0.0	0.0	13.08	121.42
26	0.041	8.55	11.5	23.52	0.176	0.74	0.26	0.0	0.5	54.0	54.0	54.0	5.75	18.67
27	0.044	3.97	11.5	23.52	0.189	0.35	0.43	0.0	0.5	54.0	54.0	54.0	5.75	18.67
28	0.042	8.74	11.5	23.52	0.180	0.76	0.49	0.0	0.5	54.0	54.0	54.0	5.75	18.67
29	0.051	15.28	19.5	59.37	0.085	0.78	0.27	0.0	0.0	0.0	11.0	65.0	1.67	54.50
30	0.104	12.95	19.5	59.37	0.175	0.66	0.42	0.0	0.0	0.0	11.0	65.0	1.67	54.50
31	0.105	16.30	19.5	59.37	0.178	0.84	0.49	0.0	0.0	0.0	11.0	65.0	1.67	54.50
32	0.279	39.50	54.5	578.57	0.048	0.72	0.11	0.0	0.0	0.0	0.0	0.0	3.50	167.00
33	0.228	23.92	54.5	578.57	0.039	0.44	0.31	0.0	0.0	0.0	0.0	0.0	3.50	167.00
34	0.195	35.72	54.5	578.57	0.034	0.66	0.41	0.0	0.0	0.0	0.0	0.0	3.50	167.00
35	0.108	14.31	29.0	361.82	0.030	0.49	0.02	0.0	0.0	0.0	0.0	0.0	1.42	307.33
36	0.208	14.95	29.0	361.82	0.058	0.52	0.40	0.0	0.0	0.0	0.0	0.0	1.42	307.33
37	0.126	16.78	29.0	361.82	0.035	0.58	0.43	0.0	0.0	0.0	0.0	0.0	1.42	307.33
38	6.405	453.29	551.0	11,126.44	0.058	0.82	0.36	0.0	45.0	106.0	106.0	106.0	86.83	21.92

Table 1. *Cont.*

Event No.	Unit Area Soil Loss	Unit Area Runoff Volume	Precp.	EI_{30}	Soil Loss/EI_{30}	C	NAVSMC	Pp_{12}	Pp_{24}	Pp_{48}	Pp_{72}	Pp_{120}	Event Duration	TILR
	kg/m^2	mm^3/mm^2	mm	MJ-mm/ha-h	(kg/m^2)/(MJ-mm/ha-h)	-	m^3/m^3	mm	mm	mm	mm	mm	h	h
39	1.536	101.54	192.5	3204.80	0.048	0.53	0.40	0.0	0.0	0.0	18.0	538.0	58.50	52.58
40	1.099	51.77	62.0	1631.76	0.067	0.83	0.41	0.0	0.0	8.0	9.0	13.5	1.25	307.50
41	0.362	1.22	23.5	234.68	0.154	0.05	0.39	0.0	0.0	0.0	0.0	4.0	1.17	450.67
42	0.613	50.35	170.5	1311.07	0.047	0.30	0.42	0.0	0.0	3.5	6.0	46.5	55.92	107.27
43	0.413	18.75	24.5	211.19	0.196	0.77	0.43	0.0	16.0	95.5	140.5	172.5	2.25	20.17
44	0.012	0.85	6.0	7.96	0.152	0.14	0.41	0.0	1.5	2.5	2.5	12.0	22.00	91.83
45	0.058	11.14	12.5	61.42	0.095	0.89	0.43	0.0	0.5	6.0	16.5	19.0	0.75	49.25
46	0.131	20.47	58.0	325.25	0.040	0.35	0.42	0.0	0.0	0.0	14.0	31.5	61.00	72.08
47	0.717	97.78	198.8	1785.67	0.040	0.49	0.44	0.0	0.5	17.5	44.0	58.0	81.67	22.83
48	0.009	0.05	5.6	7.72	0.116	0.01	0.43	0.0	4.0	4.2	11.6	176.6	6.67	16.58
49	0.105	9.42	11.6	65.44	0.161	0.81	0.41	0.0	0.0	0.4	6.0	10.0	0.83	47.00

Table 2. Summary of Pearson correlation coefficients of field experiment results.

	Unit Area Soil Loss (kg/m^2)	Unit Area Runoff Volume (mm^3/mm^2)	Precp. (mm)	NAVSMC (m^3/m^3)	Pp_{12} (mm)	Pp_{24} (mm)	Pp_{48} (mm)	Pp_{72} (mm)	Pp_{120} (mm)	Duration (h)	TILR (h)
Unit area soil loss	1.0										
Unit area runoff volume	0.90	1.0									
Precp.	0.92	0.81	1.0								
NAVSMC	−0.02	0.03	−0.03	1.0							
Pp_{12}	0.22	0.04	0.32	−0.06	1.0						
Pp_{24}	0.53	0.52	0.32	−0.06	−0.06	1.0					
Pp_{48}	0.61	0.46	0.60	−0.01	0.31	0.51	1.0				
Pp_{72}	0.53	0.38	0.55	−0.02	0.31	0.40	0.94	1.0			
Pp_{120}	0.24	0.19	0.27	0.07	0.25	0.09	0.22	0.29	1.0		
Duration	0.65	0.56	0.82	0.09	0.53	0.06	0.44	0.45	0.34	1.0	
TILR	−0.23	−0.20	−0.26	−0.13	−0.28	−0.26	−0.44	−0.50	−0.41	−0.38	1.00

Note: the data shaded with a gray background have a p-value greater than 0.05.

The time range x chosen to calculate the accumulated rainfall (Ppx) prior to an erosion event affects the Pearson correlation coefficients. The effective erosion events collected in this study seldom had any rain within the 12 h time range before their occurrence. Hence, most of the erosion events have zero rainfall in the Pp_{12} column shown in Table 1. However, Pp_{120} includes rainfall events that may not be effective enough to initiate erosion, resulting in a higher amount of rainfall but not the equivalent soil loss. If a 72 h time range were chosen, the number of ineffective rainfall events included would be less, so the correlation would change from weakly positive to moderately positive, whereas Pp_{24} excluded the soil weakening or soil recovery effects of storms occurring 24 to 36 h prior to the current effective storm on the soil, which depended on the nature of the storm and how far apart two adjacent storms were. We found that Pp_{48} is the best parameter capable of including the antecedent storms at the experiment site and achieves the best performance in the correlation analysis.

The effect of the time interval between two adjacent events (TILR) on the antecedent soil moisture content is somewhat similar to that of the accumulated rainfall prior to an erosion event (Ppx). The shorter the TILR, the lower the chance that the soil dries naturally. In general, any ineffective rainfall such as drizzle prior to an effective erosion event may alter the soil wetness. However, the negative weak correlation between the TILR and unit area soil loss (Table 2) indicates that all storm events collected in this study fail to support TILR's contribution to soil erosion.

Duration is another rainfall characteristic considered in this study, and the correlation analysis shows that it is strongly positively correlated with unit area soil loss (Table 2) as expected. However, the duration of a rainfall event masks the contribution of the antecedent soil moisture content.

The amount of soil loss is also affected by the Rainfall–Runoff Erosivity Index (EI_{30}), which is a known fact, and EI_{30} is the foundation for describing the rainfall and runoff erosivity for most soil loss estimation models. It is relatively easy to run replicate experiments and obtain similar results using a rainfall simulator in a laboratory setting, in which the rainfall intensity, soil moisture content, and duration of recess between experiments can be controlled. However, the variables controlling rainfall characteristics and soil properties are constantly changing in field experiments under natural rainfall. Therefore, we divide the unit area soil loss by the storm's Rainfall–Runoff Erosivity Index (EI_{30}), which is denoted as Soil Loss/EI_{30} hereafter. Soil Loss/EI_{30} is the unit area soil loss per Rainfall–Runoff Erosivity. It quantifies how erodible soil becomes, which is also known as soil erodibility.

We plot Soil Loss/EI_{30} with respect to Pp_{48} but exclude data where Pp_{48} equals zero, i.e., where no rainfall occurred 48 h before the erosion event. The resultant plot that grouped four durations (Dur) is shown in Figure 5. The rationale for grouping data by rainfall event duration is that event duration affects the total soil loss, as well as EI_{30}. The effect of event durations cannot be determined in laboratory rainfall simulation studies with constant rainfall intensities and durations.

Grouping data by event duration makes the data appear scattered. Fitting trends to each data group helps visualize the general trend that a field experiment under natural rainfall conditions tends to present. The trends shown in Figure 5 indicate that the total rainfall that occurs 48 h (Pp_{48}) prior to an effective erosion event has the lowest coefficient of determination (R^2 = 0.1051) when the duration of an effective erosion event exceeds 55 h. When the duration of an effective erosion event is either 3~7 h or 10~30 h, R^2 becomes 0.7737 or 0.5832, respectively. Both trends follow a similar pattern.

We further conduct a Pearson correlation analysis on Soil Loss/EI_{30} with respect to the normalized antecedent volumetric soil moisture content (NAVSMC), Pp_{48}, and rainfall duration by first excluding cases where Pp_{48} equal zero, and the results are shown in Table 3 and Figure 6. Through a two-tailed test, it was found that all the correlation coefficients listed in Table 3 have a p-value of less than 0.05.

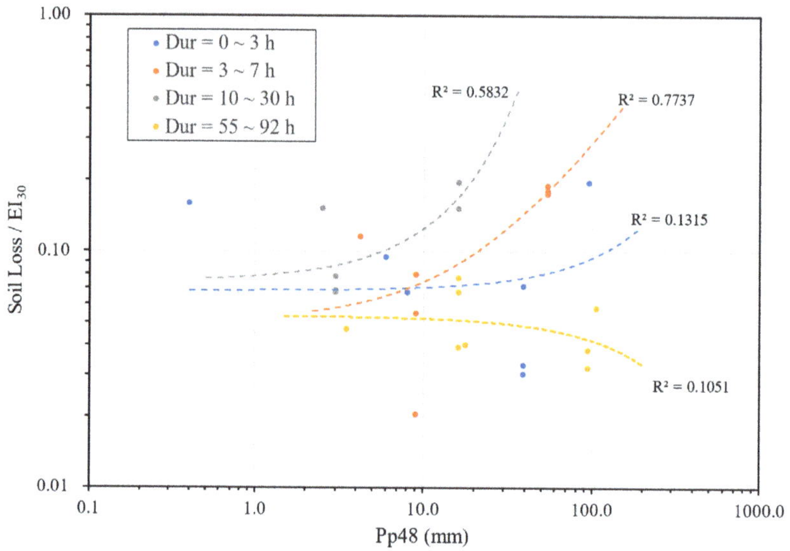

Figure 5. Trends in unit area soil loss per EI_{30} with respect to Pp_{48}, grouped by rainfall durations.

Table 3. Pearson correlation coefficients with respect to Soil Loss/EI_{30}.

Event with Durations	0~3 h	3~7 h	10~30 h	55~92 h	0~7 h	0~30 h	All Durations
NAVSMC	0.256	0.538	−0.086	0.296	0.386	0.295	0.230
Pp_{48}	0.319	0.880	0.759	−0.323	0.532	0.423	0.064
Duration	−0.325	0.808	−0.903	−0.508	0.402	−0.034	−0.466

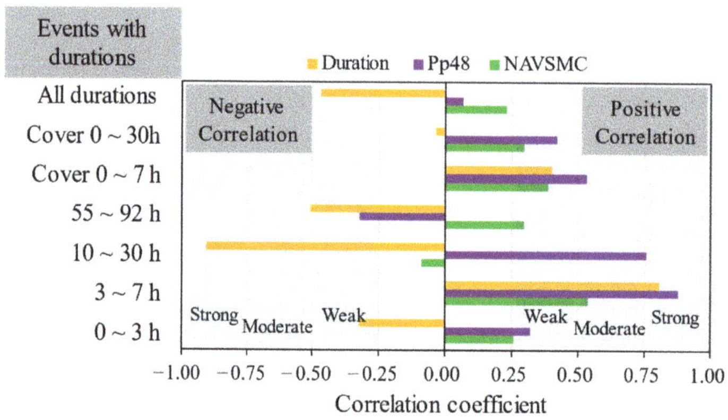

Figure 6. Pearson correlation coefficients with respect to Soil Loss/EI_{30}.

From Table 3 and Figure 6, we find that the normalized antecedent volumetric soil moisture content (NAVSMC) is moderately correlated with Soil Loss/EI_{30} when the duration of an effective erosion event is within 3 to 7 h, whereas Pp_{48} and rainfall duration are strongly correlated with Soil Loss/EI_{30}. In order to identify the conditions under which the antecedent volumetric soil moisture content affects Soil Loss/EI_{30}, we only selected data with $Pp_{12} \neq 0$ to eliminate the effect of the accumulated rainfall prior to the erosion event and generated a scatter plot on Soil Loss/EI_{30} against the NAVSMC; the result is shown in Figure 7.

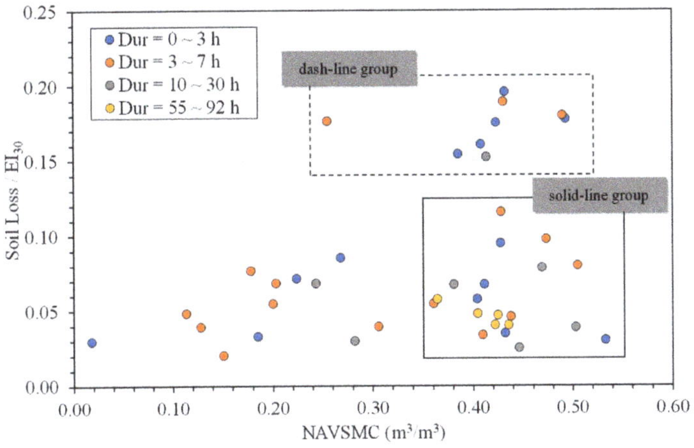

Figure 7. Soil Loss/EI_{30} vs. normalized antecedent volumetric soil moisture content (NAVSMC).

Regardless of the dispersion of the data in Figure 7, we found that the data points were segregated into three distinct clusters: a solid-line cluster, a dashed-line cluster, and an ungrouped cluster. We chose to delineate our data because each cluster possessed different characteristics in the identification process used to identify the common characteristics of the data points.

The ungrouped cluster had two characteristics: (1) NAVSMC < 0.35 and (2) Soil Loss/EI_{30} < 0.1. The remaining data were further delineated into solid- and dashed-line groups because both groups had an NAVSMC of > 0.30 but considerably different results in terms of Soil Loss/EI_{30}. In addition, the solid- and dashed-line groups demonstrated different soil erodibility under a high antecedent soil moisture content, while the ungrouped data points demonstrated a generally low soil erodibility under a low antecedent soil moisture content. Through further analysis, we found that the rainfall conditions between the solid- and dashed-line groups were noticeably different, and these rainfall conditions suggested that we had to exclude the ungrouped data from the solid-line group.

The data points in the dashed-line group in Figure 7 are Events # 26, 27, 28, 30, 31, 41, 43, 45, and 50, and the data points in the solid-line group are Events # 3, 6, 9, 18, 19, 21, 22, 24, 25, 34, 36, 37, 38, 39, 40, 42, 46, 47, 48, and 49 (Table 1). We found that the dashed-line group shared the common characteristics of a short duration, a short TILR, a low rainfall amount, a low average rain intensity, and a low EI_{30} value. The solid-line group exhibited opposite characteristics to the dashed-line group.

We further found that rainfall events with an NAVSMC greater than 0.36 in Figure 7 (data of the solid-line group) were associated with the common characteristics of stronger storms. The average rainfall duration was 22.0 h, approximately 3.5 time longer than that of the dashed-line group (6.23 h). The average rainfall amount was 103.1 mm, almost five times higher than that of the dashed-line group (19.75 mm), and the average EI_{30} was 1570.14 MJ-mm/ha-h, 20 times higher than that of the dashed-line group (78.7 MJ-mm/ha-h).

Furthermore, the average NAVSMC of these events was 0.41, slightly higher than the overall average of 0.35, while the average time interval between two adjacent events (TILR) was about 40.5 h, roughly one-third of the overall average of 104.33 h, and the average rain intensity was about 8.3 mm/h. According to the rain intensity classification of the World Meteorological Organization (2018) [34], these events fell in the lower boundary of the heavy rain category, with an average intensity between 7.6 and 50 mm/h.

Two scatter plots were subsequently drawn, as shown in Figures 8 and 9. Rainfall events with Soil Loss/EI_{30} greater than 0.15, i.e., the dashed-line group in Figure 7, were excluded from the plot in Figure 8, whereas the solid-line group in Figure 7 was excluded from the plot in Figure 9. The rationale for constructing these two figures was that the data of the solid-line group were associated with stronger storms, as previously stated, and the data of the dashed-line group were not.

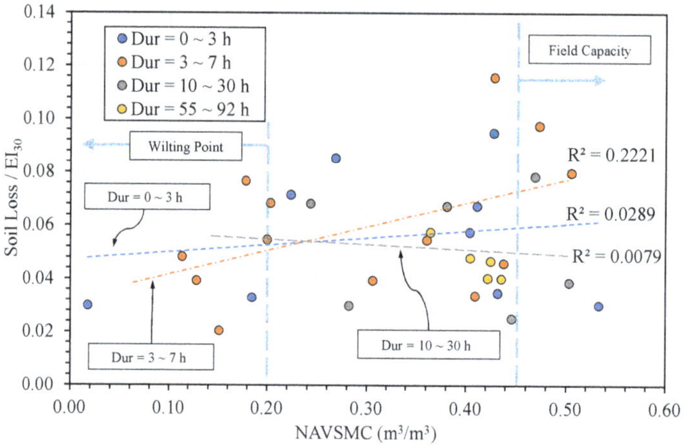

Figure 8. Soil Loss/EI_{30} vs. NAVSMC (excluding dashed-line group in Figure 7).

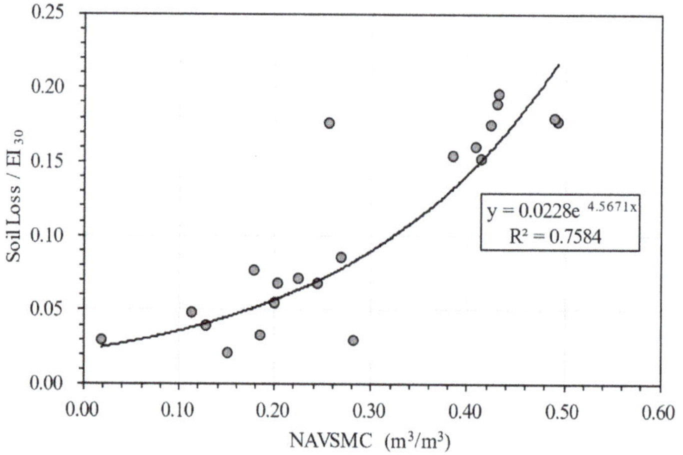

Figure 9. Soil Loss/EI_{30} vs. NAVSMC (excluding solid-line group in Figure 7).

The reasons for fitting the linear regression lines through the three duration groups in Figure 8 are as follows: (1) it is easy to see the general trends, even if the coefficients of determination are low, and (2) it is difficult to isolate the nonlinear effects of NAVSMC, if any, on soil vulnerability to erosion or soil erodibility, as this study was conducted under natural rain conditions. From Figure 8, we found that 66.7% of the events occurred with NAVSMCs within the wilting point and field capacity. The dispersion of the soil erodibility data expanded and covered a wider range when the NAVSMC exceeded 0.36, which is clearly illustrated in both Figures 7 and 8.

The best-fit curve shown in Figure 9, which excludes the solid-line group associated with stronger storms in Figure 7, exhibits a high coefficient of determination with an

$R^2 = 0.7584$, from which we conclude that (1) the impact of antecedent soil moisture content on soil erosion is conditional and (2) the impact of antecedent soil moisture content on soil erosion must be considered for rainfall events with a short duration, short TILR, lower average rain intensity, and low EI_{30}.

To verify our conclusion, a Pearson correlation analysis was again conducted. The correlation coefficients on Soil Loss/EI_{30} with respect to the NAVSMC and other parameters associated with the antecedent rainfall amount (Ppx), event duration, and time interval between two adjacent events (TILR) are summarized in Table 4. In Table 4, the data cells shaded with a gray background have a *p*-value greater than 0.05, and all data cells without a gray background have a *p*-value of less than 0.05.

Table 4. Pearson correlation coefficients of Soil Loss/EI_{30} with respect to NAVSMC and other rainfall parameters.

Total Event Counts	Duration Range (h)	NAVSMC	Pp_{12}	Pp_{24}	Pp_{48}	Pp_{72}	Pp_{120}	Duration (h)	TILR (h)
49	0~91.5	0.259	0.035	−0.125	0.164	0.179	0.030	−0.302	−0.170
39	0~30	0.318	0.342	−0.155	0.451	0.476	0.283	−0.015	−0.323
30	0~10	0.438	-	−0.215	0.465	0.495	0.172	0.140	−0.344
26	0~5	0.435	-	−0.163	0.242	0.379	0.174	−0.433	−0.199
23	0~4	0.473	-	−0.196	0.222	0.359	0.447	−0.430	−0.226
14	0~3	0.419	-	−0.347	0.131	0.312	0.343	−0.186	−0.218

Note: The data shaded with a gray background have a *p*-value greater than 0.05.

The results in Table 4 indicate that the best correlation between Soil Loss/EI_{30} and NAVSMC falls in the 0~4 h rainfall duration range. In contrast, events with a duration exceeding 10 h exhibit relatively lower correlations for most of the considered parameters. Therefore, we suggest that the impact of antecedent soil moisture content on soil erosion can be ignored for events with a duration exceeding 10 h.

4. Conclusions

This study was conducted under natural rainfall for two consecutive seasons at the plot scale to investigate the impact of the antecedent soil moisture content on the interrill erosion of clay soil. This study provides evidence that addresses the contradictory or inconclusive conclusions in previous studies on the antecedent soil moisture issue. The key points of the findings are summarized as follows:

1. Antecedent soil moisture content has an effect on unit area soil loss per EI_{30} (Soil Loss/EI_{30}) in rainfall events with short durations, low rainfall amounts, and low Rainfall–Runoff Erosivity. When the average rainfall intensity is between 7.6 and 50 mm/h, which is classified into the heavy rain category by the World Meteorological Organization (2018), the impact of antecedent volumetric soil moisture content is negligible.
2. There is a positive correlation between antecedent soil moisture content and unit area soil loss per EI_{30} (Soil Loss/EI_{30}) for rainfall events with a 0~4 h duration. When the event duration exceeds 10 h, the impact of antecedent soil moisture on soil erosion can be ignored.
3. The results of this field study on clay soil suggest that accumulative rain falling within 48 h (Pp_{48}) prior to an effective erosion event is strongly correlated with unit area soil loss per Rainfall–Runoff Erosivity (Soil Loss/EI_{30}). When the duration of an effective erosion event is less than 10 h, Pp_{48} makes a noticeable contribution to soil loss. However, when Pp_{48} is less than 20 mm, it does not affect soil loss in events with a 0~3 h duration.

Additional effort is required to continuously log the changes in antecedent soil moisture content. However, Pp_{48} can be extracted from rainfall records without much difficulty. In addition, to obtain representative values, soil moisture logging often requires additional post-processes that may overwhelm process-based soil loss simulation users. Hence, replacing the antecedent soil moisture content parameter with Pp_{48} seems to be a suitable alternative approach. Future studies are still needed, particularly on less clayey soils, to fully grasp the impact of antecedent soil moisture content on soil erosion.

Author Contributions: Conceptualization, Y.-D.C. and C.-C.W.; methodology, C.-C.W.; validation, Y.-D.C. and C.-C.W.; formal analysis, Y.-D.C. and C.-C.W.; investigation, Y.-D.C.; resources, C.-C.W.; data curation, Y.-D.C. and C.-C.W.; writing—original draft preparation, Y.-D.C.; writing—review and editing, C.-C.W.; visualization, Y.-D.C. and C.-C.W.; supervision, C.-C.W. All authors have read and agreed to the published version of the manuscript.

Funding: This study was self-funded without any external financial support.

Data Availability Statement: The original contributions presented in the study are included in the article, and further inquiries can be directed to the corresponding authors.

Acknowledgments: The authors wish to express their gratitude to those who helped prepare the field plots and assisted in collecting data.

Conflicts of Interest: The authors declare no conflicts of interest.

References

1. Bennett, H.H. *Soil Conservation*, 1st ed.; McGraw-Hill Book Company: New York, NY, USA, 1939.
2. Kirkby, M.J.; Morgan, R.P.C. (Eds.) *Soil Erosion*; Wiley: Chichester, UK; New York, NY, USA, 1980; ISBN 978-0-471-27802-3.
3. Wang, Y.; Liu, G.; Zhao, Z.; Wu, C.; Yu, B. Using Soil Erosion to Locate Nonpoint Source Pollution Risks in Coastal Zones: A Case Study in the Yellow River Delta, China. *Environ. Pollut.* **2021**, *283*, 117117. [CrossRef] [PubMed]
4. Borah, D.K.; Bera, M. Watershed-Scale Hydrologic and Nonpoint-Source Pollution Models: Review of Mathematical Bases. *Trans. Am. Soc. Agric. Eng.* **2003**, *46*, 1553–1566. [CrossRef]
5. An, J.; Zheng, F.L.; Han, Y. Effects of Rainstorm Patterns on Runoff and Sediment Yield Processes. *Soil Sci.* **2014**, *179*, 293–303. [CrossRef]
6. Wischmeier, W.H.; Smith, D.D. *Predicting Rainfall Erosion Losses: A Guide to Conservation Planning*; Department of Agriculture, Science and Education Administration: Washington, DC, USA, 1978.
7. Alewell, C.; Borrelli, P.; Meusburger, K.; Panagos, P. Using the USLE: Chances, Challenges and Limitations of Soil Erosion Modelling. *Int. Soil Water Conserv. Res.* **2019**, *7*, 203–225. [CrossRef]
8. Renard, K.G.; Foster, G.R.; Weesies, G.A.; McCool, D.K.; Yoder, D.C. (Eds.) *Predicting Soil Erosion by Water: A Guide to Conservation Planning with the Revised Universal Soil Loss Equation (RUSLE)*; Agricultural Handbook; USDA, Agricultural Research Service: Washington, DC, USA, 1997; ISBN 978-0-16-048938-9.
9. Knisel, W.G. *CREAMS: A Field Scale Model for Chemicals, Runoff, and Erosion from Agricultural Management Systems*; Conservation Research Report; Department of Agriculture, Science and Education Administration: Washington, DC, USA, 1980.
10. Young, R.; Onstad, C.; Bosch, D.; Anderson, W. AGNPS: A Non-Point-Source Pollution Model for Evaluating Agricultural Watersheds. *J. Soil Water Conserv.* **1989**, *44*, 168–173.
11. Laflen, J.M.; Elliot, W.J.; Flanagan, D.C.; Meyer, C.R.; Nearing, M.A. WEPP-Predicting Water Erosion Using a Process-Based Model. *J. Soil Water Conserv.* **1997**, *52*, 96–102.
12. Flanagan, D.; Gilley, J.; Franti, T. Water Erosion Prediction Project (WEPP): Development History, Model Capabilities, and Future Enhancements. *Trans. ASABE* **2007**, *50*, 1603–1612. [CrossRef]
13. Arnold, J.G.; Fohrer, N. SWAT2000: Current Capabilities and Research Opportunities in Applied Watershed Modelling. *Hydrol. Process.* **2005**, *19*, 563–572. [CrossRef]
14. Wischmeier, W.H.; Johnson, C.B.; Cross, B. Soil Erodibility Nomograph for Farmland and Construction Sites. *J. Soil Water Conserv.* **1971**, *26*, 5189.
15. Knapen, A.; Poesen, J.; Govers, G.; Gyssels, G.; Nachtergaele, J. Resistance of Soils to Concentrated Flow Erosion: A Review. *Earth Sci. Rev.* **2007**, *80*, 75–109. [CrossRef]
16. Kemper, W.D.; Rosenau, R.C. Soil Cohesion as Affected by Time and Water Content. *Soil Sci. Soc. Am. J.* **1984**, *48*, 1001–1006. [CrossRef]
17. Shainberg, I.; Goldstein, D.; Levy, G.J. Rill Erosion Dependence on Soil Water Content, Aging, and Temperature. *Soil Sci. Soc. Am. J.* **1996**, *60*, 916–922. [CrossRef]
18. Le Bissonnais, Y.; Bruand, A.; Jamagne, M. Laboratory Experimental Study of Soil Crusting: Relation between Aggregate Breakdown Mechanisms and Crust Structure. *Catena* **1989**, *16*, 377–392. [CrossRef]
19. Holz, D.J.; Williard, K.W.J.; Edwards, P.J.; Schoonover, J.E. Soil Erosion in Humid Regions: A Review. *J. Contemp. Water Res. Educ.* **2015**, *154*, 48–59. [CrossRef]

20. Wei, L.; Zhang, B.; Wang, M. Effects of Antecedent Soil Moisture on Runoff and Soil Erosion in Alley Cropping Systems. *Agric. Water Manag.* **2007**, *94*, 54–62. [CrossRef]
21. Auerswald, K.; Mutchler, C.K.; McGregor, K.C. The Influence of Tillage-Induced Differences in Surface Moisture Content on Soil Erosion. *Soil Tillage Res.* **1994**, *32*, 41–50. [CrossRef]
22. Fortesa, J.; Latron, J.; García-Comendador, J.; Company, J.; Estrany, J. Runoff and Soil Moisture as Driving Factors in Suspended Sediment Transport of a Small Mid-Mountain Mediterranean Catchment. *Geomorphology* **2020**, *368*, 107349. [CrossRef]
23. Massari, C.; Pellet, V.; Tramblay, Y.; Crow, W.T.; Gründemann, G.J.; Hascoetf, T.; Penna, D.; Modanesi, S.; Brocca, L.; Camici, S.; et al. On the Relation between Antecedent Basin Conditions and Runoff Coefficient for European Floods. *J. Hydrol.* **2023**, *625*, 130012. [CrossRef]
24. Jadidoleslam, N.; Hornbuckle, B.; Krajewski, W.; Mantilla, R.; Cosh, M. Analyzing Effects of Crops on SMAP Satellite-Based Soil Moisture Using a Rainfall-Runoff Model in the U.S. Corn Belt. *IEEE J. Sel. Top. Appl. Earth Obs. Remote Sens.* **2021**, *15*, 247–260. [CrossRef]
25. Schoener, G.; Stone, M.C.; Thomas, C. Comparison of Seven Simple Loss Models for Runoff Prediction at the Plot, Hillslope and Catchment Scale in the Semiarid Southwestern U.S. *J. Hydrol.* **2021**, *598*, 126490. [CrossRef]
26. Luk, S.-H.; Hamilton, H. Experimental Effects of Antecedent Moisture and Soil Strength on Rainwash Erosion of Two Luvisols, Ontario. *Geoderma* **1986**, *37*, 29–43. [CrossRef]
27. Wang, L.; Zheng, F.; Hu, W.; Zhang, X.J.; Shi, H. Interactive Effects of Rainfall Intensity, Kinetic Energy and Antecedent Soil Moisture Regime on Splash Erosion in the Ultisol Region of South China. *Catena* **2023**, *222*, 106863. [CrossRef]
28. Castillo, V.; Gomezplaza, A.; Martinezmena, M. The Role of Antecedent Soil Water Content in the Runoff Response of Semiarid Catchments: A Simulation Approach. *J. Hydrol.* **2003**, *284*, 114–130. [CrossRef]
29. Truman, C.C.; Bradford, J.M.; Ferris, J.E. Antecedent Water Content and Rainfall Energy Influence on Soil Aggregate Breakdown. *Soil Sci. Soc. Am. J.* **1990**, *54*, 1385–1392. [CrossRef]
30. Armenise, E.; Simmons, R.W.; Ahn, S.; Garbout, A.; Doerr, S.H.; Mooney, S.J.; Sturrock, C.J.; Ritz, K. Soil Seal Development under Simulated Rainfall: Structural, Physical and Hydrological Dynamics. *J. Hydrol.* **2018**, *556*, 211–219. [CrossRef] [PubMed]
31. Auerswald, K. Influence of Initial Moisture and Time since Tillage on Surface Structure Breakdown and Erosion of a Loessial Soil. *Catena Suppl.* **1993**, *24*, 93–101.
32. Ma, R.-M.; Li, Z.-X.; Cai, C.-F.; Wang, J.-G. The Dynamic Response of Splash Erosion to Aggregate Mechanical Breakdown through Rainfall Simulation Events in Ultisols (Subtropical China). *Catena* **2014**, *121*, 279–287. [CrossRef]
33. Wu, C.-C.; Wang, A.-B. Drop Size Characteristics and Erosive Kinetic Energy of Natural Rainstorms in Pingtung Laopi Area. *Chin. Soil Water Conserv.* **1996**, *27*, 151–165. (In Chinese)
34. Secretariat of the World Meteorological Organization. *Guide to Meteorological Instruments and Methods of Observation*; Secretariat of the World Meteorological Organization: Geneva, Switzerland, 2018; ISBN 978-92-63-16008-9.

MDPI AG
Grosspeteranlage 5
4052 Basel
Switzerland
Tel.: +41 61 683 77 34

Water Editorial Office
E-mail: water@mdpi.com
www.mdpi.com/journal/water

www.ingramcontent.com/pod-product-compliance
Lightning Source LLC
LaVergne TN
LVHW070451061126
851858LV00011B/217
* 9 7 8 3 7 2 5 8 8 5 2 9 9 *